World Sustainability Series

Series Editor

Due to its scope and nature, sustainable development is a matter which is very interdisciplinary, and draws from knowledge and inputs from the social sciences and environmental sciences on the one hand, but also from physical sciences and arts on the other. As such, there is a perceived need to foster integrative approaches, whereby the combination of inputs from various fields may contribute to a better understanding of what sustainability is, and means to people. But despite the need for and the relevance of integrative approaches towards sustainable development, there is a paucity of literature which address matters related to sustainability in an integrated way.

Notes on the quality assurance and peer review of this publication

Prior to publication, the works published in this book are initially assessed and reviewed by an in-house editor. If suitable for publication, manuscripts are sent for further review, which includes a combined effort by the editorial board and appointed subject experts, who provide independent peer-review. The feedback obtained in this way was communicated to authors, and with manuscripts checked upon return before finally accepted. The peer-reviewed nature of the books in the "World Sustainability Series" means that contributions to them have, over many years, been officially accepted for tenure and promotion purposes.

Arun Pratap Mishra · Amit Kumar ·
Naveen Chandra · Gajendra Singh ·
Chaitanya Baliram Pande
Editors

Threatened Medicinal Plants in the Indian Himalayan Region

Sustainability Challenges and Conservation Strategies

Springer

Editors
Arun Pratap Mishra
Department of Forestry and Remote Sensing
Earthtree Enviro Private Limited
Shillong, Meghalaya, India

Naveen Chandra
Department of Botany
Soban Singh Jeena University Campus
Almora, Uttarakhand, India

Chaitanya Baliram Pande
Indian Institute of Tropical Metrology
Pune, Maharashtra, India

Amit Kumar
Department of Habitat Ecology
Wildlife Institute of India
Dehradun, Uttarakhand, India

Gajendra Singh
Department of Forestry and Climate Change
Uttarakhand Space Application Center
Dehradun, Uttarakhand, India

ISSN 2199-7373 ISSN 2199-7381 (electronic)
World Sustainability Series
ISBN 978-3-031-73686-5 ISBN 978-3-031-73687-2 (eBook)
https://doi.org/10.1007/978-3-031-73687-2

This Springer imprint is published by the registered company Springer Nature Switzerland AG
The registered company address is: Gewerbestrasse 11, 6330 Cham, Switzerland

About the Book

The book 'Threatened Medicinal Plants in the Indian Himalayan Region—Sustainability Challenges and Conservation Strategies' provides an in-depth analysis of the major issues related to the conservation of threatened medicinal plants in the Indian Himalayan region. The book is a comprehensive resource that highlights the critical role of medicinal plants in traditional healthcare systems and identifies the significant threats that these plants face due to various anthropogenic and natural factors. The book covers ten major themes that are critical to understanding the conservation of threatened medicinal plants in the Indian Himalayan region. It provides an essential resource for researchers, conservationists, policymakers, and practitioners interested in the conservation of threatened medicinal plants in the region. The book provides an overview of the major issues related to medicinal plant conservation and suggests strategies for the sustainable management of these plants. The authors have provided a comprehensive and insightful analysis of the conservation status of medicinal plants in the region, highlighting the urgent need for concerted efforts to conserve these valuable resources.

'Threatened Medicinal Plants in the Indian Himalayan Region—Sustainability Challenges and Conservation Strategies' is more than just a collection of chapters; it is a journey through the rich tapestry of biodiversity, traditional knowledge, and the critical need for medicinal plant conservation witnessed in the Indian Himalayan Region covering Ladakh, Jammu & Kashmir, Himachal Pradesh, Uttarakhand, Sikkim and Arunachal Pradesh.

The book opens with an exploration of alpine hay meadows of Ladakh, highlighting their ecological relevance and the diverse medicinal plant diversity. Traditional land use practices, including artificial irrigation, support the growth of essential medicinal and fodder plants. This sets the stage for understanding the significance of these habitats and their contributions to local communities.

Moving forward, the book delves into the vulnerability assessment of prioritized Medicinal and Aromatic Plants (MAPs) in Paddar Valley. Through surveys and rapid mapping exercises, the study identifies key threats, categorizes the level of endangerment, and emphasizes the importance of conservation efforts to safeguard these valuable plant resources.

Shifting focus to Himachal Pradesh, the narrative unfolds the current trends in wild medicinal herb extraction. The impact of market dynamics on specific species, such as *Trillium govanianum* and *Fritillaria cirrhosa*, leading to overharvesting and endangerment, is discussed. External factors, like the MGNREGA scheme and expanding horticulture, are explored, emphasizing the need for sustainable practices.

The narrative then comprehensively reviews *Rheum* species in the Indian Himalayan Region, covering ecology, traditional uses, phytochemistry, trade, and the escalating threats faced by these high-value medicinal plants. The urgency for conservation is emphasized due to their endemicity, restricted distribution, and increasing threats.

The book proceeds with an assessment of the extent, abundance, threats, and information gap on alpine medicinal plants in the western Himalaya. Analysing past and ongoing studies, it reveals gaps in knowledge on species abundance, habitat suitability, demand and supply ratio, and climate change impact, emphasizing the necessity for more comprehensive research and management strategies.

In the Kumaun Himalaya, the documentation of 82 threatened plant species across various families serves as a comprehensive database. Aimed at aiding researchers, conservationists, and policymakers, this chapter contributes to understanding the diversity of threatened plants in the region.

The exploration then shifts to Niti Valley in the Nanda Devi Biosphere, where 72 MAPs are surveyed. The study reveals insights into medicinal plant richness, usage, and conservation status. The escalating demand for MAPs poses threats to biodiversity, emphasizing the need to protect these resources and preserve traditional knowledge.

Addressing the impact of climate change on Himalayan medicinal plants, the narrative discusses the role of plant tissue culture as a conservation strategy. The synergistic effects of plant growth-promoting microorganisms in aiding the survival and growth of micro-propagated plants in changing environmental conditions are explored.

Focusing on *Polygonatum cirrhifolium*, the narrative reviews its ecology, phytochemistry, and conservation status. The species, known for its medicinal properties, faces threats from overexploitation and habitat degradation. The chapter emphasizes collaborative conservation initiatives at regional and international levels.

The book continues with the exploration of the commercialization and conservation potential of *Swertia chirayita*. It reviews the ecological significance, traditional uses, and challenges in cultivation. The chapter advocates for sustainable practices, including genetic enhancement, to alleviate strain on natural populations.

Providing a comprehensive assessment of *Phlomoides superba*, the narrative covers taxonomic history, growth phenology, and demographic characteristics. The species faces anthropogenic threats, prompting the need for conservation efforts. The chapter also explores alternative propagation methods through tissue culture.

Utilizing species distribution modelling, the book evaluates the potential distribution of 25 MAPs in the alpine region of Kumaun Himalaya. It identifies dominant habitats and assesses habitat suitability, offering insights for reintroduction efforts and highlighting influential environmental factors.

Focusing on *Withania somnifera*, the book predicts its potential distribution in Uttarakhand using MaxEnt modeling. The study emphasizes the importance of considering multiple environmental variables for accurate predictions and provides insights for species restoration, conservation planning, and localization efforts.

The book concludes with a review of the uses of 169 plant species among rural communities in the Sikkim Himalaya, documenting traditional knowledge associated with alpine and sub-alpine vegetation. It highlights the utilitarian importance of these plants and their role in traditional healthcare practices.

The final chapter focuses on indigenous medicinal plant utilization across Arunachal Pradesh. It documents traditional knowledge from various tribes, emphasizing the reliance on plant-based remedies. The study underscores the need for conservation efforts to preserve both traditional knowledge and medicinal plant resources in the region.

As we journey through the chapters of this book, a clear story emerges one of urgency, resilience, and the delicate interplay between nature and people in the vulnerable landscapes of the Indian Himalayan Region. This collection goes beyond being a set of scientific findings; it is a heartfelt call to action. It earnestly pleads for the conservation of our botanical heritage, emphasizing that these plant legacies are crucial for ensuring sustainable futures.

Arun Pratap Mishra
Amit Kumar
Naveen Chandra
Gajendra Singh
Chaitanya Baliram Pande

Introduction

Arun Pratap Mishra, Amit Kumar, Naveen Chandra, Gajendra Singh and Chaitanya Baliram Pande

Challenges and Conservation Strategies for Threatened Medicinal Plants in the Indian Himalayan Region

The Indian Himalayan region (IHR), towering and stretching from Jammu and Kashmir in the North to Arunachal Pradesh in the east, is one of the most prominent and ecologically sensitive zones in the country. This mountain range separates the vast plains of Indian sub-continent from the frigid Tibetan Plateau. Over the years, changes in the landscape and climate are not only having an impact on the majestic mountain range hosting myriad of communities living downstream. IHR is significant for economic development and human welfare as a result of the abundance of public goods and services it offers, including food, energy, water, and medicinal plants. The region is a rich repository of biodiversity, harbouring numerous plant species with medicinal properties that have been traditionally used by local communities for centuries. It is home to around 7,000 plant species, of which approximately 2,000 are used for medicinal purposes. The traditional healthcare system of the IHR relies heavily on the use of medicinal plants, and the knowledge of their use is passed down through generations. However, in recent years, the region has experienced rapid urbanization, deforestation, and overexploitation of natural resources, leading to the depletion of many medicinal plant species. This book aims to understand the challenges facing the conservation of threatened medicinal plants in the Indian Himalayan region and explore potential conservation strategies to address these challenges.

Challenges Facing Conservation

One of the key challenges of threatened medicinal plant conservation in the Indian Himalayan region is the lack of information on the population status, available growing stock and market chain. While some plant species are well-known and actively harvested, others are less known and threatened due to decline in their population in wild. As such, there is a need for detailed research to identify medicinal plant species that are risk of extinction due to overexploitation, habitat loss, and other factors. Another challenge is the lack of effective management policies and regulations to protect medicinal plant species in the region. While some protected areas exist, they are often not well-enforced, and there are few restrictions on the harvesting and trade of medicinal plants. As such, it is a high time to enforce proactive legislation in order to ensure the sustainable harvesting and conservation of threatened plant species. Furthermore, the socio-economic factors driving the overexploitation of medicinal plant species in the region need to be better understood. Many local communities rely on the harvesting and sale of medicinal plants for their livelihoods, and as such, they may be reluctant to adopt conservation measures that limit their access to these resources. Understanding the economic incentives and disincentives for sustainable harvesting practices can help researchers develop more effective conservation strategies that balance the needs of local communities with the conservation of threatened plant species.

Anthropogenic Pressure on Medicinal Plants

In the IHR, anthropogenic pressure on medicinal plants is a significant threat to its biodiversity and traditional healthcare system. In recent years, the region has been facing an increasing anthropogenic pressure due to a variety of factors, which has put the region's biodiversity and the traditional healthcare system at risk. The anthropogenic pressure on medicinal plants in the IHR can be broadly categorized into four main categories:

Overharvesting and unsustainable collection practices: Medicinal plants are being collected from the wild in large quantities, often using unsustainable methods that hamper the plant life cycle and surround ecosystem. The high demand for medicinal plants of IHR, both nationally and internationally, has led to overharvesting of certain species, such as *Angelica glauca*, *Dactylorhiza hatagirea*, *Picrorhiza kurroa* and *Swertia chirayita* to name a few. This has not only deteriorated the population of these species in the region but has also led to the loss of genetic diversity, which is crucial for the long-term survival of the species.

Habitat loss and degradation: The IHR is undergoing rapid land-use changes due to expanding human settlements, agriculture, and infrastructure development. The conversion of natural habitats to agricultural land and the construction of roads and hydroelectric projects have resulted in the fragmentation and loss of habitats, making it difficult for threatened medicinal plant species to survive and reproduce.

Climate change: The IHR is highly vulnerable to climate change, which is affecting the distribution and phenology of medicinal plant species. The rising temperatures and changes in precipitation patterns have altered the timing of flowering, fruiting, and seed dispersal of many species, leading to a decline in their population in the region.

Lack of awareness and conservation efforts: There is a general lack of awareness and recognition of the value of medicinal plants in the IHR. The traditional knowledge of medicinal plants is not adequately documented, and the efforts to conserve medicinal plant species are minimal. There is also a lack of regulation and monitoring of the collection, trade, and use of medicinal plants, which makes it difficult to control the unsustainable harvesting practices.

Therefore, in order to address the anthropogenic pressure on medicinal plants in the IHR, several conservation strategies need to be practiced. Firstly, a comprehensive inventory of medicinal plant species, their distribution, and their population and conservation status needs to be studied. Secondly, efforts must be made to conserve and restore the habitats of medicinal plant species through specific micro-plans on habitat restoration and reforestation programs. Thirdly, sustainable harvesting practices need to be developed, which includes promoting the cultivation of medicinal plant species and developing regulations to control their trade and use. Further, there is an urgent need to address the issue through conservation efforts, sustainable harvesting practices, and promoting awareness and recognition of the value of medicinal plants. Failure to address this issue could lead to the loss of valuable medicinal plant especially threatened species and the loss of traditional knowledge that has been passed down through generations.

Conservation Strategies

Many local communities rely on the harvesting and sale of medicinal plants for their livelihoods, and as such, these communities should be actively involved in efforts to promote sustainable harvesting practices. As the book emphasizes on the importance of involving local communities in the conservation process, the strategies should involve providing education and training on sustainable harvesting techniques, as well as developing alternative livelihood options that are not reliant on the harvesting of medicinal plants. Another potential conservation strategy is the development of community-based conservation programs. These programs should involve the establishment of community-managed protected areas, where local communities are responsible for managing and conserving medicinal plant species within the area. This approach has been successful in other regions and could help promote the sustainable harvesting and conservation of threatened plant species in the Indian Himalayan region. Further, it is submitted that the conservation of threatened medicinal plants in the Indian Himalayan region is a complex and multifaceted issue that requires a comprehensive and collaborative approach. While there are many challenges facing the conservation of the threatened medicinal plant species, there is an urgent need of proactive legislation to support conservation strategies that can be employed to address these challenges. Additionally, involving local communities in the conservation process, developing community-based conservation programs, and conducting comprehensive studies of the threatened medicinal plant species, researchers can help to promote the sustainable harvesting and conservation of threatened plant species in the Indian Himalayan region.

Knowledge Gaps

While this book provides a useful overview of the challenges facing the conservation of threatened medicinal plants in the Indian Himalayan region, there are several areas that could benefit from further research. One important area is the need for more comprehensive studies including long term monitoring of the medicinal plant species found in IHR. Through conducting more comprehensive studies of the medicinal plant species found in the region, researchers will be able to identify the species that are at higher risk of extinction and hence, develop targeted conservation strategies to address such threats. Further, there is urgent need to better understand the socio-economic factors that drive the over-exploitation of medicinal plant species in the region. Many local communities rely on the harvesting and sale of medicinal plants for their livelihoods, and as such, they may be reluctant to adopt conservation measures that limit their access to these resources. Understanding the economic incentives and disincentives for sustainable harvesting practices can help researchers develop more effective conservation strategies that balance the needs of local communities with the conservation of threatened plant species. There is also a need for more research on the potential uses of threatened medicinal plant species in modern medicine. While traditional healthcare systems in the region have long relied on the use of medicinal plants, there may be potential for these species to be used in modern medicine as well. Thus, exploring the potential uses of threatened medicinal plant species in modern medicine, researchers can help to promote the conservation of these species by demonstrating their value beyond their traditional uses. IHR is particularly vulnerable to the impacts of climate change, and changes in temperature and precipitation patterns could have significant impacts on the distribution and abundance of medicinal plant species. Therefore, it is imperative to understand the impact of climate change on medicinal plant species in the region to develop strategies and promote their conservation and adaptation to changing environmental conditions.

Contents

About the Editors

Mr. Arun Pratap Mishra is a passionate environmental researcher with a masters degree in Environmental Science from Hemwati Nandan Bahuguna Garhwal (Central) University in Uttarakhand, India. His dedication to advancing ecological research and sustainability is evident in his extensive body of work, which includes 40 research articles, 10 book chapters, 3 conference papers, and 1 technical report. He also serves as the editor of four notable books that focus on Himalayan plant ecology, natural resource management, climate change in the Himalayas, and the Earth's carbon footprint, utilizing cutting edge techniques and cloud based solutions. With professional experience spanning esteemed institutions like the Uttarakhand Space Application Centre, Botanical Survey of India, Wildlife Institute of India, and RBased Services Private Limited, Mr. Mishra has developed a deep expertise in geospatial technologies. He has further solidified his credentials with a postgraduate diploma in Remote Sensing and GIS from the Indian Institute of Remote Sensing (IIRS, ISRO). Currently, Mr. Mishra holds the position of GIS Analyst at Earthtree Enviro Pvt. Ltd., Shillong, Meghalaya, where he focuses on utilizing innovative computational methods and geospatial analysis to advance ecological conservation and sustainable devel opment. His goal is to address pressing global challenges such as climate change and carbon management through data-driven, practical solutions. Mr. Mishra is always open to collaboration, believing that interdisciplinary efforts are key to driving meaningful change. His work is driven by a passion for science and its potential to create sustainable, real-world impact.

Amit Kumar is a faculty at the Wildlife Institute of India, Dehradun. Plant ecology and field botany are the main focuses of his research. His current research includes understanding vegetation patterns, plant associations, floristic diversity assessment, ethnobotany, and the threatened medicinal plants conservation in the Western Himalaya. He enjoys photographing plants, hiking, biking and playing table tennis.
e-mail: amit@wii.gov.in

Dr. Naveen Chandra is serving as a faculty in Department of Botany, Soban Singh Jeena University, Almora. He did his Post graduation in 2016 and obtained Ph.D. degree (Botany) from Kumaun University, Nainital in 2022. Prior to joining the university, he served as a Project Fellow at the Uttarakhand Space Application Centre (USAC), Dehradun, for over six years and worked on different aspect of Himalayan biodiversity. His research pursuits encompass diverse domains within the botanical sciences, including plant taxonomy and ecology, vegetation ecology, invasion ecology, high-altitude physiology, medicinal plants, ecological modelling, and the application of Remote Sensing (RS) and Geographic Information System (GIS) in biodiversity and ecological research. His scholarly contributions include 35 research papers, 7 book chapters, and 2 scientific reports. Notably, he currently serves as an editor for two books published by Springer Nature on different aspect of Himalayan ecosystem. His research activities involve comprehensive surveys of high-altitude regions in the western Himalaya, focusing on monitoring vegetation patterns and conducting long-term ecological monitoring initiatives.

Dr. Gajendra Singh is a Scientist in Forestry and Climate Change Division at Uttarakhand Space Application Centre (USAC), Department of IT, Good Governance and Science Technology, Government of Uttarakhand. He has been conducting research on Himalayan Plant Taxonomy, Forest Ecology, Spatial Phyto-resource Monitoring and Natural Resource Management. Till date, he has contributed in over 65 research papers/book chapters, 04 Books and 02 Manuals. His significant contribution is in discovering one (01) new species of flowering plant to the science, addition of five (05) species to the flora of Western Himalaya, three (03) species to Uttarakhand state and one (01) species to Himachal Pradesh. Rediscovery of one (01) species from Uttarakhand and three (03) local check list of floras.

Chaitanya Baliram Pande He completed Ph.D. in Environment Science from Sant Gadge Baba Amravati University, Amravati and M.Sc. in Geo-informatics from Amravati University in 2011. His research interests include Remote Sensing, GIS, Google Earth Engine, Machine Learning, Watershed management, Hydrogeology, Hydrological Modeling, Drought Monitoring, Land Use and Land Cover analysis, Groundwater Quality, urban planning, Hydro-geochemistry, Groundwater Modelling, Geology, Hyperspectral Remote Sensing, Remote Sensing and GIS application in natural resources management, watershed management and Environmental Monitoring and assessment subjects. Dr. Pande has more than 13 years of teaching, research and industrial experience. So far, he has published above 120+ international and national research papers, 3 textbooks, 7+ edited books, 19 conferences papers and 30 book chapters published in Scopus index. His publications cited more than 4489 citations with H-Index-42 and I-10 Index-90. He also acts as a reviewer for several scientific international journals of the International reputes with editorial board member in Journal of American Journal of Agricultural and Biological Sciences. He has been included in the prestigious 2024 Stanford University list of the world's top 2% scientists.

Plant Diversity in the Hay Meadows of Ladakh, Indian Trans-Himalaya: Focus on Ethno-medicinal and Fodder Species

Aimon Bushra, Amit Kumar, Gautam Talukdar, Hitendra Padalia, Jikmat Stanzin, Mohd. Raza, and Gopal Singh Rawat

Abstract Hay meadows are among the most species-rich habitats in the Indian Himalayan Region (IHR), as well as in other parts of the world. However, their ecological relevance, species diversity and conservation significance remain largely underexplored. During recent floristic explorations in western Ladakh, we surveyed a few semi-natural or artificially irrigated hay meadows in the Suru, Drass, Aryan and Wakha valleys of Kargil district. The traditional land use practices such as diverting the glacial melt streams into stone-built walls or canals, allow growth of numerous plant species that are valuable medicinal and fodder plants, along with graminoids. Medicinal plants such as *Aconitum heterophyllum, Dactylorhiza hatagirea, Delphinium cashmerianum, Gentianella moorcroftiana, Heracleum pinnatum* and *Podophyllm hexandrumim* and fodder plants such as *Astragalus thomsonii, Cicer microphyllum, Medicago falcata, Trifolium pratense, Oxytropis hypoglottoides, Poa alpina and Prangos pabularia,* are harvested from these hay meadows by the locals. These plants are a valuable commodity for the local people due to low-cost source of livestock-feed and their healing properties. The present study documents the plant diversity observed in the mountain hay meadows of western Ladakh along with their ethno-medicinal and nutritional values.

A. Bushra (✉) · A. Kumar · G. Talukdar · Mohd. Raza · G. S. Rawat
Wildlife Institute of India, Dehradun 248001, Uttarakhand, India
e-mail: aimonbushra16@gmail.com; aimon@wii.gov.in

H. Padalia
Indian Institute of Remote Sensing, Indian Space Research Organization, Dehradun, Uttarakhand, India

J. Stanzin
GB Pant National Institute of Himalayan Environment, Ladakh Regional Centre, Leh 194101, Ladakh, India

Mohd. Raza
Snow Leopard Conservancy India Trust, Leh, Ladakh, India

G. S. Rawat
Uttarakhand State Council for Science and Technology, Dehradun, Uttarakhand, India

A. P. Mishra et al. (eds.), *Threatened Medicinal Plants in the Indian Himalayan Region*, World Sustainability Series, https://doi.org/10.1007/978-3-031-73687-2_1

Keywords Agroecology · Cold desert · Grassland biodiversity · Medicinal plants · Traditional practices

1 Introduction

Hay meadows are semi-natural grasslands or pastures that are left to grow over the spring and summer before being harvested for hay. The vegetation of hay meadows is dominated by forbs and grasses, but unlike other pastures, they are protected from grazing during growing season (Kumar et al. 2023; Mishra et al. 2023). The meadows are maintained on well-drained and relatively more fertile soils. They are different from the natural rangelands and treeless vegetation in terms of management approaches and management intensity (Mucina et al. 2016; Rodríguez-Rojo et al., 2017).

The hay meadows are traditionally managed across temperate regions of Europe (also referred to as neutral grasslands) and adjacent parts of Asia (Mucina et al. 2016). Once wide spread across Europe such meadows have restricted distribution in transhumance pastoral areas of Norway (Bele et al. 2024), Carpathian mountains (Wehn et al. 2018; Janišová et al. 2023), Romania (Ivaşcu et al. 2016), Switzerland (Schwab et al. 2002) and Sweden (Dahlström et al. 2013). These meadows are categorized as endangered habitats in Europe due to their tremendous reduction in quantity and quality (European Red List of Habitats—Grasslands Habitat Group 2016). Such reduction has been attributed to less scientific attention to traditional management practices, agriculture mechanization and intensification over the past 100 years (Janišová et al. 2023; Bele et al. 2024).

In Tibet, maintaining hay meadows as a land-use practice has evolved following certain land policies resulting in sedantarization of pastoral communities (Ning et al. 2002). Some herders sow pastures for hay and have individual or exclusive rights over the field (Miller 1990). In the Indian Himalayan Region (IHR), agro-pastoral communities have developed hay meadow system for optimizing resource use from limited land-holdings. These hay meadows are maintained by the local communities to meet their winter forage requirements. Such hay meadows are locally called *Ghasnis* or *Phats* (Himachal Pradesh), *Mangs* (Uttarakhand), and *Lu-ungs* (Ladakh).

In Trans-Himalayan region of Ladakh, hay meadows are more common in the west, where subsistence farming is prominent than in eastern Ladakh where pastoral production is major land use practice. In western areas such as Kargil, water spillage from irrigation canals and restricted livestock grazing provide ample moisture and manure thereby allowing growth and regeneration of hay meadows. Additionally, livestock grazing follows a rotational scheme allowing growth and maintenance of alpine pastures. Hay produce are collected in autumn and stored as supplementary livestock feed in winter as extreme snow fall restricts livestock movement in alpine pastures. Moreover, the hay meadows are rich in plant diversity besides providing multiple services like controlling soil erosion, providing medicinal, fodder and food resources.

Several studies have been conducted in Ladakh to understand diversity, ethnobotany and phytogeography of plants (Stewart 1916a, 1916b; Kachroo et al. 1977; Dhar and Kachroo 1983; Murti 2001; Rawat 2007; Hartmann 2009; Srivastava and Shukla 2015; Dvorský et al. 2018). However, no research has been conducted on hay meadows, which serve as an oasis for alpine plant diversity in the region. It is worth noting that most studies have concentrated on eastern Ladakh, which is devoid of such meadows. We evaluated the significance of hay meadows in terms of their diversity, ethno-botanical and fodder values in an underexplored region of western Ladakh.

2 Study Area

The study was carried out in Kargil district of Ladakh, which forms the western part of the region, bordering Jammu & Kashmir (Fig. 1). Spanning an area of over 60,000 sq. km, the Union Territory of Ladakh is one of the most distinct regions in terms of biogeography, culture and topography. Drainage to this part of the country is through Indus river and its main tributaries, the Shayok, Chang Chenmo, Hanle, Zangskar and Suru-Drass (Warikoo 2020). Ladakh is delimited by the Karakoram Range in the northeast and Great Himalayan range in the southwest.

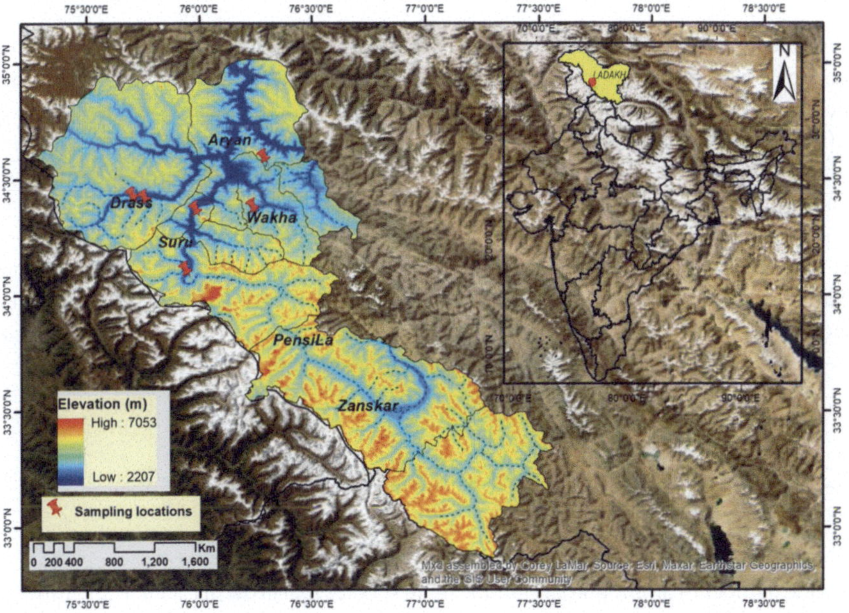

Fig. 1 Map showing the location of the study area and sampling sites; district Kargil is denoted with a red circle on the inset map

This region is characterized by sparse vegetation, generally devoid of trees and often dominated by alpine dry scrub (scrub steppe), alpine dry pastures (desert steppe) or mixed herbaceous formations and low primary productivity with a short growing season (Rawat and Adhikari 2005). Land use can be broadly divided in three production systems: nomadic, transhumance and agro-pastoralists. People in these landscapes are used to capitalize on suitable lands for food crop and livestock production. Livestock production is more intensive in areas above 4500 m (above mean sea level) elevation.

Likewise, agriculture is practiced in areas which have easy access to water and is more diverse at elevation below 3500 m. Moreover, eight ethnic groups are identified in the region, namely *Bot, Mon, Beda, Garra, Brokpa, Balti, Changpa* and *Purigpa.* Most of the ethnic groups are agro-pastoralists except the *Changpas* in eastern Ladakh who practice nomadic pastoralism. Livestock types include poultry, yak hybrids, cow, donkey, horse, mule, sheep and goats. Traditionally cultivated crop type includes barley, wheat, buckwheat, peas, lentils, foxtail millet, porso millet. Alfalfa, a traditional fodder crop is widely cultivated across the region.

3 Material and Methods

3.1 Field Sampling

The valleys were driven through, and hay fields near settlements were searched. The meadows were usually located on hills above the hamlet, and were also seen along the roads. In order to understand the plant diversity, a homogeneous one-hectare plot (100×100 m) was laid in each meadow. Within each plot, 5 sub-plots of size 2×2 m were laid on each corner and one at the center. The information on geo-coordinates, name of the species, habit of the species and their numbers were recorded in each 2×2 m plot, and the plant species were photographed for their identification. The plants were identified using relevant literature (Dhar and Kachroo 1983; Chowdhery and Wadhwa 1984; Polunin and Stainton 1984; Murti 2001; Dvorský et al. 2018; Shukla and Srivastava 2020) and online database such as Plants of World Online [https://powo.science.kew.org/] and subsequently verified with the help of experts.

3.2 Investigating Medicinal and Fodder Plants Species of the Hay Meadows

A wide number of alpine and sub-alpine plant species that grow in Ladakh, are utilized as food, fodder, fuel and for their medicinal benefits. A few important medicinal plants that have high value and significant use in the Ladakhi Amchi/Sowa-rigpa system of traditional medicines have been studied for their medicinal applications. Similarly, commonly observed forage and fodder plants of hay meadows were investigated for their dietary and fiber subsistence needs.

4 Results

4.1 Plant Species Richness in the Hay Meadows

The current study investigated hay meadows in villages across four valleys, i.e. in Suru valley (Faruna and Panikhar), Drass valley (Lhamochan and Holiyal), Wakha valley (Fokar), and Aryan valley (Chulichan). The study revealed a total of 47 unique species under 21 families. The highest number of species (n = 19) were recorded in the hay meadow of Suru valley, at Panikhar which was located along the road, nearby agricultural fields. Another site, Faruna which was located above the village, on a mountain slope and irrigated by the waters of Ghangshalmo glacier had 11 species. The hay meadows of Drass valley were also species rich, with Holiyal and Lhamochan having 15 and 12 species respectively. The meadows of Aryan and Wakha valleys were relatively low in species number. Chulichan had seven species, whereas Fokar had five species only. The number of species and families recorded at each site is provided in Fig. 2. Figure 3 shows the typical hay fields of Kargil. Of the plant species recorded in these hay meadows, one is critically endangered *Dactylorhiza hatagirea* (D.Don) Soó, two are endangered *Aconitum heterophyllum* Wall ex Royle, and *Podophyllm hexandrum* Royle while three species, namely *Aconitum violaceum* Jacquem. ex Stapf, *Physochlaena praealta* (Decne.) Miers, and *Ferula jaeschkeana* Vatke are vulnerable as per the IUCN. The elevation range, vegetation structure and dominant species of each site are provided in Table 1.

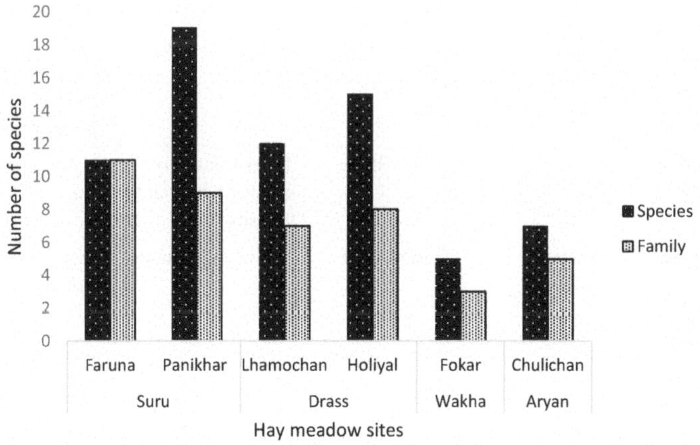

Fig. 2 Number of species and families recorded in different hay meadows of Kargil

Fig. 3 Hay fields at **a** Faruna and **b** Panikhar, villages of Kargil

Table 1 Topography, vegetation characteristics and commonly occurring species in the hay meadows

S. no.	Valley	Hay meadow site	Elevation (m)	Vegetation structure	Dominant species
1	Suru	Faruna	2990–3000	Shrubs, herbs and grasses	*Epilobium angustifolium, Medicago falcata, Oxytropis* spp., *Androsac* spp.
2		Panikhar	3215–3276	Herbs and grasses	*Aconitum heterophyllum, Delphinium cashmerianum, Heracleum pinnatum, Phragmites australis*
3	Drass	Lhamochan	3190–3243	Herbs and grasses	*Taraxacum officinale, Artemisia* spp., *Poa alpine, Medicago varia*
4		Holiyal	3059–3158	Herbs and grasses	*Prangos pabularia, Medicago falcata, Medicago sativa, Calamagrostis* spp.
5	Wakha	Fokar	3792–3815	Trees (planted), shrubs, herbs and grasses	*Medicago falacta, Rosa webbiana, Artemisia* spp.
6	Aryan	Chulichan	3665–3773	Herbs and grasses	*Cicer microphyllum, Medicago falacta*

4.2 *Medicinal and Fodder Plants of the Hay Meadows and Their Usage in Ladakh*

Though, the majority of the plant species harvested from the hay meadows are used for hay-making, some of the species are occasionally collected by the local inhabitants and utilized as medicines. Out of the 47 plant species recorded from the hay meadows, 19 species have been commonly reported to have some or the other medicinal applications (Table 2).

Several plant species that occur in the hay meadows, belonging to Fabaceae and Poaceae families offer good fodder because of their high protein and fiber content (Table 3).

5 Discussion

Hay meadows are considered biodiversity-rich areas holding both ecological and ethno-botanical importance. Additionally, hay meadows have received less attention in terms of their ecological importance in the region. Neither their difference from natural herbaceous meadows has been assessed. But in most of Europe studies it is revealed that such areas contain more than 60% of the native plants (Myklestad and Sætersdal 2004). In Kargil, the present study reports 47 species in different valleys with varying species counts. Besides, a trend between plant diversity, moisture, and elevation in the hay meadows of the region was also observed. Compared to other valleys, Suru and Drass receive more precipitation in the form of snow and rain, which increases the concentration of glaciers and water. Maximum species were recorded in Suru and Drass valley compared to other valleys of the region. This discrepancy can be attributed to higher precipitation and their location at mid-elevations, as indicated by Namgail et al. (2012) and Klimešová et al. (2011) in Ladakh, Singh et al. (2007) in Himachal Pradesh, and Sekar et al. (2023) in Uttarakhand. Such an increase in plant species richness at mid-elevation is also reported by Austrheim (2002) in Norway. Additionally, topographical differences (Rawat and Adhikari 2005; Marini et al. 2007) and the age or development history of hay meadows (Homburger and Hofer 2012) affect species composition and diversity. It is noteworthy that agriculture practice and acreage under agriculture are more profound in Drass and Suru valley, thus hay meadows are widespread in these areas. However, these observations need to be statistically tested and verified in future studies to understand the effect of such environmental variables on plant diversity in the hay meadows of the region. Furthermore, in Kargil, Behera et al. (2014) found higher herb diversity in agro-forest areas compared to alpine scrubs and pastures. Similarly, Ghimire et al. (2006) report that in contrast to alpine meadows, pastures with controlled grazing and more precipitation were richer in medicinal plants in Nepal. This illustrates the significance of customs and culture employed in the management of these hay meadows which harbour local biodiversity.

Table 2 Medicinal plants recorded from hay meadows and their use in Kargil, Ladakh

Scientific name	Local name	Family	Part used	Medicinal uses
Aconitum heterophyllum Wall ex Royle	Buma-karpo	Ranunculaceae	Roots	Used for treating jaundice, stomach ache, arthritis, and body inflammation pain
Aconitum violaceum Jacquem. ex Stapf	Buma-nakpo	Ranunculaceae	Roots	Used in treatment of asthma, cough, joint pain, stomach pain, as well as neural disorders, and cardiac diseases
Artemisia sieversiana Ehrh.	Kha-khaz/ Khamchu	Asteraceae	Aerial parts i.e. leaves, stem, flowers and fruits	Used for detumescence, hemostasis and reduce kidney inflammation/pain and urinary tract burning sensation
Bunium persicum (Boiss.) B.Fedtsch	Zeur	Apiaceae	Aerial parts	Treating abdominal pain, relieving flatulence, cold, headache, stomachache, joint pain, tuberculosis, fever etc.
Carum carvii L.	Ko-snyot	Apiaceae	Fruits and seeds	Treating gastrointestinal including dyspepsia, and various spasmodic conditions, bloating, diarrhea and also or improving eye vision
Cicer microphyllum Benth.	Sari	Fabaceae	Roots	Treating asthma and enlargement of liver
Dactylorhiza hatagirea (D.Don) Soó	Angu-lakhpa/ Wang bolakpa	Orchidaceae	Roots	Used by the locals for treating diarrhoea, dysentery, chronic fever and kidney problems

(continued)

Table 2 (continued)

Scientific name	Local name	Family	Part used	Medicinal uses
Datura stramonium L.	Datur/Dhadura	Solanaceae	Flowers, fruits and seeds	Used to alleviate stomach and intestinal pain caused by worm infestation, toothache, and fever caused by inflammation
Delphinium cashmerianum Royle	Charkyang	Ranunculaceae	Aerial parts	Used as cardiac and respiratory depressant; used in local medicine for destroying maggots in wounds of livestock (especially sheep); treating dysentery, diarrhoea with bleeding, inflammation, and wounds
Epilobium angustifolium L.	Punar	Onagraceae	Flowers	Treating stomachache
Ferula jaeschkeana Vatke	Prangos	Apiaceae	Roots	Used as antihistaminic, antiseptic; and to cure a variety of ailments such as chest pain, tumors, ulcers, and chronic wounds
Gentianella moorcroftiana (Wall. ex Griseb.) Airy Shaw	Chu-bitik	Gentianaceae	Aerial parts	Used to cure jaundice
Heracleum pinnatum C.B.Clarke	Spru/Spisho	Apiaceae	Stem, fruits	Used as digestive, and to cure abdominal cramps caused by intestinal worms
Mentha longifolia (L.) L.	Phololing	Lamiaceae	Leaves	It is used for the relief of congestion, edema, indigestion, asthma and body pain

(continued)

Table 2 (continued)

Scientific name	Local name	Family	Part used	Medicinal uses
Pedicularis punctata Decne.	Lugru-marpo	Orobanchaceae	Flowers	Used for their antioxidant, anti-inflammatory, antimicrobial, anticancer, and antispasmodic properties
Podophyllm hexandrum Royle	Denmo-kushu	Berberidaceae	Fruits and roots	Used for joint pain, skin allergies and blood purification
Rosa webbiana Wall. ex Royle	Sia-marpo	Rosaceae	Flowers and fruit	Used for treating skin diseases, gastrointestinal and respiratory disorders
Taraxacum officinale F.H.Wigg.	Hand/Khurma	Asteraceae	Whole plant	Used as anti-fungal, diuretic, liver tonic and in fever
Verbascum Thapsus L.	Monmoshing	Scrophulariaceae	Flowers and seeds	For treating fever and skin allergies

Of the 47 recorded species, more than 19 species are used in traditional medicines while 11 species have fodder value. Communication with herders in Drass and Suru valley revealed that species such as *Prangos pabularia*, *Cicer microphyllum*, *Astragalus* spp., *Epilobium angustifolium*, *Lindelofia stylosa*, and *Medicago falcata* are considered important fodder plants and are used as supplementary livestock feed in winter. Similar results were also found by Tiwari et al. (2016) in Ladakh and in northern areas of Pakistan by Khan (2003) where *Cicer microphyllum* and *Medicago sativa* is fed to livestock in colder winter months. This indicates that Traditional Ecological Knowledge (TEK) among local agro-pastoralists, as held among *Amchis* practitioners, is still intact in the region. Thus, this study emphasizes on preserving this knowledge system as otherwise loss of TEK and policies undermining traditional practices have led to decline in hay meadow in most of European region (Dahlström et al. 2013; Wehn et al. 2018).

Moreover, conservation of plant diversity along with their habitats in Ladakh is critical for multiple reasons such as climate change (Pant et al. 2018; Kattel 2022), longer growing seasons (Dolezal et al. 2016) and rapid glacier retreat (Schmidt and Nüsser 2017). Likewise, there has been a shift in the employment avenues among the local people of Ladakh, with younger generations opting for modern education and moving out jeopardizing traditional medicine system (Angmo et al. 2017).

Table 3 Important fodder plants occurring in the hay meadows

S. no.	Scientific name	Common name	Family	Indigenous uses
1	*Medicago falcata* L.	Alfa-alfa, Yellow clover	Fabaceae	Important fodder species, preferred by livestock, also stocked for winters
2	*Astragalus thomsonii* Podlech	Thomas milkvetch	Fabaceae	Fodder species
3	*Trifolium pratense* L.	Red clover	Fabaceae	Important fodder species, stocked for winter in dried form
4	*Oxytropis hypoglottoides* (Baker) Ali	Hypoglottis locoweed	Fabaceae	Palatable during flowering, used as forage
5	*Taraxacum officinale* F.H.Wigg.	Common dandelion	Fabaceae	Leaves and young shoots of plant are utilized in preparation of local food
6	*Lotus corniculatus* L.	Common bird's foot trefoil	Asteraceae	Important fodder species, stocked for winter in dried form
7	*Geranium himalayense* Klotzsch	Himalayan Crane's-bill	Geraniaceae	Fodder species; Flowers have aesthetic values
8	*Agrostis gigantea* Roth	Redtop	Poaceae	A good fodder and forage species, important for soil erosion control
9	*Poa alpina* L.	Alpine meadow grass	Poaceae	An excellent forage and fodder species
10	*Phragmites australis* (Cav.) Trin. ex Steud.	Common reed	Poaceae	Invasive species, but a strong soil binder and fodder alternative
12	*Prangos pabularia*	Prangos	Apiaceae	Important fodder plant

6 Conclusion

Keeping in view the aforementioned challenges, it is evident that the hay meadows are critical plant habitats and their age-old management practices are of prime significance for their conservation in Ladakh. The grassland management plans should assess their current status across the region. Their conservation and management should be prioritized as these meadows not only provide surplus fodder for the livestock in wintertime but also act as reservoirs of some of the highly threatened plant species. Given the variety of species recorded from the studied hay meadows, it is highly possible that these areas attract good numbers of birds and insects and other faunal species. The significance of such agro-ecological strategies, which emerged as a result of the challenges faced by the people of mountains, should be well studied and preserved. However, a larger study would be required to draw any conclusions on the comparative status of plant diversity and environmental factors across different valleys. Nonetheless, hay meadows are species-rich habitats that should be prioritized for conservation and management.

Acknowledgements The authors would like to acknowledge the Director and Dean Wildlife Institute of India (WII) and the Chief Wildlife Warden, Union Territory of Ladakh for facilitating the expeditions and granting permission to work in Ladakh. We also would like to thank the National Mission on Himalayan Studies (NMHS), Ministry of Environment, Forest & Climate Change (Govt. of India) for providing financial support for the project.

References

Angmo K, Adhikari BS, Rawat GS (2017) Sowa-Rigpa: a healthcare practice in Trans-Himalayan region of Ladakh, India. SDRP J Plant Sci 2(1):45–52

Austrheim G (2002) Plant diversity patterns in semi-natural grasslands along an elevational gradient in southern Norway. Plant Ecol 161(2):193–205. https://doi.org/10.1023/A:1020315718720

Behera MD, Matin S, Roy PS (2014) Biodiversity of Kargil cold desert in the Ladakh Himalaya. Integr Obs Assess 253–274. https://doi.org/10.1007/978-4-431-54783-9_13

Bele B, Svalheim E, Grenne SN, Norderhaug A (2024) Semi-natural hay meadows and traditional ecological knowledge (TEK) in Norway—what can we learn from written sources? J Nat Conserv 77(1431):126510. https://doi.org/10.1016/j.jnc.2023.126510

Chowdhery HJ, Wadhwa BM (1984) Flora of Himachal Pradesh—analysis, vol 1–3. Botanical Survey of India, Howrah

Dahlström A, Iuga AM, Lennartsson T (2013) Managing biodiversity rich hay meadows in the EU: a comparison of Swedish and Romanian grasslands. Environ Conserv 40(2):194–205. https://doi.org/10.1017/S0376892912000458

Dhar and Kachroo, 1983 Dhar U, Kachroo P (1983) Alpine flora of Kashmir Himalaya. Scientific Publishers, Jodhpur, 132 pp

Dolezal J, Dvorský M, Kopecky M, Liancourt P, Hiiesalu I, Macek M et al (2016) Vegetation dynamics at the upper elevational limit of vascular plants in Himalaya. Sci Rep 6(1):srep24881

Dvorský M, Klimeš L, Doležal J, Wild J, Dickoré WB (2018) A field guide to the flora of Ladakh. Academia, Praha, 148 pp

European Red List of Habitats—Grasslands Habitat Group (2016) Mountain hay meadow. https://forum.eionet.europa.eu/european-red-list-habitats/library/terrestrial-habitats/e.-grasslands/e2.3-mountain-hay-meadow/download/en/1/E2.3%20Mountain%20hay%20meadow.pdf

Ghimire SK, Mckey D, Aumeeruddy-Thomas Y (2006) Himalayan medicinal plant diversity in an ecologically complex high altitude anthropogenic landscape, Dolpo, Nepal. Environ Conserv 33(2):128–140. https://doi.org/10.1017/S0376892906002943

Hartmann H (2009) A summarizing report on the phytosociological and floristical explorations (1976–1997) in Ladakh (India). Switzerland

Homburger H, Hofer G (2012) Diversity change of mountain hay meadows in the Swiss Alps. Basic Appl Ecol 13(2):132–138. https://doi.org/10.1016/j.baae.2011.12.004

Ivaşcu C, Öllerer K, Rákosy L (2016) The traditional perceptions of hay and hay-meadow management in a historical village from Maramureş County, Romania. Martor 21:39–51

Janišová M, Bojko I, Ivaşcu CM, Iuga A, Biro A, Magnes M (2023) Grazing hay meadows: history, distribution, and ecological context. Appl Veg Sci e12723. https://doi.org/10.1111/avsc.12723

Kachroo P, Sapru BL, Dhar U (1977) Flora of Ladakh: an ecological and taxonomical appraisal. Bishen Singh Mahendra Pal Singh, Dehradun, 172pp

Kattel GR (2022) Climate warming in the Himalayas threatens biodiversity, ecosystem functioning and ecosystem services in the 21st century: is there a better solution? Biodivers Conserv 31(8–9):2017–2044

Khan AG (2003) NASSD background paper: rangelands and livestock. IUCN Pakistan, Northern Areas Progamme, Gilgit. https://www.iucn.org/sites/dev/files/import/downloads/bp_liv_range.pdf

Klimešová J, Doležal J, Dvorský M, de Bello F, Klimeš L (2011) Clonal growth forms in Eastern Ladakh, Western Himalayas: classification and habitat preferences. Folia Geobot 46(2–3):191–217. https://doi.org/10.1007/s12224-010-9076-3

Kumar A, Adhikari BS, Mishra AP (2023) Status of threatened medicinal and Biosphere Reserve, Uttarakhand. A report (TR No./2022/35) submitted to International Union for Conservation of Nature—India, p 103

Marini L, Scotton M, Klimek S, Isselstein J, Pecile A (2007) Effects of local factors on plant species richness and composition of Alpine meadows. Agr Ecosyst Environ 119(3–4):281–288. https://doi.org/10.1016/j.agee.2006.07.015

Miller DJ (1990) Grasslands of the Tibetan Plateau. Rangelands 12(3):159–163

Mishra AP, Kumar A, Yadav SN (2023) Ecology and conservation of threatened medicinal plants in the Trans-Himalayan region of Nanda Devi Biosphere Reserve, Western Himalaya. Trees For People 14:100451. https://doi.org/10.1016/j.tfp.2023.100451

Mucina L, Bültmann H, Dierßen K, Theurillat JP, Raus T, Čarni A et al (2016) Vegetation of Europe: hierarchical floristic classification system of vascular plant, bryophyte, lichen, and algal communities. Appl Veg Sci 19(Suppl. 1):3–264

Murti SK (2001) Flora of cold deserts of Western Himalaya, India, vol 1. Botanical Survey of India, Kolkata, p 452

Myklestad Å, Sætersdal M (2004) The importance of traditional meadow management techniques for conservation of vascular plant species richness in Norway. Biol Conserv 118(2):133–139. https://doi.org/10.1016/j.biocon.2003.07.016

Namgail T, Rawat GS, Mishra C, van Wieren SE, Prins HHT (2012) Biomass and diversity of dry alpine plant communities along altitudinal gradients in the Himalayas. J Plant Res 125(1):93–101. https://doi.org/10.1007/s10265-011-0430-1

Ning W, Jian L, Zhaoli Y (2002) Grazing intensity on the plant diversity of alpine meadow in the eastern Tibetan plateau. Rangifer 24(4):9

Pant GB, Kumar PP, Revadekar JV, Singh N (2018) Climate change in the Himalayas. Springer, Cham, Switzerland

Polunin O, Stainton A (1984) Flowers of the Himalaya. Oxford University Press, Delhi

Rawat GS (2007) Alpine vegetation of the Western Himalaya: species diversity, community structure, dynamics and aspects of conservation. D.Sc. thesis, Kumaun University, Nainital, 239 pp

Rawat GS, Adhikari BS (2005) Floristics and distribution of plant communities across moisture and topographic gradients in Tso Kar basin, Changthang plateau, eastern Ladakh. Arct Antarct Alp Res 37(4):539–544. https://doi.org/10.1657/1523-0430(2005)037[0539:FADOPC]2.0.CO;2

Rodríguez-Rojo MP, Jiménez-Alfaro B, Jandt U, Bruelheide H, Rodwell JS, Schaminée JH et al (2017) Diversity of lowland hay meadows and pastures in Western and Central Europe. Appl Veg Sci 20(4):702–719

Schmidt S, Nüsser M (2017) Changes of high altitude glaciers in the Trans-Himalaya of Ladakh over the past five decades (1969–2016). Geosciences 7(2):27

Schwab A, Dubois D, Fried PM, Edwards PJ (2002) Estimating the biodiversity of hay meadows in north-eastern Switzerland on the basis of vegetation structure. Agr Ecosyst Environ 93(1–3):197–209. https://doi.org/10.1016/S0167-8809(01)00347-4

Sekar KC, Thapliyal N, Pandey A, Joshi B, Mukherjee S, Bhojak P, Bisht M, Bhatt D, Singh S, Bahukhandi A (2023) Plant species diversity and density patterns along altitude gradient covering high-altitude alpine regions of west Himalaya, India. Geol Ecol Landsc 1–15. https://doi.org/10.1080/24749508.2022.2163606

Shukla AN, Srivastava SK (2020) Flora of Ladakh: an annotated inventory of flowering plants. In: Dar GH, Khuroo AA (eds) Biodiversity of the Himalaya: Jammu and Kashmir state, vol 18. Springer, Singapore, pp 673–730

Singh KN, Lal B, Singh RD, Todaria NP, Ahuja PS (2007) Species richness, distribution pattern and conservation status of higher plants in the Spiti cold desert of trans-Himalaya, India. Int J Biodivers Sci Manag 3(4):223–233. https://doi.org/10.1080/17451590709618176

Srivastava SK, Shukla AN (2015) Flora of cold desert Western Himalaya, vol 2 Dicotyledons. Botanical Survey of India, Kolkata

Stewart RR (1916a) Flora of Ladak, Western Tibet. I. Discussion of the flora. Bull Torrey Bot Club 43(11):571–590

Stewart RR (1916b) Flora of Ladak, Western Tibet. II. List of Ladak plants. Bull Torrey Bot Club 43(12):625–650

Tiwari J, Pareek K, Raghuvanshi MS, Kumar P, Roy MM (2016) Fodder production system-a major challenge in cold arid region of Ladakh, India. MOJ Ecol Environ Sci 1(1):22–28. https://doi.org/10.15406/mojes.2016.01.00005

Warikoo K (2020) Ladakh: India's Gateway to Central Asia. Strat Anal 44(3):177–192

Wehn S, Burton R, Riley M, Johansen L, Anders K, Rønningen K (2018) Land use policy adaptive biodiversity management of semi-natural hay meadows: the case of West-Norway. Land Use Policy 72:259–269. https://doi.org/10.1016/j.landusepol.2017.12.063

Vulnerability Assessment of Alpine Medicinal and Aromatic Plants in Paddar Valley of North Western Himalaya

Anil Thakar, Ajaz Ansari, and Deeksha Dave

Abstract The Himalayan region harbour a unique habitat for a number of high-value medicinal plants. Owing to inaccessibility of medical facility, inhabitants of this region are highly dependent on medicinal plants. In the current study, a total of 12 highly utilized and commercially exploited threatened Medicinal and Aromatic Plants species (MAPs) species were prioritized for ecological status and vulnerability assessment in Paddar valley out of the 72 alpine MAPs found in the region. To gather the information, a semi-structured questionnaire survey was conducted in 40 representative sites across the alpine landscape of Paddar valley based on the plant parts used, occurrence locality, area visited by pastoral communities and socio-economic aspects. Field surveys were carried out using a Rapid Mapping Exercise in alpine meadows which involved dividing the landscape in heterogenous habitats and laying transects of 500 m, and within each transect, establishing four plots of 1 m × 1 m at 50 m intervals resulting in a total of 44 plots in each transect. The transects was laid along an elevational gradient to assess the species population. To elucidate the intensity of threats, a Rapid Threat Assessment approach was employed using 8 vulnerability criteria such as plant parts used, life forms, habitat preferences, endemism, population size, patterns of use, trade and overall threat status. The results of the threat assessment indicated that *Picrorhiza kurroa* Royle ex Benth, *Dactylorhiza hatagirea* (D.Don) Soó, *Aconitum heterophyllum* Wall. Ex Royle and *Fritillaria cirrhosa* D.Don were among the most exploited species. Furthermore, it was observed that roots and rhizomes were the most commonly utilized parts of the MAPs, accounting for 80% of usage, followed by inflorescence and fruits at 20%. Six species were categorized as endangered (EN), five as critically endangered (CR) one as vulnerable (VU) as per CAMP and IUCN status. This study provides valuable insights into the status of selected Medicinal and Aromatic Plants species found in the alpine region of the Paddar Valley, highlighting the species that are currently under

A. Thakar (✉) · D. Dave
School of Interdisciplinary and Transdisciplinary Studies, Indira Gandhi National Open University, New Delhi, India
e-mail: anilenv0@gmail.com

A. Ansari
Department of Environmental Sciences, University of Jammu, Jammu, Jammu and Kashmir, India

A. P. Mishra et al. (eds.), *Threatened Medicinal Plants in the Indian Himalayan Region*, World Sustainability Series, https://doi.org/10.1007/978-3-031-73687-2_2

significant threat. The findings underscore the importance of conservation efforts to safeguard these valuable plant resources and the ecosystems they inhabit.

Keywords Alpine meadows · Medicinal plants · Rapid mapping exercise · Threat assessment · Vulnerability

1 Introduction

The high-altitude Himalayan regions showcase remarkable diversity concerning topography, flora, fauna, and culture (Chandra et al. 2021; Mishra et al. 2023a). These areas endure prolonged and severe winters, lasting up to 6–8 months, with precipitation contributing to glacier formation, which serves as a critical water source for major rivers throughout the year. These challenging environmental conditions have led to the evolution of highly adapted and resilient flora, fauna, and human communities in these areas (Gulzar et al. 2022). The local populations heavily rely on natural resources found in high altitudes for their daily sustenance and fuel requirements (Rather et al. 2022; Rana and Rawat 2017). Due to the varied habitats found in mountain ecosystems, they harbour numerous rare, and significant flora, including various medicinal and aromatic plants (MAPs). Currently, more than 90% of plant species utilized in the pharmaceutical industry are sourced from wild habitats, with a significant proportion originating from the subalpine and alpine regions of the Himalayas (Hamid et al. 2020; Kumar et al. 2023). In India, the pharmaceutical sector relies on around 280 medicinal and aromatic plant species (MAPs), out of which 175 are found in the Indian Himalayan region (Mathela et al. 2021; Rai et al. 2017). In the past, there was limited knowledge and documentation regarding the traditional Tibetan Medicinal System or Amchi system of medicine i.e., Sowa-Rigpa (Mathela et al. 2021; Thakur et al. 2023). However, due to globalization and improved road accessibility, this situation has now changed. Some areas are no longer isolated from global societies, leading to significant cultural and social transformations but the remote areas at high altitudes often lack proper access to medical facilities and livelihood opportunities, compelling the local inhabitants to depend on medicinal plants for healthcare (Körner 2016; Bargali et al. 2023). However, the escalating demand for trade, unscientific harvesting practices, and grazing pressure on natural populations of MAPs have led to a substantial increase in their exploitation (Shaheen et al. 2015; Chandrasekar et al. 2023; Bisht et al. 2022). The habitats of these plants frequently coincide with heavily grazed alpine meadows, causing limitations on their population size and distribution due to the impact of grazing animals, resulting in the depletion of medicinal herbs (Ganie et al. 2019; Mishra et al. 2023b). Moreover, when collected from the wild, approximately 70% of the harvesting practices involve destructive methods (Malik et al. 2015).

Over the past two decades, the continuous exploitation of MAPs from the wild and the loss of their habitats have resulted in the decline of many valuable species, leading to a gradual reduction in their regeneration potential and overall diversity in the region. Such depletion and loss of medicinal plant diversity directly impact the health and livelihood of the local communities who depend on them (Rawal and Dhar 1997; Rawat et al. 2021). Due to the challenging climatic conditions, rugged terrain and remote location of Paddar Valley, there has been limited botanical research conducted in this region. While some recent ethnoveterinary studies have been carried out by Singh et al. (2022) but there remains a significant gap in understanding of the population and ecological status of alpine medicinal plants. The current study aimed to determine the vulnerability status and prioritization of high-value MAPs found in the alpine region. This prioritization was established by considering factors such as local usage, trade demand, data from secondary databases, community perceptions, and rarity of these MAPs. Additionally, this research aimed to fill the knowledge gap by assessing the ecological characteristics and the level of threat faced by these MAPs in the Paddar Valley of District Kishtwar, J&K.

2 Material and Method

2.1 Study Area

The alpine landscape of Paddar Valley in J&K also Known as Sapphire valley is a rich treasure of medicinally important plants (Fig. 1). The total geographic area of Paddar valley is 3300 km^2 where the subalpine-alpine zone ranges from 2600–4500 m followed by nival zone (>4500 m). A total of 40 alpine meadows sites has been selected for sampling based on the semi-structured questionnaire of locals inhabiting in the watershed of sub valleys such as Panji dhar, Bheda dhar, Japan dhar, Gandhari, Ungayee, Kabban, Bujaz-Hagshu meadows, Dharlang, Bharnaz and Buzun valley. The valley is drained by Chanderbhaga river flowing from Pangi valley and its tributaries like Bhot nallah, Kabban-Sansari nallah and many streams. This remote region is nestled amidst rugged mountains and steep cliffs, experiences extreme cold, particularly during the winter months, with heavy snowfall and sub-zero temperatures (min. $-30°$ in winters and $-1°$ in summers). The area remains covered with snow from November to mid-June and the short growing season with significant day-night temperature fluctuations further contribute to the harsh environmental conditions for the flora found in the region. The landscape in the alpine area is characterized by rocky terrain with sparse vegetation and glacial features which constitutes of diverse physiognomic units that contribute to the richness of MAPs. These units include stunted forests, alpine scrub, alpine meadows and pioneer communities found on scree slopes and moraines. Considering factors such as altitude, aspect, moisture availability, and the duration of the growing season, six distinct types of alpine meadows such as tall forb meadows, mixed herbaceous meadows or short forb meadows, matted shrub or shrubbery meadows, *Danthonia* grasslands, *Kobresia* sedge meadows, and cushionoid vegetation have been identified in the Western Himalaya (Rawat 2005).

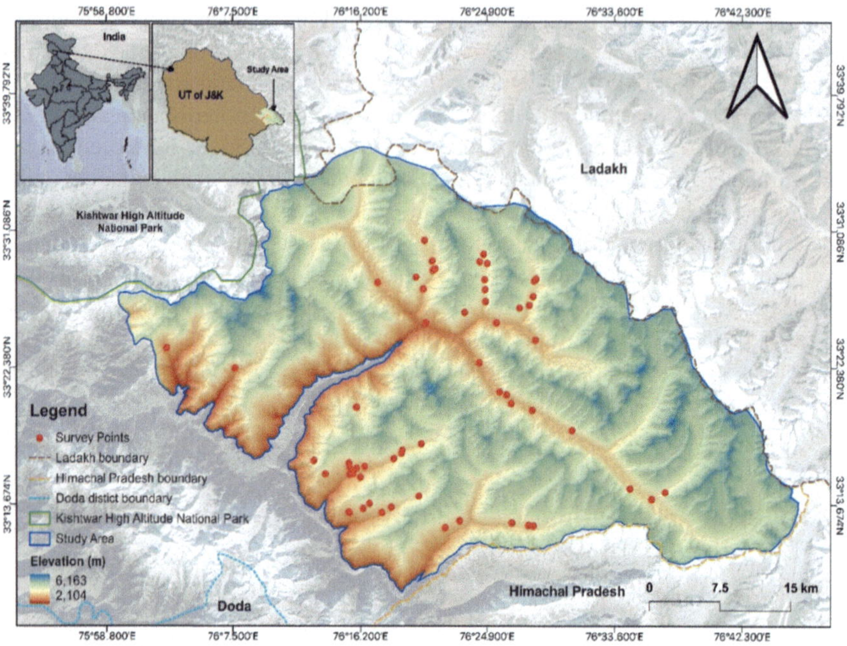

Fig. 1 Map showing sampling sites in the study area

The lives of the local inhabitants residing in high altitude villages are closely intertwined with the alpine meadows, which is evident from numerous folklores and traditional tales (Tali et al. 2015). These meadows hold significant religious importance, as many religious sites are situated in and around them, attracting regular visits from the local population as well as from other areas (Kala 2000, 2006). A substantial population of migratory and indigenous pastoral communities utilizes alpine meadows for seasonal grazing of their livestock. The augmentation in livestock numbers and excessive utilization of specific pastures has resulted in the deterioration of alpine ecosystems.

2.2 *Questionnaire Survey and Selection of Medicinal Plant Species*

A semi-structured questionnaire survey was employed to evaluate the ecological status and prioritization of MAPs taking into consideration of details about the plant parts used, occurrence locations, and socioeconomic significance (Ved et al. 1998). Local residents were requested to enlist the MAPs that were highly utilized and had greater demand both within the village and in the market. MAPs listed by over

80% of the respondents were considered as high-value species. A total of 75 knowledgeable individuals (comprising 60% male and 40% female, aged between 29 and 64 years) directly or indirectly engaged in the collection, utilization, and trade of MAPs were interviewed across the major sub valleys. The interviewees encompassed mainly traditional healers known as *Amchis, Vaidyas* (Ayurvedic practitioners), and villagers representing various communities residing in the high-altitude region of Paddar Valley possessing traditional herbal medicine expertise. Pertinent information concerning trade practices, government policies, revenue generation, and species prioritized for collection were gathered from local organizations, namely Buddhist Association Paddar, and the Jammu and Kashmir Forest Department.

Through a comprehensive analysis of questionnaire surveys and literature reviews, considering factors such as occurrence patterns, rarity, and conservation status as documented in Conservation Assessment and Management Plans (CAMP) and International Union for the Conservation of Nature (IUCN) reports, we prioritized and selected a total of 12 species in this study. Among these, six were categorized as "endangered (EN)", five as "critically endangered (CR)" one as "vulnerable (VU)".

2.3 Population and Vulnerability Assessment of MAPs

The population of MAPs was estimated using a Rapid Mapping Exercise (RME) technique (Rawat et al. 2004). To assess the major habitat types for the species a 500-m long transect was established with 10 plots (each 5-m circle) at 50-m intervals in 40 alpine meadows of Paddar Valley. Within each 5 m circular plot, four quadrats measuring 1 × 1 m were placed in the north, east, west, and south directions to evaluate the MAPs population. In each 1 × 1 m quadrat, all the MAPs present were counted and approximately 30–40 plots were laid in each transect, with the actual number depending on the accessibility of the area being studied. The purpose of this method was to efficiently and systematically estimate the abundance of MAPs in their respective habitats.

Whereas for the vulnerability assessment the threat assessment of MAP's was conducted using a rapid threat assessment (RTA) approach, which is a modified method of rapid vulnerability assessment (Cunningham 1996; Shreshtha and Shrestha 2012). The assessment involved eight vulnerability criteria for each species, including plant parts used, life forms, habitat, endemism, population size, use pattern, trade, and threat status (Table 1). Local users were interviewed to gather information on the plant parts used (criteria 1), and this data was validated using secondary literature. Species that utilized roots and rhizomes were considered more vulnerable since the entire plant needed to be destroyed to obtain these parts. The life forms of various MAPs were assessed based on existing literature (criterion 2). Long-lived perennial species were considered more vulnerable as they heavily rely on their underground parts for survival, and the destruction of these parts could significantly impact their populations. Habitat assessments (criterion 3) were conducted through literature reviews and field surveys. Species occurring in fragile habitats with minimal human interference, such as gravelly and rocky slopes, were deemed more vulnerable to natural and anthropogenic disturbances (Pandey et al. 2018; Singh and Rawat 2011). Conversely, species growing in areas with substantial human interference, like grassy slopes, shrubberies, and meadows, were considered less vulnerable due to their better adaptation. Information on nativity and endemism (criterion 4) was obtained from regional flora references (Dhar and Kachroo 1983). Population trend data (criterion 5) was determined by the species' availability across meadows. High frequency values (100%) specify widespread occurrence, while lower percentages indicated the species occurred in fewer places. Density values were also considered to assess distribution patterns with sparse and thin distribution less than 1 and greater than 1, respectively.

Extraction trends of medicinal plants (criterion 6) were evaluated by interviewing local people and healers. The current threat status (criterion 7) was based on assessments by various organizations like IUCN, CAMP, and the Forest Department of J&K. Values for use patterns (criterion 8) were assigned based on information from local inhabitants, medicinal plant harvesters, and available literature. By aggregating scores from related categories for each species, a final threat score was obtained. The species were then categorized into four threat levels: Very high, high, medium, and lower threat based on their respective final scores (≥25, 20–24, 15–19, and <15).

3 Results

In the alpine region of Paddar valley, we identified 12 highly utilized and commercially valuable medicinal and aromatic plant species (Table 2). The species were prioritized based on their threat evaluation as per local demand, trade value and plant parts used. All the species were herbs and roots were the mostly used parts which are being extracted uncontrollably. All species are categorised under threatened category (Table 3) as per CAMP (2003).

3.1 Population Status and Habitat Preference of MAPs

Fritillaria cirrhosa D.Don and *Picrorhiza kurroa* Royle ex Benth. with density 1.1 and 0.5 individuals/m^2 respectively were highly extracted species for their use and high demand signifying a high risk of extinction in its natural habitat. *Dactylorhiza hatagirea* (D.Don) Soó exhibited the highest density at 1.9 individuals/m^2 followed by *Sinopodophyllum hexandrum* (Royle) T.S.Ying and *Jurinea macrocephala* Benth. at 1.8 and 1.5 individuals/m^2, respectively. On the other hand, *Saussurea atkinsonii* C.B.Clarke and *Picrorhiza kurroa* Royle ex Benth. had the lowest densities with only 0.3 and 0.5 individuals/m^2 respectively. Mixed herbaceous and grassy slopes are the most preferred habitat of these plant species due to the accessibility they are easily extracted by the people (Rawat and Negi 2021). Vulnerability assessment

Table 1 Parameters used for vulnerability criteria, categories, status and scores assigned for plant species identified by the participants in study area

S. no.	Parameter	Categories	Status	
			Ranking score	Categorical percentage (%)
1	Plant parts used	Whole plant/rhizome/root	4	27
		Bark, stem	3	8
		Flower, inflorescence fruits, seeds	2	22
		Leaves	1	43
2	Life forms	Long-lived perennial	4	52
		Short-lived perennial	3	13
		Biennial	2	31
		Annual	1	4
3	Habitat	Stony/gravel/soil/rocky slopes	4	24
		Moist, marshy, glacial moraine land	3	26
		Along the stream, agriculture land, shrubberies	2	11
		Grassland/meadows/pastures	1	39
4	Nativity and endemism	Native/endemic to India	4	1
		Native/endemic to Himalaya	3	58
		Native/endemic to Himalaya and nearby countries	2	41
		Cosmopolitan	1	0
5	Population size	Few places sparse	4	21
		Few places thin	3	22
		Many places thick or thin	2	23
		Everywhere thick or thin	1	34
6	Extraction status	Commercial + self-use	4	36
		Commercial	3	26
		Self-use	2	38
		No use	1	0
7	Threat status	Status in three categories or more	4	41
		Status in two categories	3	50
		Status in one category	2	9
		Not assigned	1	0
8	Use patterns	Local people + local exchange + trade	4	32
		Local people + trade	3	26
		Local people + local exchange	2	22
		Local people	1	20

Table 2 Vulnerability scores of high-altitude medicinal plants from alpine region of Paddar Valley

S. no.	Species name	P1	P2	P3	P4	P5	P6	P7	P8	Total score	Category
1	*Aconitum heterophyllum* Wall. ex Royle	4	1	3	3	4	1	3	3	22	III
2	*Dactylorhiza hatagirea* (D.Don) Soó	4	3	3	3	2	4	2	3	24	III
3	*Fritillaria cirrhosa* D.Don	4	3	1	3	4	4	2	4	25	IV
4	*Gentiana kurroo* Royle	4	3	2	4	4	2	2	1	22	III
5	*Jurinea macrocephala* Benth.	4	3	4	2	1	4	1	3	22	III
6	*Lagotis cashmeriana* (Royle) Rupr.	4	3	1	3	2	2	2	2	19	II
7	*Picrorhiza kurroa* Royle ex Benth.	4	3	1	3	3	4	3	4	25	IV
8	*Sinopodophyllum hexandrum* Royle	4	1	2	3	1	2	3	1	17	II
9	*Rheum australe* D.Don	4	3	4	3	4	4	2	3	27	IV
10	*Saussurea atkinsonii* C.B.Clarke	4	2	1	2	4	2	3	2	20	III
11	*Dolomiaea costus* (Falc.) Kasana & A.K. Pandey	4	3	2	2	3	4	4	4	26	IV
12	*Valeriana jatamansi* Jones ex Roxb.	4	3	2	2	2	3	2	2	20	III

P1 = Plant parts used, P2 = Life forms, P3 = Habitat, P4 = Nativity and Endemism/Distribution, P5 = Population, P6 = Extraction trend, P7 = Threat and P8 = Use pattern

of these shows that *Dactylorhiza hatagirea, Fritillaria cirrhosa, Picrorhiza kurroa, Rheum australe* D.Don and *Dolomiaea costus* (Falc.) Kasana & A.K.Pandey were the most vulnerable species (Table 2). *D. hatageria* was found at the lower elevation in subalpine zone while *P. kurroa* being at the higher elevation along gradient of 2800–4100 m (Fig. 2). The findings underscore the importance of conservation efforts and management plans to safeguard these valuable plant species and their ecosystems in the face of growing exploitation and demand (CAMP 2003).

Table 3 Status of the threatened plant species found and identified by informants in study area

S. no.	Species name	Elevation at which species is found (m)	Parts extracted	Density (individuals/ m^2)	Status		
					CAMP	IUCN	CITES
1	*Aconitum heterophyllum* Wall. ex Royle	2940	Root	1.3	CR	EN	II
2	*Dactylorhiza hatagirea* (D.Don) Soó	3150	Rhizome	1.9	CR	EN	
3	*Fritillaria cirrhosa* D.Don	3400	Bulb	1.1	EN	VU	
4	*Gentiana kurroo* Royle	4000	Root, rhizome	0.9	CR	CR	
5	*Jurinea macrocephala* Benth.	2950	Roots, rhizome	1.5	EN	EN	II
6	*Lagotis cashmeriana* (Royle) Rupr.	3800	Root	1.0	VU	EN	
7	*Picrorhiza kurroa* Royle ex Benth.	4100	Rhizome	0.5	EN	EN	II
8	*Sinopodophyllum hexandrum* Royle	2800	Rhizome, roots	1.8	EN	EN	II
9	*Rheum australe* D.Don	3600	Root, leaves, bark	0.9	EN		II
10	*Saussurea atkinsonii* C.B.Clarke	3100	Root	0.3	EN		
11	*Dolomiaea costus* (Falc.) Kasana & A.K. Pandey	3000	Root	0.6	CR	CR	I
12	*Valeriana jatamansi* Jones ex Roxb.	3690	Roots	1.0	CR		II

3.2 Trade Value of MAPs

The collection and trade of MAPs have been longstanding practices for supplementing household income. In the remote, high-altitude regions of the Himalayas, where access to healthcare and medical facilities is limited, communities rely on wild MAPs to treat various illness. 75 key respondents were interviewed who were involved in MAPs collection in alpine areas to explore extraction trends in the region (criteria 6).

The majority of respondents indicated that they were extracting and using plants like *Aconitum heterophyllum, Dactylorhiza hatagirea, Fritillaria cirrhosa, Picrorhiza kurroa, Rheum australe* and *Dolomiaea costus* for both commercial purposes and personal use. These species were mainly used to treat abdominal

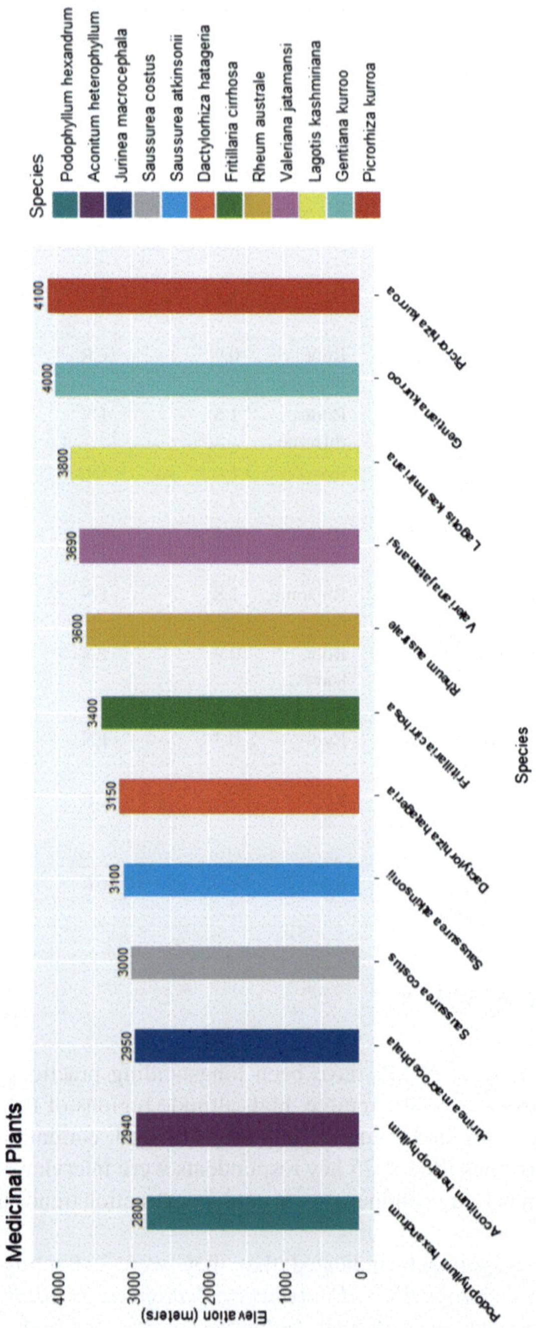

Fig. 2 Elevation gradient of MAPs found in the study area

pain, fever, cough, asthma and joint pain. On the other hand, species like *Fritillaria cirrhosa, Dolomiaea costus* and *Picrorhiza kurroa* were primarily extracted for economic gain through trade due to their high economic value. The trade is normally done at local level by the people and selling to nearby valley Sural bhatori of Pangi, HP. Among the 12 species studied, it was reported that each household collected approximately 8–15 kg of *P. kurroa*, 0.2–0.8 kg of *A. heterophyllum* and 10–20 kg of *F. cirrhosa* annually from the wild for local trade only. Other species are collected for self-use only and owing to their lower value in market they are extracted in lesser quantity. In India, the annual turnover of herbal medicine is estimated to be 177,000 metric tons, which involves the active trade of 960 plant species (Ved and Goraya 2008). This heightened demand exerts significant pressure on natural populations leading to a decline in their abundance and placing them at various levels of threat (Mathela et al. 2021; Uniyal et al. 2006).

3.3 Vulnerability Assessment of MAPs

Fritillaria cirrhosa, Picrorhiza kurroa, Rheum australe and *Dolomiaea costus* were the most vulnerable species (Table 2) on the basis of their vulnerability scores. The scores are allotted to each species on the basis of various (8) parameters/criteria. The various categories based on the vulnerability scores ranges from I to IV.

Category I: The most vulnerable category comprises species with a vulnerability score of ≥25. The plants in this category are characterized by low population density with restricted distribution, primarily found in a few meadows. Four species fall into this category: *Fritillaria cirrhosa* (1.1 Individuals/m^2), *Picrorhiza kurroa* (0.5 Individuals/m^2), *Rheum australe* (0.9 Individuals/m^2) and *Dolomiaea costus* (0.6 Individuals/m^2). These species are at a significant risk of extinction in their natural habitats due to uncontrolled extraction owing to their controlled occurrence. Hagshu, Bujaz, Tun, Chandi padhar, Panjidhar and Gudanag are suitable habitats for good population of these species.

Category II: These species are characterized by limited or sparse populations and currently there is relatively low demand of them. A total of six species had vulnerability score of 20–24 viz., *Aconitum heterophyllum* (1.3 Individuals/m^2), *Dactylorhiza hatagirea* (1.9 Individuals/m^2), *Gentiana kurroo* Royle (0.9 Individuals/m^2), *Jurinea macrocephala* (1.5 Individuals/m^2), *Saussurea atkinsonii* (0.3 Individuals/m^2) and *Valeriana jatamansi* Jones ex Roxb. (1.0 Individuals/m^2). The species are highly vulnerable but lower as compare to category 1 species. Kabban, Hagyote, Ligri, Bhedadhar, Hezul, Lossaini and Sumcham are good habitats for the existence of these species.

Category III and Category IV: The species falling in the category of III and IV have clumped and widely distributed population respectively. Two species viz., *Sinopodophyllum hexandrum* Royle (1.8 Individuals/m^2) and *Lagotis cashmeriana*

(Royle) Rupr. (1.0 Individuals/m^2) having vulnerability score between 15–19 falls under the category III. These species have gregarious population and are in medium anthropogenic pressure. None of the species falls under the category IV having score less than 15.

4 Discussion

In the alpine region of J&K threat assessment of 94 high valued medicinal plants was carried out by Ganie et al. (2019). Out of these 94 species, 12 high valued species of medicinal and aromatic plants were studies in alpine region of Paddar valley for their ecological status and vulnerability assessment. Among these, five species are categorised as Critically Endangered (CR), six species are in endangered category of CAMP and *Lagotis cashmeriana* (Royle) Rupr. falls under vulnerable category (Table 3) of IUCN and CITES. The vulnerability scores indicated that the utilization of plant parts (accounting for 21% of the vulnerability), the geographical distribution (14%), the habitat conditions (10%), and the pattern of extraction (13%) were the primary threat factors for alpine MAPs. In the western Himalaya parts of the plant used determines the maximum vulnerability of MAP's (Chandra et al. 2021; Kumar et al. 2021a, b). *Fritillaria cirrhosa, Picrorhiza kurroa, Rheum australe* and *Dolomiaea costus* were the most vulnerable species with vulnerability score ≥25 similarly to the vulnerability criteria in alpine areas of Uttarakhand state (Chandra et al. 2021). *F. cirrhosa* had the maximum demand by the locals owing to its high market price. A total of 12 threatened species were identified in the study area similarly 20 species were categorized as threatened among 50 species in alpine region of western Himalaya in Uttarakhand (Chandra et al. 2021). *L. kashmiriana* and *S. hexandrum* have high pressure and shows gregarious distribution (score between 15–19) in the Panjidhar, Hagshu, Gandhari, Chiring nallah. Whereas none of the species have low pressure and wide distribution (Score < 15) depicting the declining population of medicinal plants in north western Himalaya (Dhar et al. 2000; Tali et al. 2015). Through the vulnerability assessment we found that 39% of species were short-lived perennial, 58% are endemic to Himalaya, and 31% species having thick or thin population in the Paddar valley. There are about six major habitat types in alpine region where 24% MAPs were growing in gravel/soil or rocky slopes and 39% in Grassland/herbaceous meadows (Table 1). Excessive extraction of *R. australe, A. heterophyllum* and *F. cirrhosa* was made by locals as they are restricted to the Kabban, Dharlang, Barnaz, Bhazun meadows and are ultimately traded to the Himachal Pradesh through Killar-Pangi highway. Few adjoining meadows like Dachan, Kiar, Ananth, Kibber Nallah in Kishtwar High Altitude National Park are also known for their availability (Tali et al. 2015; Rather et al. 2022). Due to their high demand these species are being exploited continuously leading to their declining population in western Himalaya (Kumar et al. 2022; Sharma et al. 2021). *Picrorhiza kurroa* and *Fritillaria cirrhosa* were the most traded species and it is estimated that about 2000–2500 kg both are collectively traded off from the Kishtwar district which is lower than the tradable quantity in Pithoragarh district of Uttarakhand (Garbyal et al. 2005). Kumar et al. (2021a, b) executed the similar survey of MAPs in Lahaul and Spiti landscape of Himachal Pradesh and laid stress on the fragile habitat of these threatened species. Given the absence of a dedicated conservation policy for MAPs in the whole Indian Himalayan region, it becomes crucial to prioritize the protection and monitoring of these ecologically rich areas in North Western Himalaya also where significant populations of MAPs exists. Unfortunately, the increasing number of livestock and excessive use of certain pastures such as Warwan and Marwah have resulted in the degradation of the pristine alpine habitats (Ganie et al. 2019). Moreover, human activities in these areas, such as wood collection for fuel and NTFP extraction, recreational activities like camping, trekking and even cultivation on steep slopes, further exacerbates the situation (Chandra et al. 2021). The cumulative impact of these anthropogenic activities poses a severe threat to the fragile habitats of the alpine region and if left unchecked, the consequences may become irreversible. Thus, considering multiple threats to the flora of Himalaya hotspot we need to carefully address these challenges to safeguard the ecological integrity of these vital ecosystems (Mehta et al. 2021). These sites hold great importance in terms of botanical diversity, and safeguarding them is of paramount significance to preserve the unique and abundant plant species present in these regions.

5 Conclusion

The ecological assessment of high valued MAPs in the study area reveals a concerning scenario. This study has identified several MAP species facing critical threat, with factors such as plant part utilization, geographical distribution, habitat conditions and extraction patterns emerging as significant threats. Particularly alarming is the high demand for species like *Fritillaria cirrhosa* which has led to extensive extraction and placed immense pressure on their populations. Furthermore, anthropogenic activities, including wood collection, non-timber forest product extraction, recreational pursuits, and cultivation on steep slopes compound these threats jeopardizing the fragile alpine habitats. The lack of dedicated conservation policies in the Indian Himalayan region underscores the urgent need for prioritizing the protection and monitoring of these ecologically rich areas. The cumulative impact of these human activities poses a grave and potentially irreversible danger to these unique ecosystems. Therefore, this study underlines the paramount importance of implementing robust conservation measures to safeguard the diverse and abundant plant species present in these alpine regions with the ultimate goal of preserving their ecological integrity and biodiversity.

References

Bargali H, Kumar A, Singh P (2023) Plant studies in Uttarakhand, Western Himalaya—a comprehensive review. Trees For People 8:100203. https://doi.org/10.1016/j.tfp.2022.100203

Bisht A, Pandey A, Bhatt ID, Pandey V (2022) Propagation of *Mahonia jaunsarensis*: an endemic medicinal shrub of Western Himalaya, India. Plant Biosyst Int J Deal all Aspects Plant Biol 156(4):1050–1055. https://doi.org/10.1080/11263504.2022.2073397

Conservation Assessment and Management Prioritization (CAMP) (2003) For the medicinal plants of Jammu & Kashmir, Himachal Pradesh & Uttaranchal. In: Ved DK, Kinhal GA, Ravikumar K, Prabhakaran V, Ghate U, Shankar RV, Indresha JH (eds) FRLHT, Bangalore, India, p 206

Chandra N, Singh G, Lingwal S, Bisht MPS, Tewari LM, Joshi VC (2021) Ecological status of alpine medicinal and aromatic plants of Western Himalaya. J Herbs Spices Med Plants. https://doi.org/10.1080/10496475.2021.1963903

Chandrasekar K, Thapliyal N, Pandey A, Joshi B, Mukherjee S, Bhojak P, Bisht M, Bhatt D, Singh S, Bahukhandi A (2023) Plant species diversity and density patterns along altitude gradient covering high-altitude alpine regions of west Himalaya, India. Geol Ecol Landsc. https://doi.org/10.1080/24749508.2022.2163606

Cunningham AB (1996) People, park and plant use: recommendations for multiple use zones and development alternatives around Bwindi Impenetrable National Park, Uganda; People and Plants Working Paper 4. UNESCO, Paris

Dhar U, Kachroo P (1983) Alpine flora of Kashmir Himalaya. Scientific Publishers, Jodhpur, India

Dhar U, Rawal RS, Upreti J (2000) Setting priorities for conservation of medicinal plants—a case study in the Indian Himalaya. Biol Conserv 95:57–65. https://doi.org/10.1016/S0006-3207(00)00010-0

Ganie AH, Tali BA, Khuroo AA, Reshid ZA, Nawchoo IA (2019) Impact assessment of anthropogenic threats to high-valued medicinal plants of Kashmir Himalaya, India. J Nat Conserv 50:125715. https://doi.org/10.1016/j.jnc.2019.125715

Garbyal SS, Aggarwal KK, Babu CR (2005) Traditionally used medicinal plants in Dharchula Himalays of Pithoragarh District, Uttaranchal. Indian J Tradit Knowl 4:199–207

Gulzar A, Hamid M, Dar FA, Wani SA, Malik AH, Kamili AN, Singh CP, Khuroo AA (2022) Patterns of floristic and functional diversity in two treeline ecotone sites of Kashmir Himalaya. Environ Monit Assess 194:420

Hamid M, Khuroo AA, Malik AH (2020) Assessment of alpine summit flora in Kashmir Himalaya and its implications for long-term monitoring of climate change impacts. J Mt Sci 17:1974–1988. https://doi.org/10.1007/s11629-019-5924-7

Kala CP (2000) Status and conservation of rare and endangered medicinal plant in the Indian trans-Himalaya. Biol Conserv 93:371–379. https://doi.org/10.1016/S0006-3207(99)00128-7

Kala CP (2006) Medicinal plants of the high-altitude cold desert in India: diversity, distribution and traditional uses. Int J Biodivers Sci Manag 2(1):43–56

Körner C (2016) Plant adaptation to cold climates. F1000Research 5:2769. https://doi.org/10.12688/f1000research.9107.1

Kumar A, Sathyakumar S, Goraya GS, Gupta AK, Adhikari BS, Rawat GS (2021a) Assessment of medicinal and aromatic plant species on their collection, usage, demand, markets, price trends and life cycle in Lahaul and Pangi landscape, Himachal Pradesh. A report submitted to Himachal Pradesh Forest Department and United Nations Development Programme, p 141

Kumar R, Arya D, Chandra Sekar K, Bisht M (2021b) Threat assessment and prioritization of high-value medicinal plants in Pindari valley, Nanda Devi Biosphere Reserve, India. Curr World Environ 16:1–14

Kumar A, Kant R, Dogra KS, Chandra Sekar K, Kharkwal K, Dash SS (2022) Diversity of threatened plants and conservation efforts in Great Himalayan National Park, Himachal Pradesh, Western Himalaya. Indian For 147:1172–1175. https://doi.org/10.36808/if/2021/v147i12/166614

Kumar A, Adhikari BS, Mishra AP (2023) Status of threatened medicinal and Biosphere Reserve, Uttarakhand. A report (TR No./2022/35) submitted to International Union for Conservation of Nature—India, p 103

Malik ZA, Bhat JA, Ballabha R, Bussmann RW, Bhatt AB (2015) Ethnomedicinal plants traditionally used in health care practices by inhabitants of Western Himalaya. J Ethnopharmacol 172:133–144. https://doi.org/10.1016/j.jep.2015.06.002

Mathela M, Kumar A, Sharma M, Goraya GS (2021) Hue and cry for *Fritillaria cirrhosa* D.Don, a threatened medicinal plant in the Western Himalaya. Discov Sustain 2:38. https://doi.org/10.1007/s43621-021-00048-5

Mehta P, Bisht K, Chandra Sekar K (2021) Diversity of threatened medicinal plants of Indian Himalayan Region. Plant Biosyst Int J Deal Aspects Plant Biol 155:1121–1132

Mishra AP, Kumar A, Yadav SN (2023a) Ecology and conservation of threatened medicinal plants in the Trans-Himalayan region of Nanda Devi Biosphere Reserve, Western Himalaya. Trees For People 14:100451. https://doi.org/10.1016/j.tfp.2023.100451

Mishra AP, Chandra N, Mandy JJ, Dwivedi SK, Alruzuq A, Pande CB (2023b) Analytical Hierarchy Process (AHP) based on the spatial assessment of an endangered alpine medicinal herb Aconitum heterophyllum in the Western Himalayan Environment. In: Climate change impacts on natural resources, ecosystems and agricultural systems. Springer International Publishing, Cham, pp 579–594. https://doi.org/10.1007/978-3-031-19059-9_25

Pandey A, Sekar KC, Joshi B, Rawal RS (2018) Threat assessment of high-value medicinal plants of cold desert areas in Johar Valley, Kailash Sacred Landscape, India. Plant Biosyst Int J Deal All Aspects Plant Biol. https://doi.org/10.1080/11263504.2018.1448010

Rai ID, Singh G, Rawat GS (2017) Plants of Kedarnath Wildlife Sanctuary, Western Himalaya—a field guide. Bishen Singh Mahendra Pal Singh, Dehradun, p 393

Rana SK, Rawat GS (2017) Database of Himalayan plants based on published floras during a century. Data 2(36):1–9

Rather ZA, Ahmad R, Dar TH, Khuroo AA (2022) Ensemble modelling enables identification of suitable sites for habitat restoration of threatened biodiversity under climate change: a case study of Himalayan Trillium. Ecol Eng 176. https://doi.org/10.1016/j.ecoleng.2021.106534

Rawal RS, Dhar U (1997) Sensitivity of timberline flora in Kumaun Himalaya, India—conservation implications. Arct Alp Res 29:112–121

Rawat GS, Adhikari BS, Chandola SA (2004) Manual for rapid inventory and mapping of medicinal and aromatic plants in Uttarakhand. Medicinal and Aromatic Plant Board, Govt. of Uttarakhand

Rawat N, Mishra AP, Chandra N, Painuly V, Negi GS, Bisht MPS, Purohit S (2021) Maxent modeling for predicting the potential distribution of Valeriana jatamansi jones. In: Chakrata forest division of Garhwal Himalaya. Indian J Ecol 48(2):367–372

Rawat B, Negi AS (2021) Plant diversity patterns along environmental gradients in Nanda Devi Biosphere Reserve, West Himalaya. Trop Ecol 62:61–70. https://doi.org/10.1007/s42965-020-00122-5

Rawat GS (2005) Alpine meadows of Uttarakhand: ecology, land use and status of medicinal and aromatic plants. Bishen Singh Mahendra Pal Singh, Dehra Dun

Shaheen H, Mashwani ZUR, Dar ME (2015) Spatial patterns and diversity of alpine vegetation across Langer-Shandur Valley, Hindukush Himalayas. Curr Sci 108:1534–1539

Sharma R, Kaur S, Uniyal SK (2021) Population and vulnerability assessment of high value medicinal plants in the Alpine regions of western Himalaya. J Appl Res Med Aromat Plants 26:100353. https://doi.org/10.1016/j.jarmap.2021.100353

Shrestha N, Shrestha KK (2012) Vulnerability assessment of high-valued medicinal plants in Langtang National Park, Central Nepal. Biodiversity 13:24–36. https://doi.org/10.1080/14888386.2012.666715

Singh K, Kumar B, Kumar P, Lone JF, Sharma YP, Gairola S (2022) Documentation of ethnoveterinary knowledge: harnessing potential phytotherapy in high mountainous areas of Paddar, District Kishtwar (India). Ethnobot Res Appl 24:19. https://doi.org/10.32859/era.24.19.1-22

Singh G, Rawat GS (2011) Ethnomedicinal survey of Kedarnath wildlife sanctuary in Western Himalaya, India. Indian J Fundam Appl Life Sci 1:35–46

Tali BA, Ganie AH, Nawchoo IA, Wani AA, Reshi ZA (2015) Assessment of threat status of selected endemic medicinal plants using IUCN regional guidelines: a case study from Kashmir Himalaya. J Nat Conserv 23:80–89. https://doi.org/10.1016/j.jnc.2014.06.004

Thakur S, Rai ID, Singh B, Dutt HC, Musarella CM (2023) Predicting the suitable habitats of Elwendia persica in the Indian Himalayan Region (IHR). Plant Biosyst Int J Deal All Aspects Plant Biol 157(4):769–778. https://doi.org/10.1080/11263504.2023.2204090

Uniyal SK, Kumar A, Lal B, Singh RD (2006) Quantitative assessment and traditional uses of high value medicinal plants in Chhota Bhangal area of Himachal Pradesh, Western Himalaya. Curr Sci 91:1238–1241

Ved DK, Mudappa A, Shankar D (1998) Regulating export of endangered medicinal species need for scientific rigour. Curr Sci 75:341–344

Ved DK, Goraya GS (2008) Demand and supply of medicinal plants in India; BSMPS, D. Dun, and FRLHT, Bangalore, India

Current Trends in Commercial Extraction of Wild Medicinal Herbs from the Western Himalaya and Management Implications

G. S. Rawat, G. S. Goraya, V. Jishtu, and Amit Kumar

Abstract The chapter reviews the current trends in the demand and extraction of wild medicinal herbs in the Western Himalaya. Using the case studies from the state of Himachal Pradesh, it provides information on highly traded species, current extraction pressure driven by market dynamics, and impact on the ecology and threat status of such species. The study reveals a shift in demand, with certain species like *Trillium govanianum* and *Fritillaria cirrhosa* experiencing a sudden surge in trade, primarily driven by illegal exports to China. The attractive market value has led to overharvesting, endangering the wild populations. Hotspots of extraction and changing trends in the number of collectors are discussed, emphasizing the need for evolving better management strategies based on regional CAMP workshops.

Keywords Current extraction · Market dynamics · Threatened medicinal plant · Traded species · Threat status

1 Introduction

World over human societies are increasingly using medicinal and aromatic plants for health care and as nutraceuticals owing to preference for chemical free natural products as they are safer to use and are cheaper (Nishteswar 2014). A large number of indigenous ethnic communities in developing countries also earn a significant

G. S. Rawat (✉)
Uttarakhand State Council for Science and Technology, Vigyan Dham, Dehradun, Uttarakhand, India
e-mail: gsrawat59@gmail.com

G. S. Goraya
Govt. of Himachal Pradesh, Shimla, Himachal Pradesh, India

V. Jishtu
Himalayan Forest Research Institute, Panthaghati, Shimla, Himachal Pradesh, India

A. Kumar
Wildlife Institute of India, Chandrabani, Dehradun, Uttarakhand 248001, India

A. P. Mishra et al. (eds.), *Threatened Medicinal Plants in the Indian Himalayan Region*, World Sustainability Series, https://doi.org/10.1007/978-3-031-73687-2_3

proportion of their cash income through sale of wild medicinal plants as several species have very high demand from the herbal industries (Ved and Goraya 2007). Nearly 25% of the modern medicines are derived directly or indirectly from the plants and nearly 80% of all synthetic drugs are derived from medicinal plants (Ramakrishnappa 2002; Bauer and Brönstrup 2014). It has been established that many drugs that are used in modern medicine have come from folk-use and use of plants in indigenous cultures. However, availability of many medicinal plant species has declined rapidly during recent years due to over extraction (Ved and Goraya 2007; Kumar et al. 2023; Mishra et al. 2023). It is seen that nearly 90% of medicinal plant species used by the industries are collected from the wild and most of the wild harvest is unsustainable owing to over harvest of roots, bark, stem and whole plants. Premature and destructive harvest hampers natural regeneration causing rapid decline of wild populations (Dhar et al. 2000; Badola and Aitken 2003).

India is among the topmost countries in the trade of medicinal plants with a list of 1178 species forming source of 1622 botanical raw drugs (Goraya and Ved 2017). In India over 2,00,000 tons of raw material is used annually by over 1.5 million practitioners of the Indian systems of medicine with estimated 25,000 effective plant-based formulations, many of which are commercially manufactured in about 8600 drug manufacturing units.

The Himalayan region is home to many of the traded medicinal plant species; and in addition to the trade, local folk-medicine systems also pose severe threat to the wild population of medicinal plants (Badola and Aitken 2003; Tali et al. 2015). It is estimated that in the Indian Himalayan Region (IHR), 1748 plant species are of medicinal value (1685 angiosperms, 12 gymnosperms and 51 pteridophytes); of which 1020 are herbs, 335 shrubs and 330 are trees (Samant et al. 1998). The Western Himalaya is particularly rich in diversity of high altitude wild medicinal plants (Kumar et al. 2009; Kumar and Hamal 2011; Gupta et al. 2013; Kumar and Sharma 2013; Gairola et al. 2014).

Most of the high value medicinal and aromatic plants (MAPs) found in the IHR come from the sub-alpine and alpine areas. Traditionally the people in the higher altitudes of the Himalaya used the wild medicinal herbs for their own use or for a limited barter system in exchange of commodities from the lower areas. With the change in socio-economic conditions and demand for cash income coupled with the coming into high demand certain medicinal herbs by herbal industries, the rural people in the high altitudes have shifted their attention towards their commercial harvest. Declining population of MAPs in the wild due to overexploitation has raised concern during recent years. Unsustainable harvest in the form of uprooting the entire plant, extraction before the seeds are shed and extraction of immature parts not only causes decline of populations but also leads to local extinction of species already under stress or having small population size. In some cases, the population size of a species can inherently be low or it might have become low due to its overexploitation and habitat degradation. Sometimes low habitat diversity results in low population of species and at other times anthropogenic pressures in the form of grazing, trampling and extraction result in the decline of their population. Increased growth of unpalatable/weedy species also limits their growth and consequently results in their low population (Rawat 2007). Other harmful effects of over harvest are soil erosion and habitat degradation. There are cases where faulty or improper harvesting has either restricted the distribution of species or it has completely resulted in its depletion in a particular locality.

Several authors including Dhar et al. (2000), Kala (2005) and Jishtu et al. (2021) have underlined the need for threat assessment and conservation prioritization of commercially traded medicinal plants. Attempts have been made to assess the conservation status of medicinal plant species of western Himalayan region, with CAMP (Conservation Assessment and Management Prioritisation) Workshops held at Kullu in 1998, Shimla in 2003 and 2010, and Leh in 2021. Consolidated outcome of the three CAMP Workshops (Kullu and Shimla) organized to assess conservation status of medicinal plants of Himachal Pradesh alone, 66 of the 87 medicinal plant species taken up for assessment were categorized as Red-listed, 35 of these being 'Critically Endangered' and 'Endangered', with harvest for trade and habitat degradation being the major threats to their wild populations (Goraya et al. 2012).

Goraya and Ved (2017) have worked out a comprehensive inventory of medicinal plants in trade in India. While attempting to group species sourced from different regions in the country, they have flagged 60 traded medicinal plant species sourced from the IHR which need priority conservation action. Information about the specific location of wild harvest of medicinal plant species in trade is however scanty, and is limited to only a few localities in the Western Himalaya. Major hotspots of wild medicinal plants in the region are alpine areas of Pithoragarh, Chamoli, Uttarkashi, Kullu, Kangra, Lahaul-Spiti, Kishtwar and Anantnag districts. *Amchis*, the herbal healers of Ladakh, collect medicinal plants mostly from moist valleys of Zanskar and partly from parts of the Great Himalayan range in Himachal Pradesh and Kashmir. Among the highly traded MAPs from the region are Atis or Ativisha (*Aconitum heterophyllum*), Colchicum (*Colchicum luteum*), Maida (*Polygonatum verticillatum*), Mushkbala (*Valeriana jatamansi*), Jatamansi (*Nardostachys jatamansi*), Pushkarmool (*Inula racemosa*), Kutki (*Picrorhiza kurroa*), Dhup (*Dolomiaea macrocephala* Syn. *Jurinea dolomiaea*), Kuth (*Dolomiaea costus* Syn. *Saussurea costus*) and Salampanja (*Dactylorhiza hatagirea*).

A rapid survey of market and local villages in and around Great Himalayan National Park (GHNP) conducted by the authors revealed that nearly two decades ago the eco-development zone of this park generated revenue to the tune of Rs. 2.04 crore from wild medicinal plants, which were collected mainly from the alpine areas. In terms of quantity collected, lichens ranked highest i.e., up to 105 tons followed by *Dolomiaea macrocephala* (82 tons) and *Dioscorea deltoidea* (45 tons). Lichens, morel mushroom (*Morchela esculenta*) and *Viola* spp. have relatively low biomass per plant and hence, more individuals were collected for per unit weight of these species. The highest price of medicinal plants in the local market was for morel mushroom (Rs. 2500/- per kg), followed by *Dactylorhiza hatagirea* (Rs. 850/- per kg) and *Aconitum heterophyllum* (Rs. 800/- per kg). Even though roots of *Dolomiaea macrocephala* fetched relatively lower rates in the local market (Rs. 75/- per kg), the number of people engaged in its extraction was only second to the morel mushroom collectors. A similar study by Tandon (1997) brings out that about 70% of 1600 odd households in the GHNP conservation area were involved in the wild harvest of medicinal herbs, and that trade of thus harvested medicinal herbs contributed an average of Rs. 14,115/- to the rural household incomes of the area.

Extraction of MAPs in parts of GHNP Conservation Area was being carried out till its final notification on the basis of 'rights' given to the local people as per the Anderson's Settlement Report. Not surprisingly, most of the herb collectors of the area claimed to be the 'right holders', and these rights continue to be transmitted to the succeeding generations. The studies on the impact of herb collection on the habitat and on the ecological status of herbs showed that there was enormous stress on some habitats and pressure on some species. Interviews with the herb collectors also indicate a progressive decline in the quality and quantity of herbs. For example, Dhoop (*Dolomiaea macrocephala*) roots according to the local people used to be as thick as a person's forearm some 10 years ago and today it is difficult to get a root even as thick as a finger of a person. Under the original settlement plan by Anderson in 1886 the collection of medicinal plants was permitted only for 2 months in a year from 15th August. By the beginning of 1960s there was a significant expansion of commercial market for the MAPs giving local people a major new source of cash income. Steady increase in market forces led to uncontrolled extraction through the summer months (April to November) as against the original system of rights established in Anderson's Settlement which was restricted to a collecting season of only two months from 15 August to 15 October. The irresistible pull of the booming market prompted the collectors to expand herb harvesting to the entire season, from April to November, forming the major reason for the decline of herbs in this area.

Rana and Rawat (2019) estimated the density of a few highly traded MAPs in Kishtwar region of Jammu and Kashmir and found that there was a large variation in population sizes of high value medicinal plants, ranging from 6.5 $\pm$ 2.9 SD to 2221 $\pm$ 1055 SD per m^2. Further, these authors observed that all critically endangered species had very small population size with <50 individuals, among which *Lilium polyphyllum* and *Gentiana kurroo* were the rarest with a mean population size of 6.5 $\pm$ 2.9 SD and 6.7 $\pm$ 1.5 SD respectively. Population size of other Red-listed species ranged between 18.9 $\pm$ 13.5 SD for *Paris polyphylla* and 385.2 $\pm$ 196.1 SD for *Ephedra gerardiana*. These authors recommended that in order to ensure sustainable use of these valuable resources and maintain their viable populations in the long run, it would be necessary to follow an integrative approach based on *in-situ* and *ex-situ* conservation strategies.

In this paper we provide current status of commercial trade of selected high value species and its impact on their populations in the wild with special reference to the state of Himachal Pradesh.

2 Current Trends in Extraction of Wild Medicinal Herbs in Himachal Pradesh

Some of the wild harvested medicinal herbs in high trade from Himachal Pradesh include *Aconitum heterophyllum, Fritillaria cirrhosa, Picrorhiza kurroa, Sinopodophyllum hexandrum, Rheum moorcroftiana, Trillium govanianum* and *Valeriana jatamansi* (Goraya and Ved 2017; Kumar et al. 2021a), *Elwendia persica syn. Bunium persicum* (Sharma et al. 2022). Besides, Chharila (mainly the shield lichen, *Parmelia* sp.) is collected in large quantities from the high-altitude forests and sold to the traders at Mandi. Reportedly the unsorted lichens are further graded at larger collection centres at Amritsar or Kharibawli in , Delhi and eventually traded for dyes, as spice etc. Interactions with local traders reveal that wild harvest of medicinal plants has always been dependent upon the prevailing market dynamics. For example, the large-scale trade of *Dioscorea deltoidea* roots, so common upto late 1980s with an average annual trade volume of 3500 qtls., is presently but a trickle. Similar is the case of *Dolomiaea macrocephala*, the annual trade volume of which has come down from an average of 5000 qtls. per year till late 1990s to less than 100 qtls. per year at present.

According to a leading trader in medicinal herbs based at Mandi in Himachal Pradesh, presently the state is witnessing high demand for only two species viz., Nag-chattri (*Trillium govanianum*) and Van lahsun (*Fritillaria cirrhosa)* which are sold at Rs. 1200–1500/kg and Rs. 15,000–18,000/kg respectively at Mandi market. It also comes out that there is no demand for these herbs from the national pharmaceutical industries and all the raw material is traded illegally for export to China. The usual trade route for clandestine trade of these two species used to be via Pakistan. However, trade disruption between India and Pakistan has impacted trade via this route and now most of this export happens through Nepal, that too mainly on foot/mules. That the demand for Nag-chhatri and Van lahsun is a recent phenomenon becomes amply clear from the fact that wild harvest and trade of these two species did not figure in the surveys carried out during the mid-1990s by the present authors and Tandon (1997). Further, perusal of the CAMP Workshop reports on medicinal plants from Himachal Pradesh reveals that trade of these two species did not get red-flagged even in the CAMP Workshops held at Kullu in 1998, and at Shimla in 2003 (Ved et al. 2003) and 2010 (Goraya et al. 2012). Whereas *Fritillaria cirrhosa*, one of the 'ashtavarga' group of herbs of ayurvedic importance, did get assessed due to its traditional usage in the Indian Systems of Medicine, no abnormal trade of *Trillium govanianum* had come to notice even upto the last CAMP Workshop.

From the above, it comes out that extraction and trade of *Trillium govanianum* for its rhizomes, and *Fritillaria cirrhosa* for its bulbs, started somewhere towards the end of 2009 and witnessed a sudden spurt thereafter in Himachal Pradesh, Jammu & Kashmir and Uttarakhand. The pull to collect these two species from the wild became very high with the prices offered to the wild gatherers shooting up from Rs. 800 to 2600 per kg dried rhizomes of Nag-chhatri, and from Rs. 1200 to 6000 per kg dried bulbs of Van lahsun over the next 2–3 years. An idea about the level of destructive harvest could be had from the fact that it requires uprooting of an average of 2000 plants to make 1 kg of dried material in respect of each of these species (Goraya and Ved 2017). Recently, Jishtu et al. (2021) estimated the annual trade (200–500 MT) of *T. govanianum* from remote Dodra-Kwar with the price ranging from Rs. 2000–2500 per kg during 2014–15. As per this study, 28 temperate species were under trade and many species among them are categorized and banned due to their trade or high conservation concern at national and international level by various agencies such as IUCN, CITES, CAMP.

The attractive market value of these species led to their collection, even from protected areas such as the GHNP, and from regions that were off-limits for collection in a given year. This resulted in the confiscation of significant amounts of illegally gathered material. The temptation of substantial profits motivates individuals to persist in their relentless harvesting of these two herbs, leading to the near eradication of their wild populations. According to a local herb collector from Tirthan Valley, high price of Van lahsun and Nag-chhatri at Mandi market had resulted in heavy extraction of these species. According to him, early harvest i.e., soon after the snow-melt in alpine areas and competition among the collectors may have led to low availability of Van lahsun and Nag-chhatri in the area. As per his conservative guess, availability of Van lahsun and Nag-chhatri in the alpine areas had declined by 85 and 75%, respectively. An idea about the drastically shrinking availability of Van lahsun can be had from the reports by Kumar et al. (2021a, b) and Mathela et al. (2021), which brings out that the wild gatherers in Pangi area of Himachal Pradesh were recently offered a price as high as Rs. 18,000 per kg for its dried bulbs.

It is quite intriguing to see the current state of banaksha (*Viola* serpens) as a decade ago, banaksha had a rate of Rs. 700–800 per kg, which was still profitable to the collectors. However, today the price is in the range of Rs. 7500–8500 per kg, but the collectors make no more than Rs.400 per day for their meticulous collections. Therefore, today there is not much collection being made, while the demand for banaksha is very high.

It has been observed that the majority of the herb collection involved the use of entire plants, amounting to its destructive harvest, and people share that the availability of some of the species has become scarce. It was also noted that some plant species, though in less frequent use, still remained very important as these were needed for very specific treatment. The people expressed a general fear that the increasing scarcity of many plants might lead to the loss of traditional knowledge about the medicinal uses of such species. The sudden decrease in the availability of medicinal plant species such as *Aconitum heterophyllum, Paris polyphylla, Dactylorhiza hatagirea, Sinopodophyllum hexandrum, Rhododendron campanulatum, Swertia chirayita* and *Zanthoxylum armatum* was reported by the local people during surveys in the North-West region of the country (Goraya and Ved 2017).

In a random walk of 600 m (by two m on either side) along an alpine meadow of GHNPCA in June 2023, the first and second authors could count only 40 individuals of *Fritillaria cirrhosa.* Whereas, the area was teaming with wild populations of *Fritillaria cirrhosa* and *Trillium govanianum* upto early 2010s when these species had just come under commercial trade (Plate 1).

Fritillaria cirrhosa

Trillium govanianum

Plate 1 Rich populations of *Fritillaria cirrhosa* and *Trillium govanianum* in GHNP (May 2013)

Alongside the trend of increasing harvest of medicinal herbs in the State, a parallel trend of many people staying away from such wild collection is also visible due mainly to profitable expansion of horticulture, especially apple, in the state. For example, the outer Seraj area of Kullu, bordering the north-western region of the GHNPCA, has witnessed a drastic decline in extraction of medicinal herbs from wild. The major reason is the rise in the spread of apple cultivation in this region. The apple has provided a boost to the local economy here, and the number of herb collectors has been declining since 1990's.

Another factor that is curtailing the number of collectors going to the forest is the MGNREGA scheme of the government, which provides for hundred days of assured work to the registered low-income populace near their homes. As per one herb collector, reduction in the number of people engaged in herding is another reason of reducing footfall of local herb gatherers in the forests. Earlier, the herders would spend the entire day roaming with their herds in the forests and collect a number of herbs, which they would subsequently store and sell to the village level procurers. With habitats shrinking, people find it difficult to herd their livestock and have opted to stall feeding.

3 Hotspots of Extraction in Himachal Pradesh

In Chamba district, the hotspots for medicinal herbs collection are Hegh, Chobia dhar, and Kugti pass areas. In the Churah Forest Division, the hotspots for collecting the herbs are, Makhan dhar, Ahenpukhar dhar of Sahoo and Devkothi, Sanwaal and Bhanera areas of around Bairagarh (Tissa). A review of trade data collected from Chamba Forest Division in Western Himachal Pradesh reveals that prior to 2019–20 there were around 80–85 permit holders for the collection of wild medicinal herbs. However, after the COVID-19 pandemic, there has been a sharp decline in the number of wild gatherers (Table 1). One of the plausible reasons in this case seems to be moving out of the young generation to the more lucrative options in cities and towns.

The hotspots of medicinal plant extraction in the GHNP and the surrounding areas include, the villages of Malana and Tosh in the adjoining Parvati valley, and Rela and Deori in the Sainj valley, and the villages in the Sarchi, Bandal, Pekhari, Kalwari and Plaich of Tirthan valley. In the main Beas valley, the hotspots for plant collections are the villages in the upper Manali valley.

As per the information received from the local market and villagers, rate of collection of MAPs is decreasing in the area. Number of collectors and amount of collection varies with the demand of plants in the market. Based on the information gathered from local market/village surveys and field observations, the current price trends and quantum of collection including market rates during 1997–1998 of MAPs in the buffer zone of GHNP are provided in Tables 2 and 3, respectively.

As per data obtained from Banjar Forest Division for 2020–21, the significant NTFP exported was mehndi/Chharila (890 Quintals), khanor (538 Quintals), karu (113 Quintals) and pathanbel (109 Quintal). Some species like mushkbala (70 kg), dhoop (50 kg), bankakdi (46 kg) and dori (*Potentilla sp.*) (25 kg) were also being traded in quantities less than 100 Quintals per annum. The major species traded and their market rates in Kullu region of Himachal Pradesh are provided in Table 4.

In Pangi region of Lahaul & Spiti, the hotspots for herb collection are Udan Batori, Sural Batori, Karru Dhar (Shor area), Twan Dhar, Chasak Dhar and Sechu Dhar. The major species traded and their market rates in Pangi region of Himachal Pradesh are provided in Table 5.

In Rohru area, the hotspots for herb collection are Chanshal slopes, Muraldanda area, Dodra Kwar, Ransar Valley and Dulgi Dhar. In the adjoining Rampur Forest Division, the hotspots for collecting medicinal herbs are Kashpat area, Sarahan, Daranghati, Kut, Sarpara and Phancha. In Kinnaur, the hot spots are Rupi, Bhabha, Bari Kanda, Sangla and Kilba Kanda.

In Mandi, the information with the forest department pertains mainly to Chir, Kail and deodar cones. The medicinal plant exported in good quantities are tejpatta (*Cinnamomum tamala*) and Pathanbel (*Bergenia ciliata*). The hotspots of herb

Table 1 Recent trends in the number of herb collectors (official permit holders) in District Chamba, Himachal Pradesh

S.no	Year	No. of permit holders
1	2015–16	85
2	2016–17	82
3	2017–18	78
4	2018–19	75
5	2019–20	84
6	2020–21	55
7	2021–22	32
8	2022–23	30
9	2023–24	23

Table 2 Current market rates of wild medicinal herbs collected from Sainj and Tirthan area of Great Himalayan National Park

Species	Local name	Price (Rs. per kg)
Morchella esculenta	Gucchi	8000–15,000
Viola serpens	Banaksha	1500–2000
Parmelia sp.	Mehndi/Chalora	120–200
Nardostachys jatamansi	Balchadd	400–6000
Picrorhiza kurroa	Karu	1000–1500
Sinopodophyllum hexandrum	Bankakadi	90–200
Angelica glauca	Chora	100–150
Jurinea dolomiaea	Dhoop	150–200
Acorus calamus	Bach	70–120
Aconitum heterophyllum	Atish/Meetha Patish	3500–6000
Dactylorhiza hatagirea	Hathpanja	3500–6500
Fritillaria cirrhosa	Jangli lahsun	10,000–15,000
Polygonatum verticillatum	Meda	100–150
Salvia moocroftiana	Thuth	60–150
Bergenia ciliata	Pathanbel	25–50
Valeriana jatamansi	Muashak bala/Nihanu	100–150

collection here are Kotala, Baggi, Malwara, Drang, Piun, Cheltigarh and Shivadhar. Notably, the number of permit holders in Mandi has increased from 10 (5 years ago) to 18 (2023).

4 Management Implications

Wild harvest of MAPs is subject to following the in-place legal and regulatory framework viz., no harvest from protected areas, harvest as per approved four-year extraction cycle, and no harvest of notified species. The above analysis, however, reveals that the high value medicinal plants continue to be harvested from the wild irrespective of the legal status of the area, regulatory mechanism in place, and threat status of the species. It is pertinent to mention that the Ministry of Environment and Forests, Government of India has, under Section 38 of the Biological Diversity Act, 2002, notified eight plant species namely *Aconitum deinorrhizum, A. heterophyllum, A. violaceum, Eremostachys superba, Jasminum parkeri, Nardostachys grandiflora, Dactylorhiza hatagirea* and *Taxus wallichiana*, prohibiting their collection from the wild (Anonymous 2009) without specific permission. Six of these eight species are routinely collected as raw drugs. It is obvious that neither the industries and the traders, nor the poor collectors, are concerned about the conservation status of the species, and collection of many of these species continues unabated. The alpine and

Table 3 Commercial extraction of MPs in the buffer zone of GHNP during 1997–98 through village surveys by Rawat (2007)

S. no	Name of the species	No. of collectors/Yr	Amount collected (kg/yr)	Price (Rs./kg)	* Total Income/Yr (In lakh)
1	*Aconitum heterophyllum*	345	1050	800	8.40
2	*Aconitum violaceum*	842	5100	750	38.25
3	*Acorus calamus*	65	158	50	0.07
4	*Ainsliea aptera*	95	252	60	0.15
5	*Angelica glauca*	368	4675	45	2.1
6	*Dactylorhiza hatagirea*	462	1285	850	10.92
7	*Diplazium maximum*	38	1085	5	0.05
8	*Dioscorea deltoidea*	583	45,000	30	13.50
9	*Hypericum patulum*	15	500	75	0.37
10	*Jurinea dolomiaea*	880	82,800	75	62.10
11	*Morchella esculenta*	1224	1092	2500	25.12
12	*Nardostachys jatamansi*	125	320	100	0.32
13	*Parmelia* spp.	658	105,280	24	25.27
14	*Picrorhiza kurrooa*	355	8350	85	7.09
15	*Sinopodophyllum hexandrum*	215	450	65	0.30
16	*Polygonum amplexicaule*	335	4800	22	1.05
17	*Taxus wallichiana*	35	9750	12	1.17
18	*Thymus linearis*	15	25	40	0.001
19	*Valeriana hardwickii*	125	3055	100	3.05
20	*Valeriana jatamansi*	448	4652	90	4.18
21	*Viola serpens*	336	1170	120	1.40
	Total Income from the MPs = Rs. 204.91 lakhs				

*Approximate revenue earned from the sale of each species in the local market

Table 4 Major species traded and their approximate market prices prevalent in Kullu region of Himachal Pradesh

S. no	Botanical name	Local name	Price (Rs. per kg)
1	*Sinopodophyllum hexandrum*	Ban Kakdi	700–800
2	*Trillium govanianum*	Nag chatri	1400–1600
3	*Fritillaria cirrhosa*	Van lahsun	15,000–18,000
4	*Picrorhiza kurroa*	Karru	1200–1500
5	*Dactylorhiza hatagirea*	Hatpanja	6000–7000
6	*Aconitum heterophyllum*	Atish	4000–5000
7	*Dolomiaea macrocephala*	Dhoop	300–500
8	Lichens	Chaloora	Not available

Table 5 Major species traded and their approximate market prices prevalent in Pangi region of Himachal Pradesh

S.no	Botanical Name	Local name	Price (Rs. per kg)
1	*Sinopodophyllum hexandrum*	Ban Kakdi	1000–1200
2	*Aconitum heterophyllum*	Atish	2000–2500
3	*Aconitum violaceum*	Meetha Patish	1000–1100
4	*Fritillaria cirrhosa*	Van lahsun	11,000–12,000
5	*Picrorhiza kurroa*	Karru	3000–4000
6	*Dactylorhiza hatagirea*	Hatpanja	2000–3000
7	*Dolomiaea macrocephala*	Dhoop	500–600
8	Lichens	Mehndi/Chaloora	1200–1500
9	*Arnebia benthamii*	Ratanjot	500–600
10	*Angelica glauca*	Chora	1200–1300
11	**Dolomiaea costus*	Kuth	2500–3000
12	*Eremurus himalaicus*		1000–1100
13	*Carum carvi*	Gurnu	500–600
14	*Elwendia persica (Bunium persicum)*	Kala Zeera	3500–3600

* Govaerts et al. (2021)

sub-alpine areas forming the major source of the medicinal plant species in high trade are too vast, porous and remote to protect and patrol on a regular basis with the given strength of regulatory staff. It is noteworthy that *N. grandiflora* DC. was hitherto considered as homotypic synonym of *N. jatamansi* (D. Don) DC. where as the two are totally distinct species and the former is not traded. This needs to be corrected in the official notifications.

Under these circumstances what can be done to control illegal harvest of high value medicinal plants and ensure conservation of the species in perpetuity? The foregoing analysis reveals that engagement of economically backward families in alternate means of livelihoods and keeping strict vigil on illegal trade of wild medicinal plants or other forest products may be effective in reducing the pressure on populations of wild medicinal plants in the region. Goraya and Ved (2017), however, argue that weaning away the local communities from wild gathering of medicinal herbs through their engagement in alternative vocations may not help protect the wild populations of medicinal herbs, as wild harvest is a function of demand, and that industries would find other ways to get the material collected from the wild to fulfil their needs. They opine that shift from the present unorganized and individualistic wild collections to a community based organized wild collection system working on time tested harvest and post-harvest handling protocols might be more sustainable. Kumar et al. (2021a, b) suggest that Biodiversity Management Committees (BMC) would be very useful in organizing the local communities and handling sustainable management of medicinal herbs. The authors too support active engagement of local communities in sustainable management of medicinal herbs, and believe that strengthening the BMCs being set up under the Biological Diversity Act 2002 through establishing these committees at every remote village level and empowering these BMCs with full responsibility for conservation and sustainable harvest and for equitable benefit sharing amongst stakeholders would be a highly desirable action.

It has been established that *in-situ* conservation is the most viable system of managing wild populations of priority medicinal plant species. Impact of the *in situ* conservation action can be enhanced by strengthening wild populations of the species under stress through plantations (Goraya and Ved 2017). In view of the increasing demand of medicinal herbs, several authors have come to suggest cultivation of commercially traded species towards their *ex-situ* conservation measure (Goraya et al. 2012; Phondani 2016; Goraya and Ved 2017; Kumar et al. 2021a, b). National level programs to promote cultivation of threatened medicinal plants have also been put in place by the National Medicinal Plants Board, Government of India. This, however, needs working out of farm-scale agro-techniques for priority species, and extension of these programs to the farmers and the herbal industry managements. Precious little seems to have been achieved in this area so far; agro-techniques for all these species are either not available or are not farmer-friendly, and the industry continues to buy raw material through agents and brokers exploiting the medicinal plant growers. The authors believe that research institutions dealing with this aspect need to come up with farm-scale protocols for cultivation techniques and economics of cultivating high value species, and set up demonstration fields; and the national agencies should support programs to develop farmer friendly market platforms.

The medicinal plants for which rhizomes and roots are harvested are particularly vulnerable to local extinction. Whereas, there is a case for putting in place appropriate regulatory mechanism to check pre-mature and over harvest of such species, viz. of *Fritillaria cirrhosa, Trillium govanianum, Aconitum heterophyllum* and *Dactylorhiza hatagirea*, there is an equally urgent need to revisit the hotspots of wild harvest of medicinal plants listed in this chapter, to establish baselines on the population densities of highly traded medicinal plants, and to subsequently monitor these populations at regular periodicity by the state management authorities in close coordination with research institutions and community-based organizations dealing with the subject.

Emerging trends in trade in high value MAPs under harvesting stress, and shifts in the hotspots of extraction, calls for in-depth geo-spatial analysis of the suitable habitats available for the traded species, assessment of their populations, and assessment of their threat status through CAMP workshops organized using IUCN's threat categories for formulation of priority conservation strategies (https://envis.frlht.org/newsletters/sep2006.htm). The last such workshop was conducted way back in 2010 at Shimla. Since then, many species such as *T. govanianum* which was not considered in the CAMP, Shimla for assessment at the state level should be assessed in the future workshops, as its global status has been changed to Endangered by IUCN (Jishtu et al. 2021). Therefore, it is imperative that regular, periodic CAMP Workshops be organized in order to ascertain the threat status and any changes in that status for high value medicinal plants that are traded from the area.

References

Anonymous (2009) Notification dated March 17, 2009 of the Ministry of Environment and Forests, Government of India, under Section 38 of the Biological Diversity Act, 2002; published vide S. No. 481 in the Gazette of India extraordinary on 19 March 2009

Badola HK, Aitken S (2003) The Himalayas of India: a treasury of medicinal plants under siege. Biodiversity 4:3–13

Bauer A, Brönstrup M (2014) Industrial natural product chemistry for drug discovery and development. Nat Prod Rep 31(1):35–60

Dhar U, Rawal RS, Upreti J (2000) Setting priorities for conservation of medicinal plants: a case study in Indian Himalayas. Biol Cons 95:57–65

Gairola S, Sharma J, Bedi YS (2014) A cross-cultural analysis of Jammu, Kashmir and Ladakh (India) medicinal plant use. J Ethnopharmacol 155:925–986

Goraya GS, Ved DK (2017) Medicinal plants in India: an assessment of their demand and supply. National Medicinal Plants Board and Indian Council of Forestry Research and Education, Dehradun

Goraya GS, Jishtu V, Rawat GS, Ved DK (2012) Wild medicinal plants of Himachal Pradesh: an assessment of their conservation status and management prioritisation. An outcome of the 'CAMP' workshop held at HFRI, Shimla on 1–4 December 2010. NMPB & HPFD

Govaerts R, Nic Lughadha E, Black N, Turner R, Paton A (2021) The world checklist of vascular plants, a continuously updated resource for exploring global plant diversity. Sci Data 8:215. https://doi.org/10.1038/s41597-021-00997-6

Gupta S, Sharma O, Raina N, Sehgal S (2013) Ethno-botanical study of medicinal plants of Paddar Valley of Jammu and Kashmir, India. Afr J Tradit Complement Altern Med 10:59–65

Jishtu V, Bhondge SW, Bhushan B, Chauhan M, Chauhan A (2021) Threatened ethnomedicinal plants of Dodra-Kwar region of Himachal Pradesh, NW Himalaya. J Med Plants Stud 9(2):151–159

Kala CP (2005) Indigenous uses, population density, and conservation of threatened medicinal plants in protected areas of the Indian Himalayas. Conserv Biol 19(2):368–378

Kumar S, Hamal IA (2011) Herbal remedies used against arthritis in Kishtwar high altitude national park. Indian J Traditional Knowl 10:358–361

Kumar S, Sharma S (2013) Species diversity, uses and distribution of medicinal plants along an altitudinal gradient in Paddar valley, Northwestern Himalaya. Int J Med Aromatic Plants 3:2249–4340

Kumar M, Paul Y, Anand VK (2009) An ethnobotanical study of medicinal plants used by the locals in Kishtwar, Jammu and Kashmir, India. J Ethnobiol Ethnomed 13:1240–1256

Kumar A, Sathyakumar S, Goraya GS, Rawat GS (2021a) Assessment of medicinal and aromatic plant species on their collection, usage, demand, markets, price trends and life cycle in Lahaul and Pangi landscape, Himachal Pradesh. Technical Report No. 2021/14. Wildlife Institute of India

Kumar A, Sathyakumar S, Goraya GS, Gupta AK, Adhikari BS, Rawat GS (2021b) Sustainable harvesting and cultivation protocols of threatened medicinal and aromatic plants of the Western Himalaya. Submitted to Himachal Pradesh Forest Department and United Nations Development Programme, 64 pp

Kumar A, Adhikari BS, Mishra AP (2023) Status of threatened medicinal and biosphere reserve, Uttarakhand. A report (TR No./2022/35) submitted to International Union for Conservation of Nature—India, p 103

Mathela M, Sharma M, Kumar A, Goraya GS (2021) Hue and cry for *Fritillaria cirrhosa* D.Don, a threatened medicinal plant in the Western Himalaya. Discover Sustain 2:38. https://doi.org/10.1007/s43621-021-00048-5

Mishra AP, Kumar A, Yadav SN (2023) Ecology and conservation of threatened medicinal plants in the Trans-Himalayan region of Nanda Devi Biosphere Reserve, Western Himalaya. Trees, For People 14:100451. https://doi.org/10.1016/j.tfp.2023.100451

Nishteswar K (2014) Depleting medicinal plant resources: a threat for survival of Ayurveda. Ayu 35(4):349–350

Phondani P (2016) Promoting medicinal plants cultivation as a tool for biodiversity conservation and livelihood enhancement in Indian Himalaya. J Asia-Pac Biodivers 9(1):39–46

Ramakrishnappa A (2002) Biodiversity and the ecosystem approach in agriculture, forestry and fisheries. FAO, Rome

Rana SK, Rawat GS (2019) Ethnobotanical survey of medicinal plants for assessment of their conservation status in Chenab Valley, J&K. In: Proceedings of 1st Himalayan researchers consortium, vol 1, pp 63–70

Rawat GS (2007) Alpine vegetation of the Western Himalaya: species diversity, community structure, dynamics and aspects of conservation. D.Sc. Thesis. Department of Botany, Kumaun University, Nainital

Samant SS, Dhar U, Palni LMS (1998) Medicinal plants of Indian Himalaya: diversity, distribution, potential values. Gyanodaya Prakashan, Nainital, Uttarakhand

Sharma M, Mathela M, Sharma R, Bargali H, Goraya GS, Kumar A (2022) Counting the cost: high demand puts *Bunium persicum* (Boiss.) B.Fedtsch. in jeopardy. J Threatened Taxa 14(1):20530–20533. https://doi.org/10.11609/jott.6831.14.1.20530-20533

Tali BA, Ganie AH, Nawchoo IA, Wani AA, Reshi ZA (2015) Assessment of threat status of selected endemic medicinal plants using IUCN regional guidelines: a case study from Kashmir Himalaya. J Nat Conserv 23:80–89

Tandon V (1997). Status of collection, conservation, trade and potential for growth in sustainable use of major medicinal plant species found in the Great Himalayan National Park and Its Environs in Kullu District of Himachal Pradesh. A Report. FRLHT

Ved DK, Goraya GS (2007) Demand and supply of medicinal plants in India, Bishen Singh, Mahendra Pal Singh, Dehradun and FRLHT, Bangalore, India

Ved DK, Kinhal GA, Ravikumar K, Prabhakaran V, Ghate U, Sankar RV, Indresha JH (2003) Conservation assessment & management prioritisation for the medicinal plants of Jammu & Kashmir, Himachal Pradesh & Uttaranchal. Foundation for Revitalisation of Local Health Traditions, Bangalore, India

In-Depth Assessment of Rhubarb, *Rheum* L.: A High-Value Medicinal Plant Species in the Indian Himalayan Region

Manisha Mathela, Amit Kumar, and Arun Pratap Mishra

Abstract The genus *Rheum* L. commonly known as 'Rhubarb' is a native of Europe, tropical to temperate Asia, and comprises of 60 species distributed worldwide. In the India Himalayan Region (IHR) eight species of *Rheum* have been identified. High medicinal efficacy of the diverse bioactive secondary metabolite has made *Rheum* one of the most sought-after genus in IHR. The communication comprehensively reviews the current knowledge and research gaps concerning the *Rheum* species. The review covers various aspects such as ecology, traditional medicinal uses, phytochemistry, trade, and threats the species face in the IHR. The distribution of *Rheum* species in IHR varies by state, Uttarakhand has the highest number of species (6), followed by Sikkim (5), Himachal Pradesh, Jammu and Kashmir, and Ladakh (4 species each), and Arunachal Pradesh (2). Since antiquity, the species have been used for over 50 different culinary and medicinal practices in IHR. A variety of bioactive phytochemicals are responsible for the medicinal properties attributed to *Rheum* species. Owing to the exceptional medicinal and therapeutic properties and growing demands, the species are under severe threats such as premature harvesting, illegal trade, unscientific and unorganized harvesting, and other anthropogenic pressures. It has been established that most *Rheum* species are threatened as per regional assessments in IHR, raising a high conservation concern. Thus, owing to endemicity, restricted distribution, and lack of information on population status and market, in-situ conservation viz., delineation of high conservation value areas and ex-situ conservation like gene banks, germplasm, cryopreservation, botanical gardens for their sustainable utilization is suggested.

Keywords Conservation · Himalaya · Medicinal plant · *Rheum* · Environment

---

M. Mathela · A. Kumar (✉) · A. P. Mishra
Department of Habitat Ecology, Wildlife Institute of India, Dehradun 248001, Uttarakhand, India
e-mail: amit@wii.gov.in

A. P. Mishra et al. (eds.), *Threatened Medicinal Plants in the Indian Himalayan Region*, World Sustainability Series, https://doi.org/10.1007/978-3-031-73687-2_4

1 Introduction

The Indian Himalayan Region (IHR) is home to *ca.* 18,000 plant species (Banerji and Basu 2011), however, according to Singh et al. (2019) *ca.* 11,157 species of flowering plants belonging to 2359 genera under 241 families have been reported. Of these, the medicinal and aromatic plants (MAPs) accounts for 1748 species (23.4% of India) that includes 1685 species of angiosperms, 12 gymnosperms, and 51 pteridophytes (Samant et al. 1998). The high therapeutic value owing to the chemical constituents has resulted in high demand and an immense pressure on the wild population. Thus, it has resulted in inclusion of 121 species that are listed in the Red Data Book of Indian plants (Nayar and Sastry 1987–1988; 1990) and over 120 species are classified under different threat categories of IUCN (Samant and Pal 2003; Ved et al. 2005). According to Kabir et al. (2015), over 90% of the plant species used in the herbal industry are collected from the wild, majority of which comes from the sub-alpine and alpine zones of the Himalaya. India has immense wealth of herbal medicines and traditional medicinal systems that employ a large number of plant species in Ayurveda (2000 species), Siddha (1121 species), Unani (751 species) and Tibetan (337 species) (Kumar et al. 2011). According to Rajasekharan and Ganeshan (2002), an approximate revenue worth US$ 800 is being generated by medicinal plants and herbal medicines, globally. There has been an increase in awareness of the use of herbals among the masses, a tremendous rise in the herbal industry, and rapid demand for wild medicinal plants in different medicinal systems and abroad. This has resulted in an increment in the demand for natural and local products, where local traditional knowledge is the prime resource for all ethnobotanical investigations resulting in an over-exploitation of MAPs, which are directly retrieved in situ (Badola and Aitken 2003; Kumar et al. 2021a, b, 2023; Mishra et al. 2023). The locals, however, have been using MAPs as a primary source of health care since antiquity, the local herb sellers and traditional healers locally known as *amchis, hakims, vaidyas, lamas,* and *bongthing*, administer herbs to people based on the ethnobotanical heirloom (Farooq et al. 2014; Tamang et al. 2017). The short growing seasons and a combination of other threats such as premature harvesting, unscientific collection practices, and over-exploitation to meet the demands of illegal trade, the wild medicinal plant species face immense threat (Kumar et al. 2016, 2021a, b). The increasing market demand results in over-extraction resulting in the further population decline of several MAPs (Mathela et al. 2020, 2021). Therefore, to address these issues, consistent records of vegetation data including distribution and population status are necessary (Clubbe et al. 2010; Bargali et al. 2021). Against this backdrop, the current communication aimed to provide a comprehensive review of (i) ecology, (ii) ethnobotany, (iii) phytochemistry including therapeutic potential, (iv) trade and threats of the *Rheum* L. species (Buckwheat family) in the IHR.

2 Study Area

The Indian Himalayan Region (IHR) (Fig. 1), lies between 36° 3′ N; 72° 31′ E as the north-western boundary and 27° 52′ N; 97° 21′ E in the east, and has a total geographical area of 530,795 km^2 (Singh 2006). According to Singh et al. (2019), the dominant families in IHR include Poaceae, Asteraceae, Orchidaceae, Fabaceae, Rosaceae, Cyperaceae, Scrophulariaceae, Rubiaceae, Lamiaceae and Arecaceae, whereas *Carex, Impatiens, Rhododendron, Primula, Pedicularis, Saxifraga, Bulbophyllum, Taraxacum, Corydalis* and *Astragalus* are the dominant genera. The difference in climatic conditions, topography, natural history, and orogeny influences phytogeography, and hence, the flora of Western Himalaya (WH) and Eastern Himalaya (EH) differ from one another. The Eastern Himalaya has been considered 'the cradle of flowering plants' which harbour an estimated 5800 species, of which 2000 (36%) are endemic (Takhtajan 1986; Mathur et al. 2014). Whereas, the Western Himalaya is estimated to support over 5000 species of flowering plants (Mudgal and Hajra 1997). In terms of the predominance of the genera, Singh et al. (2019) recorded *Poa, Potentilla, Astragulus, Saxifraga, Corydalis, Cotoneaster, Artemisia, Pedicularis, Euphorbia, Taraxcum, Polygonum, Astragalus, Gentiana, Epilobium, Pedicularis, Saussurea, Cyperus, Euphorbia, Nepeta,* and *Allium* in WH, whereas, *Rhododendron, Bulbophyllum, Impatiens, Ficus, Dendrobium, Primula, Carex, Aspetes, Eria, Coelogyne, Pedicularis, Juncus, Gentiana, Calanthe, Swertia* in the EH. Notably, the genus *Rheum*, known for its high demand due to therapeutic benefits and local consumption, is widespread over the IHR, with the exception of a few species that have a more restricted range.

3 Materials and Methods

After conducting an extensive literature review using major scientific search platforms such as Google Scholar (Page no. 1–30), Research-Gate, Cab Direct, Science-Direct, and Web of Science, the present communication assesses studies on *Rheum* species in the IHR. Keywords as search terms- or '*Rheum*' or 'Rhubarb' 'medicinal plants' or 'ethnobotany' or 'Himalaya' or 'IHR' or 'IUCN' or 'threatened species' or 'sustainable harvesting'; 'list' or 'checklist' along with the specific name of Himalayan states or species were used.

The archived databases of regional and international journals were also searched and supplemented with other relevant literature (scientific reports and articles) through cross-referencing from the resulting search. The initial search shortlisted over 250 articles. The abstracts of the research papers were also reviewed carefully and the relevant literature was retained for assessment. This assessment focuses on full-fledged studies on the *Rheum* species in the IHR.

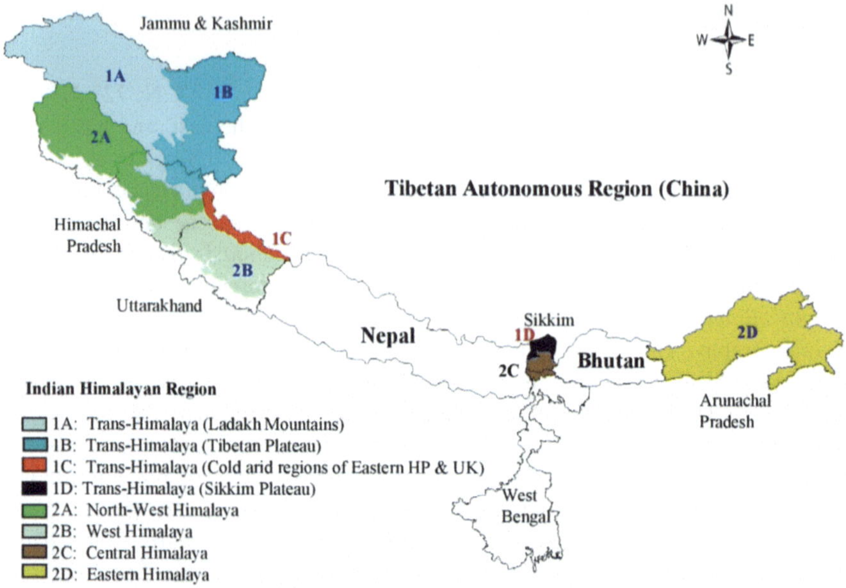

Fig. 1 Map showing the Indian Himalayan Region

4 Comprehensive Review and Analysis

The articles on the species ecology, ethnomedicinal use, phytochemistry, therapeutic potential, trade or market, and threat factors were systematically reviewed from the existing information on states from which the species were recorded. The year of publication and the nature of the study was enlisted in a tabular form. Additional information about the species such as the accepted scientific name, habit, and nativity was gathered using Plants of the World Online, Google Scholar, Research-Gate, and Shodhganga (a reservoir of Indian theses). Thus, based on the abovementioned criteria, and combination search, a total of 130 research papers were systematically screened for review. Threat categories were identified using IUCN, WPA (Amendment) 2022, and regional threat assessments such as CAMP. The screened literature for review was also discussed in detail with the subject experts and the distribution maps of the species were prepared in ArcMap version 10.8. Further, to gain insights into the present knowledge and research gaps, the current review has been categorized into major sections such as ecology, ethnobotany, phytochemistry, therapeutic potential, trade, and threats.

5 Results

The genus described by Carl Linnaeus in 1753 (Species Plantarum 1: 371) is one of the endangered, endemic, and high use value MAP species found in the Himalayan region. The Indian sub-continent harbors ten species of *Rheum* (Hooker 1897; Santapau and Henry 1973), of which eight species have been reported in India (Srivastava and Shukla 2015). These species namely, *Rheum acuminatum* Hooker. f. & Thompson, *R. australe* D.Don, *R. globulossum* Gage, *R. moorcroftianum* Royle., *R. nobile* Hook. f. & Thompson, *R. spiciforme* Royle, *R. tibeticum* Maxim. ex. Hook. f., and *R. webbianum* Royle are exclusively distributed in the Indian Himalayan states viz., Jammu & Kashmir, Himachal Pradesh, Uttarakhand, Sikkim and Arunachal Pradesh and Union territory of Ladakh (Fig. 2). Of the eight species, *R. australe, R. acuminatum, R. moorcroftianum, R. spiciforme, R. tibeticum,* and *R. webbianum* are commonly found in the Western Himalaya whereas, *R. nobile* and *R. globulossum* are exclusively reported from the Eastern Himalaya (Tayade et al. 2012; www.plantsoftheworldonline.org) (Fig. 3). The maximum number of *Rheum* species are reported from the state of Uttarakhand (6 species) followed by Sikkim (5 species), Himachal Pradesh, Jammu & Kashmir, and Ladakh (4 species each), and Arunachal Pradesh (2 species). Notably, *R. australe* is reported to have been successfully cultivated in Assam (Agarwal et al. 2001). Detailed information on the distribution, nativity, and conservation status of the *Rheum* species found in IHR is provided in Tables 1, 2 and 3.

6 Ecology: Distribution, Habitat, and Population

According to Sun et al. (2012), buckwheat members are found throughout Asia-Temperate to parts of Asia-Tropical, covering China, India, Nepal, and Pakistan. *Rheum* is native to Europe, Asia-Tropical, and Asia-Temperate and encompasses *ca.* 60 species worldwide (Wang et al. 2005). According to Li et al. (2000), the familial area of *Rheum* is believed to be the Qinghai- Tibetan Plateau (QTP), Central and Western Asia, although, the extant species thrive in the Asian monsoon climate. The genus displays substantial evolutionarily morphological and ecological variation (Wang et al. 2005; Pandith et al. 2018).

In the IHR, the occurrence of the *Rheum* species ranges between 2800–5000 m. *R. australe* is widely distributed from the high alpine zones of Kashmir valley in the Western Himalaya to Sikkim and Arunachal Pradesh in the Eastern Himalaya between 2200–5200 m elevation. *R. moorcroftianum* shows a restricted global distribution occurring only in India and Nepal. In India, this species is confined to the alpine regions of Jammu and Kashmir and Ladakh, Himachal Pradesh, and Uttarakhand between 2000–4200 m elevation in the Western Himalaya (Khan et al. 2019). The distribution of *R. spiciforme* is limited to the alpine areas of the Western Himalaya specifically Ladakh, Jammu & Kashmir, and Uttarakhand ranging from 2000–5100 m (Ganie et al. 2014). *R. webbianum* occurs between 2000–4500 m elevation in Ladakh,

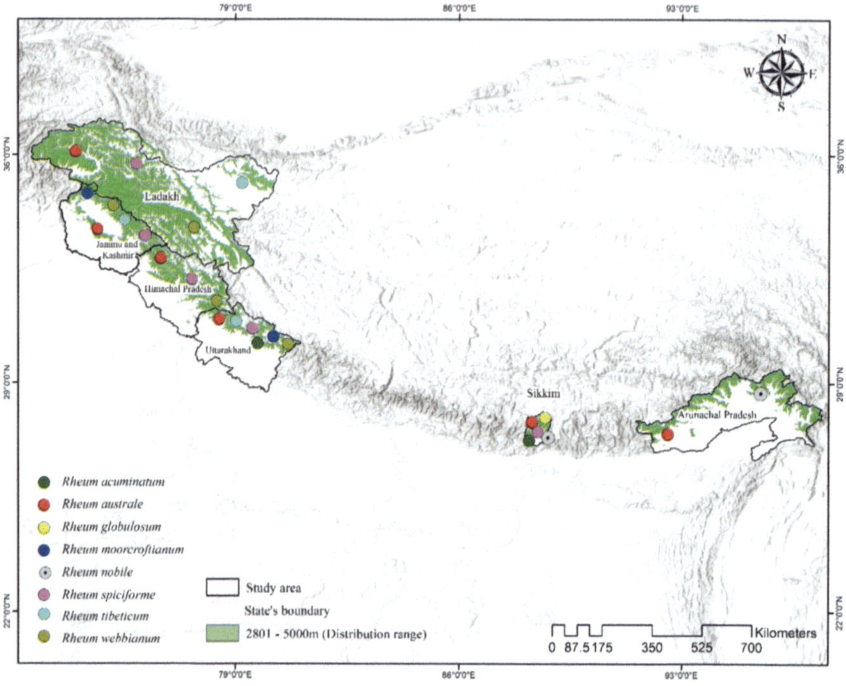

Fig. 2 Distribution of *Rheum* species in different biogeographic zones of IHR

Jammu & Kashmir, Himachal Pradesh, and Uttarakhand in the Western Himalaya (Tabin et al. 2016). *R. tibeticum* is found in rocky, arid, and steep regions of the alpine zone with an elevational range of 3628–4370 m in the Ladakh region of the Trans Himalaya, Jammu & Kashmir (Srivastava 2014) and Uttarakhand (Chandola 2009). *R. acuminatum* is found exclusively in the humus-rich soil of the alpine and sub-alpine areas ranging from 2800–4000 m (Li et al. 2000; Wang et al. 2005; Shah 2006; Rokaya et al. 2012; Srivastava 2014; Tabin et al. 2016; Kalsang 2018; Pandith et al. 2018). *R. nobile*, also known as the 'Himalayan giant glasshouse plant' is endemic to alpine regions of the Eastern Himalaya between 4000–6000 m elevation (Feng et al. 2015). Notably, *R. globulosum* is known to be distributed between 4500–5000 m elevation in parts of Sikkim (www.plantsoftheworldonline.org, www.eflora s.org), although studies pertaining to its distribution are meagre.

Owing to the fragmented nature of studies albeit regional, the information on the population status of *Rheum* species is scarce. According to Kumar et al. (2015), Niti Valley, a cold-arid region of the Nanda Devi Biosphere Reserve harbour 415 individuals per hectare of *Rheum australe*. The average density (individuals/m^2) of *Rheum australe* (*syn. R. emodi*) and *Rheum moorcroftianum* was reported to be 2.0 1.0 & 3.0 1.0 and 1.67 0.6 & 7.67 4.7, respectively in Spiti, Himachal Pradesh (Kala 2000). *R. nobile* is reported to show 20–30% of the global population in Arunachal Pradesh and Sikkim (Ved et al. 2005). Whereas, *Rheum moorcroftianum, Rheum*

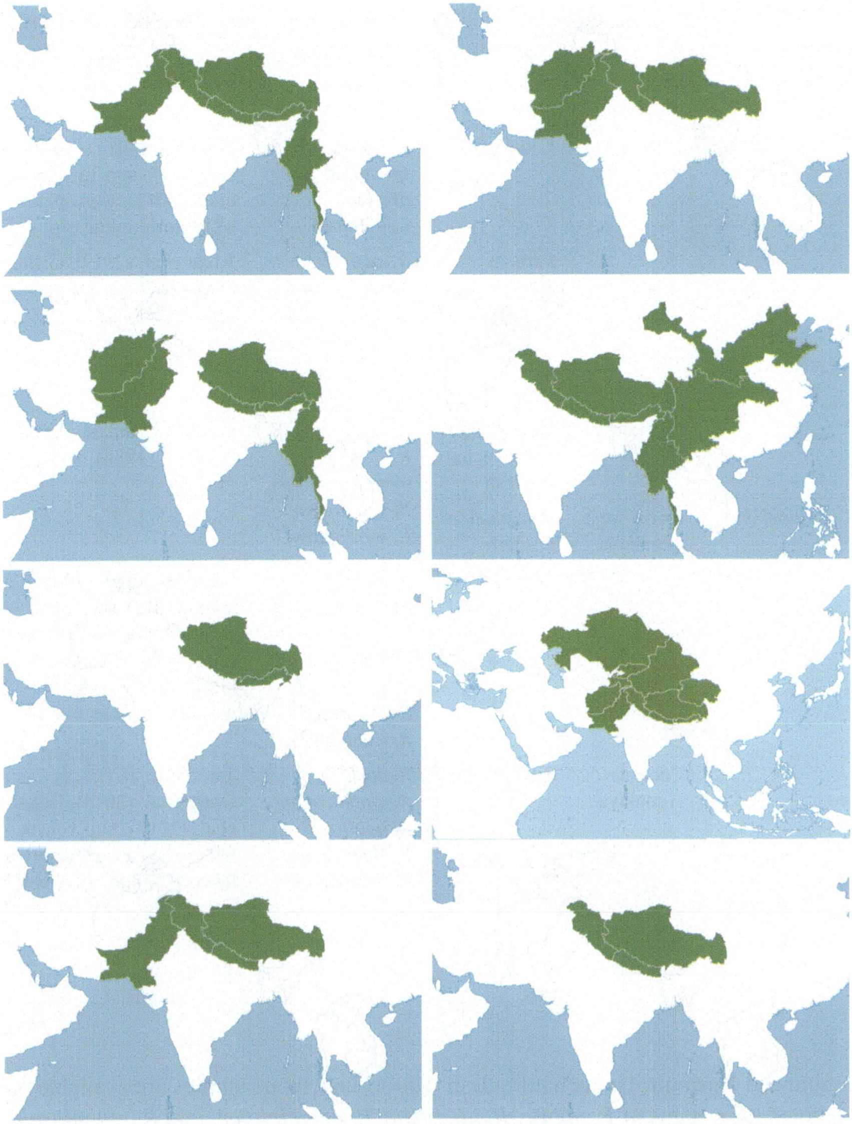

Fig. 3 Native distributional range of *Rheum* species in the world (www.plantsoftheworldonlin e.org)

australe, Rheum webbianum, and *Rheum spiciforme* account for 50, 30–40, 50–60, and 30–40% of the global population in Ladakh, respectively (Chauhan et al. 2020). According to Ved and Tandon (1998), over 20% population decline (10 years of generations) has been reported for *Rheum australe* and *Rheum webbianum* in

Table 1 Distribution of *Rheum* species in different biogeographic zones of the Indian Himalayan Region

Biogeographic zone	Biotic province	Region	Species	References
The Trans-Himalaya	Ladakh mountains	Ladakh	*R. australe, R. spiciforme, R. tibeticum, R. webbianum*	Sun et al. (2012), Stewart (1916–17) and Tayade et al. (2012), www.plantsoftheworldonline.org
		Parts of Jammu and Kashmir	*R. australe, R. moorcroftianum R. spiciforme, R. tibeticum, R. webbianum*	Khan et al. (2019), Ganie et al. (2014), www.plantsoftheworldonline.org, www.efloras.org
	Tibetan Plateau	Upper ridges of Himachal Pradesh	*R. australe, R. moorcroftianum, R. spiciforme, R. webbianum*	Rana and Samant (2011), Naithani (1984), www.efloras.org
The Himalaya	North-west Himalaya	Himachal Pradesh		
		Uttarakhand	*R. australe, R. acuminatum, R. moorcroftianum, R. spiciforme, R. tibeticum, R. webbianum*	Naithani (1984), Uniyal et al. (2007), Badoni et al. (2009), Kumar et al. (2016), www.plantsoftheworldonline.org, www.efloras.org
	Central Himalaya	Sikkim	*R. australe, R. globulosum, R. nobile R. spiciforme R acuminatum*	Tamang et al. (2017), Dutta et al. (2013), Dahal et al. (2017), http://envis.frlht.org, www.plantsoftheworldonline.org, www.efloras.org
	Eastern Himalaya	Arunachal Pradesh	*R. australe, R. nobile*	Shankar et al. (2015), Tsering and Tag (2015), www.efloras.org

Jammu and Kashmir, Himachal Pradesh. Unfortunately, no studies are available on the population status of *R. nobile, R. tibeticum, R. acuminatum, and R. globulosum.*

7 Ethno-botany

The medicinal use of *Rheum* dates back to the nineteenth and twentieth centuries (Bhatia et al. 2020). The species under this genus have been substantially used in traditional medicine, specifically in Ayurveda, Siddha, Unani, homeopathic, and *Sowa-Rigpa* system of Tibetan medicine and the Chinese system of medicine (Stewart

Table 2 Habitat and elevation of *Rheum* species in the Indian Himalayan Region

Species	Habitat	Elevation (m)	References
R. australe	Alpine zone on rocky soil, moraines, and crevices	2800–3600	Nautiyal and Nautiyal (2004), Stewar (1916–1917), Rokaya et al. (2012), www.efl oras.org
R. acuminatum	Slopes, forest, rocky, humus-rich soil of alpine and sub-alpine area of Himalaya	2800–4000	Nautiyal and Nautiyal (2004), www.eflora s.org
R. globulosum	—	4500–500	www.plantsoftheworl donline.org, www.efl oras.org
R. moorcroftianum	Slopes and river banks, mainly in the western Himalaya in the cold and dry morainic region	3500–5000	Shah (2006), Nautiyal and Nautiyal (2004)
R. nobile	Alpine zone of the Eastern Himalaya, where the soil is loamy and heavily clayey, in well-drained, yet moist areas. Found on rocky cliffs of rocks, rock ledges, and open slopes	3900–4200	Tamang et al. (2017), Tsering and Tag (2015), Kalsang (2018), Dutta et al. (2013)
R. spiciforme	Slopes and dry locations	4000–5000	www.plantsoftheworl donline.org, www.efl oras.org
R. tibeticum	Dry slopes	2500–4000 4000–4600	www.plantsoftheworl donline.org, www.efl oras.org
R. webbianum	Sub alpine areas, open slopes on moist areas	2400–4200	Tayade et al. (2012), www.efloras.org

1916–1917). In ayurvedic preparation, it has been long used as a strong laxative, stomachic-tonic, an ingredient of purgatives, and has been known for its astringent effect (Chopra et al. 1969; Thakur et al. 1989; Castleman 1995; Agarwal et al. 2001). The *Bhotia* and *Pangwal* communities of Uttarakhand and Himachal Pradesh have been using the plant locally known as chukri for many culinary and medicinal practices. The *Rheum* species are also used in the treatment of various ailments namely cuts, wounds, muscular swelling, indigestion, and tonsillitis (Tayade et al. 2012; Bano et al. 2017; Stewart 1916–1917). Further, *Rheum* species have been attributed to the treatment of various ailments in traditional as well as modern medicine that include healing diseases associated with the bladder, kidney, liver, lungs, and diseases like diabetes, relapsing fever, cancer, insect bites, diarrhea, and constipation (Mohtashami et al. 2021). The predominant plant part used includes stem and rhizomes (Kumar et al. 2021a, b). For instance, the roots of *R. australe* (the Indian rhubarb) have a well-known stomachic and cathartic use, which is darker, inferior in aroma, bitter, and finds use throughout the world (Thakur et al. 1989). The plant

Table 3 Nativity and conservation status of *Rheum* species

Species	Conservation status	% of the global population	Nativity	References
R. australe	Endangered (J&K, HP & UK) Vulnerable	30–40	Native to East Himalaya, Myanmar, Nepal, Tibet, West Himalaya	Goraya and Ved (2017), Pandith et al (2018), CAMP (1998), https://powo.science.kew.org, Goraya et al. (2013), Chauhan et al. (2020)
R. acuminatum	NA	NA	Native to China North-Central, China South-Central, East Himalaya, Myanmar, Nepal, Tibet, West Himalaya	https://powo.science.kew.org/
R. globulosum	NA	NA	Native to Central and S. Tibet to Sikkim	https://powo.science.kew.org/
R. moorcroftianum	Endangered (HP) Vulnerable (J&K) Near-threatened (UK)	50	Native to Afghanistan to W. and Central Tibet and Nepal	Goraya and Ved (2017), Goraya et al. (2013), CAMP (1998), https://powo.science.kew.org, Chauhan et al. (2020)
R. nobile	Near Threatened (AR); Vulnerable (SK)	20–30	Native to Afghanistan, East Himalaya, Pakistan, Tibet	CAMP (2005), https://powo.science.kew.org/
R. spiciforme	Vulnerable, Near-threatened	30–40	Native to Afghanistan, East Himalaya, Kazakhstan, Kirgizstan, Pakistan, Qinghai, Tadzhikistan, Tibet, West Himalaya, Xinjiang	Ved and Tandon (1998), Ved et al. (2005), Goraya et al. (2013), https://powo.science.kew.org/

(continued)

Table 3 (continued)

Species	Conservation status	% of the global population	Nativity	References
R. tibeticum	Near-threatened	NA	Native to Afghanistan, Pakistan, Tibet, West Himalaya	https://powo.science.kew.org/
R. webbianum	Vulnerable	50–60	Native to Nepal, Pakistan, Tibet, West Himalaya	Ved and Tandon (1998), Ved et al. (2003), Goraya et al. (2013), https://powo.science.kew.org, Chauhan et. al (2020)

Abbreviations used: J&K = Jammu & Kashmir, HP = Himachal Pradesh, UK = Uttarakhand, AR = Arunachal Pradesh, SK = Sikkim

forms an important constituent in a variety of medicines that holds antibiotic properties (Agarwal et al. 2001). Several authors such as (Inamori et al. 1983; Agarwal et al. 2000) reported that rhizomes of the Indian rhubarb viz. *R. australe and R. webbianum* are officially accepted and mentioned in Indian pharmacopeia, whereas *R. moorcroftianum* and *R. spiciforme* are also reported as integral constituents with many polyherbal formulations in several drugs. The roots of *R. spiciforme* show contents of free and combined anthraquinones (Huiyan et al. 2003), and the extracts of the plant also possess antifungal and antibacterial properties (Bhat et al. 2018). Based on the current review, it is revealed that the *Rheum* species have been used to cure at least 32 diseases, predominantly cuts and wounds, abdomen, arthritis, and stomach-related issues in IHR. Details on the ethnomedicinal uses and plant parts used of *Rheum* species in IHR is provided (Table 4).

8 Phyto-chemistry

The chemical constituents derived from this genus have shown extensive pharmacological activities (Zheng et al. 2013). The species are a constituent in ayurvedic preparations (Singh et al. 2017). Technological progress in recent years has made it easier to isolate anthraquinones from the plant species, whereas it was quite difficult to specify the active ingredient of Rhubarb up until the 1850s (Lee et al. 2017). About 200 bioactive compounds with six different types of skeletons such as anthraquinone, anthrone, stilbene, flavonoids, acylglucoside, and pyrone are the major constituents isolated from *Rheum* species (Lust 1974; Thakur et al. 1989; Agarwal et al. 2001; Usui et al. 2002; Dregus et al. 2003; Iwashina et al. 2004; Zhu et al. 2005; Misiti et al. 2006; Rokaya et al. 2012; Tayade et al. 2012; Zheng et al. 2013). *R. australe* due to its antimicrobial properties (Agarwal et al. 2000; Babu et al. 2003) is used

Table 4 Ethnomedicinal use of the *Rheum* species in IHR

Species	Local name	Parts used	Medicinal use	References
Rheum australe	Indian rhubarb, Himalayan rhubarb; Vern.-Dolu, Revandchini	Roots, petiole, leaves, bark	Kidney stones, gout and jaundice, stomach problems, cuts, wounds, muscular swelling, tonsillitis, and mumps Cleaning teeth and sprinkled over ulcers for quick healing	Stewart (1916–1917), Agarwal et al. (2001), Rokaya et al. (2012), Aslam et al. (2012), Bano et al. (2017)
Rheum acuminatum	Ornamental rhubarb	Roots; fruits	Herpes; digestion; diarrhea and dysentery	Rokaya et al. (2012), Tsering and Tag (2015), Dahal et al. (2017)
Rheum moorcoftianum	Vern.-Amlaparni, revand-chini, Dholu	Roots	Constipation, diarrhea, cuts, wounds, swellings, rashes, etc. fever, cuts, and wounds	Shah (2006), Kumar et al. (2016)
Rheum nobile	Vern. Tehuka (Sikkim)	Roots	Common cold, cough, and throat pain	Iwashina et al. (2004), Tsering and Tag (2015), Tamang et al. (2017)
Rheum spiciforme	*Pamb challan*, Pumbhaak, Jarochotal	Leaf, root, rhizomes	Purgative properties, wounds, cuts, fever. *Rheum* includes anti-oxidant, anti-microbial, antitumor, anti-inflammatory, anti-fungal, anti-atherosclerotic, anti-proliferative, hepatoprotective, and immuno-enhancing of treat wounds, boils, and rheumatic pain; and the leaves are used as a vegetable in the Kashmir valley	Ganie et al. (2014), Tabin et al. (2016), Bhat et al. (2018), Khan et al. (2022)

(continued)

Table 4 (continued)

Species	Local name	Parts used	Medicinal use	References
Rheum tibeticum	–	Underground parts, leaves	The underground part of *R. tibeticum* is used to treat wounds, boils, and other skin diseases, as an expectorant and appetizer. The leaves of this species are cooked and used to treat rheumatism, also as a laxative	Rao et al. (2020)
Rheum webbianum	Tatar, Tatri	Roots, stem, leaves, leaf-stalks	Indigestion, abdominal diseases, astringent, boils, purgative, wounds, and flatulence Indigestion, wounds, and gastritis Wounds of renal function disorders, hyperlipidemia, cancer and improves the memory in senile patients, of drugs combating cancer and body cholesterol. It is also highly important in treating indigestion, abdominal disorders, boils, wounds, and gastritis through traditional medicinal practices	Abe et al. (2000), Agarwal et al. (2000), Agarwal et al. (2001) Chaurasia et al. (2007), Tayade et al. (2012), Kumar et al. (2016), Tabin et al. (2016)

with the combination of different salt mordants in the coloring textile industries (Das et al. 2008). Similarly, the extracts from *R. spiciforme* cab be employed as a potent antioxidant and anti-inflammatory agent (Bhat et al. 2018). *R. australe* and *R. acuminatum* can be used interchangeably as both species almost have the same compounds (Rokaya et al. 2012). The secondary metabolites isolated from the different plant parts of *Rheum* species found in IHR species have been screened against various diseases (Table 5).

Table 5 *Rheum* species and their chief chemical compounds

Species	Secondary metabolites	References
R. australe	Chrysophanol, emodin, aloe-emodin, physcion and rhein, piceatannol and resveratrol, revandchinone-3, rhein, piceatannol, desoxyrhapontigenin, desoxyrhaponticin, gallic acid, lindleyin, isolindleyin, glucopyranoside, hydroxychromone, kaempferol-3-*O*-rhamnoside, eugenol, methyl heptyl ketone, rheinal, rhaponticin-β-d-glucoside	Rokaya et al. (2012), Pandith et al. (2018), Bano et al. (2017)
R. acuminatum	Chrysophanol, emodin, aloe-emodin, physcion and rhein, piceatannol and resveratrol	Rokaya et al. (2012)
R. globulossum	Not known	
R. moorcroftianum	Phenolic antioxidant compounds (chlorogenic acid), Micro and macro elements such as Zn, Cu, Mn, Fe, Co, Na, K, Ca and Li, Anthraquinone glycosides, Physcion, chrysophanic acid, emodin, hydroxy-methyl anthraquinone, rhein, vitamin B, C, D, 2–5-Dimethylhydroxychromene	Pandey et al. (2018), Tabin et al. (2018), Rashid et al. (2014)
R. nobile	Quercetin 3-*O*-glucoside, quercetin 3-*O*-galactoside, quercetin 3-*O*-rutinoside, quercetin 3-*O*- arabinoside and quercetin 3-*O*-[6″-(3-hydroxy-3-methylglutaroyl)-glucoside]	Fei et al. (2017), Prasad and Purohit (2001), Shi et al. (2014), Iwashina et al. (2004)
R. spiciforme	Anthraquinones derivatives—(chrysarobin, aloe-emodin, emodin, physcion, chrysophanol), chrysophanol, physcion, emodin, aloe-emodin, chrysarobin, piceatannol, glucopyranoside	Khan et al. (2023), Bhat et al. (2018)
R. tibeticum	Not known	
R. webbianum	Rhein and aloin, oxalic acid, 3′,5′-Dihydroxy-4-methoxystilbene (desoxyrhapontigenin), 1,8-Dihydroxy-3-methyl-anthraquinone, 1,6,8-trihydroxy-3-methyl-anthraquinone and 1,8-dihydroxy-3-hydroxymethyl-anthraquinone	Tayade et al. (2012), Rashid et al. (2014), Tabin et al. (2018)

9 Trade, Threat, and Conservation Status

As stated by the International Union for Conservation of Nature and Natural Resources (IUCN) committee for threatened plant species, one in ten species of vascular plants is threatened or endangered due to the high exploitation rate in the commercial and international trade sector (Dar et al. 2016). Herbal-based products show long-term and resilient outcomes that have resulted in the demand for the herbal-based product gaining popularity in the global market. The rising demand and turnover of millions of US dollars owing to their therapeutic properties have increased the demand for Himalayan medicinal plant products (Olsen 1998). As per a recent estimate, India accounts for 2.5% of the export of herbals in the international market which is anticipated to raise US$ 5 trillion by 2050 (Bhattacharya et al. 2014). A total of 195,000 metric tons of consumption of raw herbal drugs has been estimated by India's herbal industry for the year 2014–2015 (Goraya and Ved 2017). Notably, *Rheum* species are one of the important high-altitude medicinal plants used locally in the treatment of diseases and are one of the sought-after ingredients in herbal and conventional drugs. Popularly known by the trade names such as *Revan chini, Dolu, Tatri, Chukri, Padamchal, and Amlaparni*, the *Rheum* species are being sold at Rs. 100–250 kg^{-1} in the Indian market with an estimation of 100–200 metric tonnes, annually (Goraya and Ved 2017). The estimated consumption of *Rheum australe* in dry weight metric tonne (DWMT) was 158.27 mt (dry weight) annually (Goraya and Ved 2017). Further, the estimated consumption by the large and medium herbal units was 76.68% and by small and very small herbal units was 23.33%. Interestingly, *R. webbianum, R. moorcroftianum,* and *R. spiciforme* are traded as a substitute for *R. australe* (Singh 1999). The estimated annual trade of rhizomes of *Rheum australe, R. moorcroftianum,* and *R. webbianum* is 1000–1500, 100–200, and <10 MT $year^{-1}$, respectively (Goraya and Ved 2017).

In IHR, the major operative threats to *Rheum* populations are unmanaged tourism, concretization, landslides, overgrazing, over-exploitation for local use, and illicit trade (Chandra et al. 2021). Subsequently, premature harvesting and illegal trade that function parallel are prime causes resulting in the decline of the wild population (Mathela et al. 2020; Kumar et al. 2021a, b; Farooq et al. 2014). *Rheum* species, owing to their medicinal and monetary values are extracted selectively, engendering further population decline from the wild. Thus, unscientific collection and high traded volume of the stout herb from the wild have resulted in the reduction of its population making it threatened in one or more states of the IHR (Singh 1999; Goraya and Ved 2017; Mathela et al. 2020). Furthermore, one of the primary causes of the decline of population of *Rheum* species is a lack of awareness on its importance (Smith and Overgaard 2003).

10 Discussion

A number of workers have studied various aspects of *Rheum* species, for example, Losina-Losinskaya (1936) recognized pollen morphologies of several *Rheum* species; Bhardwaj (2013) worked on the cultivation practices of *Rheums australe*; Song et al. (2020) worked on *Rheum nobile*; Bhatia et al. (2020) researched on *Rheum australe*; Rana et al. (2022) studied on elevational range shifts of a few *Rheum* species. The current review revealed that majority of the studies have focussed on the ethnomedicinal use and phytochemistry of *Rheum* species, however information on its abundance and population status remained underexplored. Unfortunately, no substantial information on the distribution, population status and ethnomedicinal uses are available for *R. globulosum, R. spiciforme,* and *R. tibeticum*. The wild populations of *R. moorcroftianum* and *R. webbianum* have declined from the wild in recent decades (Kumar et al. 2015), however a lack of studies conducted on other *Rheum* species specifically on *R. globulosum, R. spiciforme, R. tibeticum, R. acuminatum,* and *R. nobile* could also be a factor contributing to their unknown population status indicating a major research gap.

The *Rheum* species have been used for many culinary and medicinal practices and are a part of many traditional herbal medicinal systems. It is quite evident that the culinary and medicinal uses of different *Rheum* species are invaluable for economic sustenance and profit generation. The presence of important secondary metabolites isolated from *Rheum* species makes this one of the most sought-after genera (Agarwal et al. 2001). However, the majority of the species are collected from the wild, facing tremendous pressure from harvesting, coupled with unorganized over-extraction and premature harvesting. The presence of illegal markets further catalyzes several precedented and unprecedented threats to the natural populations from the wild. Other operative threats include road construction, illicit trade, landslides, the concretization of the catchment area, overgrazing, and unmanaged tourism that have led to the loss of its natural habitat. Despite facing tremendous pressure, it is submitted that the conservation status of the *Rheum* species needs to be assessed. Although, *R. australe* along with *R. moorcroftianum* are being cultivated in the alpine region of Garhwal Himalaya of Uttarakhand (Bhardwaj 2013), the cultivation and harvesting protocols are available only for a few species (Kumar et al. 2016, 2021a, b). The cultivated plants of *Rheum* are often considered inferior in quality as compared to the wild populations (Schippmann et al. 2002; Devkota et al. 2010), ex-situ methods of conservation such as cultivation, and developing germplasm banks of the priority species are important to sustainably manage their demand (Uniyal et al. 2000). It is worth noting that *R. nobile* takes a long time to produce a sustainable population, which makes it vulnerable, especially when the habitat is disrupted by a variety of threats (Song et al. 2020). Therefore, in-situ conservation methods that include setting aside areas around viable populations of threatened species for conservation such as Medicinal Plant Conservation Areas (MPCAs) is suggested.

11 Conclusion

Despite being widely distributed species with several ethnomedicinal uses in IHR, the information on population status of *Rheum* species in general and *R. globulosum, R. spiciforme, R. tibeticum, R. acuminatum,* and *R. nobile,* specifically remained scarce in IHR. The present review highlights the need to assess the distribution and population of *Rheum* species following IUCN standards on both regional and global scales along with an estimation of available growing stock in the wild. Although, the *Rheum* species have been substantially used in several traditional systems of medicine, it is submitted that the depleting traditional ethnobotanical knowledge needs to be documented at the regional level. Further, in order to optimize the benefits to the local communities, and to adopt the concepts of sustainable harvesting and benefit-sharing, it is important to assess the availability, market trends, and methods for collection of the *Rheum* species in high demand. A major step towards the conservation of *Rheum* species can be spreading awareness of the dwindling populations amongst the stakeholders through capacity building. It is also imperative to develop area-specific and species-specific cultivation and harvesting protocols for the conservation of *Rheum* species.

Acknowledgements The authors are thankful to the Wildlife Institute of India for Institutional Support.

References

Abe I, Seki T, Noguchi H et al (2000) Galloyl esters from rhubarb are potent inhibitors of squalene epoxidase, a key enzyme in cholesterol biosynthesis. Planta Med 66:753–756

Agarwal SK, Singh SS, Verma S et al (2000) Antifungal activity of anthraquinone derivatives from *Rheum emodi*. J Ethnopharmacol 72:43–46

Agarwal SK, Singh SS, Lakshmi V et al (2001) Chemistry and pharmacology of rhubarb (*Rheum* species)—a review. NISCAIR-CSIR 1–9

Aslam M, Dayal R, Javed K et al (2012) Pharmacognostical and phytochemical evaluation of *Rheum emodi* Wall. Int J Curr Phar 2:471

Babu KS, Srinivas PV, Praveen B et al (2003) Antimicrobial constituents from the rhizomes of *Rheum emodi*. Phytochem 62:203–207

Badola HK, Aitken S (2003) The Himalayas of India: a treasury of medicinal plants under siege. Biodiversity 4:3–13

Badoni A, Bisht C, Chauhan JS (2009) Seed age effect on germinability in seeds of *Rheum* emodi wall. ex Meissn: an endangered medicinal plant of Garhwal Himalaya. N Y Sci J 2:81–84

Banerji G, Basu S (2011) Sustainable management of the herbal wealth of the Himalayas: prioritising biodiversity for conservation and development. In: Pre-congress workshop of 1st Indian Forest Congress, HFRI

Bano H, Siddique MAA, Gupta RC et al (2017) Response of *Rheum australe* L. (rhubarb), (Polygonaceae) an endangered medicinal plant species of Kashmir Himalaya, to organic-inorganic fertilization and its impact on the active component Rhein. J Med Plant Res 11:118–128

Bargali H, Mathela M, Sharma R et al (2021) Plant studies in Himachal Pradesh, Western Himalaya: a systematic review. J Mt Sci 18:1856–1873

Bhardwaj R (2013) Studies on the development of cultivation practices of *Rheum australe* D.Don. Ph.D. thesis submitted to Dr. Yashwant Singh Parmar University of Horticulture and Forestry, Solan, Himachal Pradesh

Bhat AH, Dar KB, Sofi MA et al (2018) *Rheum spiciforme* Royle—the medicinal herb with positive modulatory effect on controlled in vitro oxidative stress. Indian J Exp Biol 56:556–564

Bhatia A, Arora S, Nagpal A et al (2020) Screening of rhizomes of *Rheum* emodi Wall. Ex. Meissen for antimutagenic potential employing Ames assay. Nucleus 63:167–177

Bhattacharya R, Reddy KRC, Mishra AK (2014) Export strategy of Ayurvedic products from India. Int J Ayurvedic Med 5:125–128

Castleman M (1995) The healing herbs: the ultimate guide to the curative power of nature's medicines. Bantam Rodale Press, pp 305–307

Chandola S (2009) Vegetational inventory of cold desert habitat of Nilang area of Jadh Ganga catchment (Uttarkashi) in Garhwal Himalaya. PhD thesis, Hemvati Nandan Bahuguna Garhwal University, Uttarakhand, India, pp 299

Chandra N, Singh G, Lingwal S et al (2021) Alpine medicinal and aromatic plants in the Western Himalaya, India: an ecological review. Indian J Ecol 48:319–331

Chauhan A, Jishtu V, Thakur L et al (2020) Medicinal plants of the trans-Himalayan Cold Desert of Ladakh—a review. Int J Sci Environ 9:239–253

Chaurasia OP, Ahmed Z, Ballabh B (2007) Ethnobotany and plants of trans-Himalaya. Satish Serial Pub House 509–520

Chopra RN, Chopra IC, Varma BS (1969) Glossary of Indian medicinal plants/supplement. Council of Scientific and Industrial Research, India

Clubbe C, Hamilton M, Corcoran M (2010) Using the global strategy for plant conservation to guide conservation implementation in the UK Overseas Territories. Kew Bull 65:509–517

Dahal S, Sharma TP, Borthakur SK (2017) Database on medicinal plants of Tamze Medicinal Plants Conservation Area (MPCA) of Sikkim Himalaya, India. NeBio J Environ Biodivers 8:45–56

Dar KB, Bhat AH, AmIN S et al (2016) Efficacy of aqueous and methanolic extracts of *Rheum spiciformis* against pathogenic bacterial and fungal strains. J Clin Diagnos Res 10:BC18

Das D, Maulik SR, Bhattacharya SC (2008) Colouration of wool and silk with *Rheum emodi*. Indian J Fibre Text Res 33:163–170

Devkota A, Dall'Acqua S, Comai S (2010) *Centella asiatica* (L.) urban from Nepal: quali-quantitative analysis of samples from several sites, and selection of high terpene containing populations for cultivation. Biochem Syst Ecol 38:12–22

Dregus M, Schmarr HG, Takahisa E (2003) Enantioselective analysis of methyl-branched alcohols and acids in Rhubarb (*Rheum rhabarbarum* L.) stalks. J Agric Food Chem 51:7086–7091

Dutta PK, Dutta BK, Sundriyal RC et al (2013) Diversity, representativeness and biotic pressure on plant species along alpine timberline of western Arunachal Pradesh in the eastern Himalaya, India. Curr Sci 701–708

Farooq U, Abaas G, Saggoo MIS et al (2014) Ethno-botany of some selected Monochlamydeae plant species from the Kashmir Himalaya, India. J Med Plant Res 8:834–839

Fei Y, Wang J, Peng B, Peng J, Hu JH, Zeng ZP et al (2017) Phenolic constituents from *Rheum nobile* and their antioxidant activity. Nat Prod Res 31(24):2842–2849

Feng T, Pucker B, Kuang T, Song B, Yang Y, Lin N, Zhang H, Moore MJ, Brockington SF, Wang Q, Deng T, Wang H, Sun H (2015)

Ganie AH, Tali BA, Khuroo AA et al (2014) *Rheum spiciforme* Royle (Polygonaceae): a new record to the flora of Kashmir Valley, India. Natl Acad Sci Lett 37:561–565

Goraya GS, Jishtu V, Rawat GS et al (2013) Wild medicinal plants of Himachal Pradesh: an assessmentof their conservation status and management prioritization (CAMP). Himachal Pradesh ForestDepartment, Shimla, Himachal Pradesh, India

Goraya GS, Ved DK (2017) Medicinal plants in India: an assessment of their demand and supply. National Medicinal Plants Board, Ministry of AYUSH, Government of India, New Delhi and Indian Council of Forestry Research and Education, Dehradun, pp 100-5

Hooker JD (1897) The flora of British India, vol 7. L. Reeve & Co. Ltd. The Oast House. Brook, Nr. Ashford, Kent, pp 842

Huiyan X, Xiaofeng Z, Huan W et al (2003) HPLC determination of anthraquinones from various parts of *R. Tanguticum*, *R. undulatum* and *R. spiciforme*. Acta Bot Sin 23:328–331

Inamori Y, Kato Y, Kubo M et al (1983) Studies on metabolites produced by *Aspergillus terreus* var. aureus. I. Chemical structures and antimicrobial activities of metabolites isolated from culture broth. Chem Pharm Bull 31:4543–4548

Iwashina T, Omori Y, Kitajima J et al (2004) Flavonoids in translucent bracts of the Himalayan *Rheum nobile* (Polygonaceae) as ultraviolet shields. J Plant Res 117:101–107

Kabir Dar A, Siddiqui MAA, Wahid-Ul H et al (2015) Threat status of *Rheum* emodi—a study in selected Cis-Himalayan Regions of Kashmir Valley Jammu and Kashmir India. Med Aromat Plant 4:183–186

Kala CP (2000) Status and conservation of rare and endangered medicinal plants in the Indian trans-Himalaya. Biol Conserv 93:371–379

Kalsang T (2018) Climate change and its effect on the medicinal plants

Khan MI, Pandith SA, Salika R (2019) *Rheum moorcroftianum* (Polygonaceae) in Kashmir Himalaya. Phytotaxa 405:269–275

Khan MI, Shahzad A, Ganie IB et al (2022) Organogenesis, direct somatic embryogenesis, and shoot proliferation of *Rheum* spiciforme Royle: an endemic and vulnerable medicinal herb from Indian Trans Himalayas. In Vitro Cell Dev Biol Plant 58:35–50

Khan MI, Bashir N, Pandith SA, Shahzad A, Barvkar VT, Pable AA et al (2023) Co-occurrence of two ascomycete endophytes as the specialized metabolite production partners in *Rheum spiciforme* Royle. Symbiosis 1–16

Kumar GP, Kumar R, Chaurasia OP et al (2011) Conservation status of medicinal plants in Ladakh: cold arid zone of trans-Himalayas. J Med Plant Res 5:685–694

Kumar A, Mitra M, Adhikari BS et al (2015) Depleting indigenous knowledge of medicinal plants in cold-arid region of Nanda Devi Biosphere Reserve, Western Himalaya. Med Aromat Plants 4:2167–2178

Kumar A, Mitra M, Adhikari BS et al (2016) Flora of Niti valley: a cold arid region of Nanda Devi Biosphere reserve, Western Himalaya, India. Check List 12:1–16

Kumar A, Sathyakumar S, Goraya GS et al (2021a) Assessment of medicinal and aromatic plant species on their collection, usage, demand, markets, price trends and life cycle in Lahaul and Pangi landscape, Himachal Pradesh. A report submitted to Himachal Pradesh Forest Department and United Nations Development Programme, pp 141

Kumar A, Sathyakumar S, Goraya GS et al (2021b) Sustainable harvesting and cultivation protocols of threatened medicinal and aromatic plants of the Western Himalaya. Submitted to Himachal Pradesh Forest Department and United Nations Development Programme

Kumar A, Adhikari BS, Mishra AP (2023) Status of threatened medicinal and Biosphere Reserve, Uttarakhand. A report (TR No./2022/35) submitted to International Union for Conservation of Nature—India, p 103

Lee MR, Hutcheon J, Dukan E et al (2017) Rhubarb (*Rheum* species): the role of Edinburgh in its cultivation and development. J R Coll Phys Edinb 47:102–109

Li Y, Liu H, Ji X et al (2000) Optimized separation of pharmacologically active anthraquinones in Rhubarb by capillary electrochromatography. Electrophoresis 21:3109–3115

Losina-Losinskaya AS (1936) The genus *Rheum* and its species. J Inst Bot Acad Sci Common Things Soveticarum Union Socialisticarum Ser 1:5–141

Lust J (1974) The herb book. Bantam Books, New York, pp 156–169

Mathela M, Bargali H, Sharma M et al (2020) Brainstorming on the future of the highly threatened medicinal plants of the Western Himalaya, India. Curr Sci 118:1485–1486

Mathela M, Kumar A, Sharma M et al (2021) Hue and cry for *Fritillaria cirrhosa* D.Don, a threatened medicinal plant in the Western Himalaya. Discov Sustain 2:38

Mathur V, Sivakumar K, Onial M et al (2014) India's fifth national report to convention on biological diversity. Ministry of Environment, Forests and Climate Change, Government of India 5(10)

Mishra AP, Kumar A, Yadav SN (2023) Ecology and conservation of threatened medicinal plants in the trans-Himalayan region of Nanda Devi Biosphere Reserve, Western Himalaya. Trees For People 14:100451. https://doi.org/10.1016/j.tfp.2023.100451

Misiti F, Sampaolese B, Mezzogori D et al (2006) Protective effect of rhubarb derivatives on amyloid beta (1–42) peptide-induced apoptosis in IMR-32 cells: a case of nutrigenomic. Brain Res Bull 71:29–36

Mohtashami L, Amiri MS, Ayati Z et al (2021) Ethnobotanical uses, phytochemistry and pharmacology of different *Rheum* species (Polygonaceae): a review. In: Pharmacological properties of plant-derived natural products and implications for human health, 2021, pp 309–352

Molur S, Walker S (1998) Conservation Assessment and Management Plan (CAMP) workshop report. Zoo Outreach Organisation and CBSG, India, Kulu, Himachal Pradesh

Mudgal V, Hajra PK (1997) Floristic diversity and conservation strategies in India. In: Hajra PK (ed) Botanical Survey of India, Ministry of Environment and Forests, vol 1, p 409

Naithani BD (1984) Flora of chamoli, vol I & II. Botanical Survey of India, Howrah, New Delhi, India

Nautiyal MC, Nautiyal BP (2004) Agrotechniques for high altitude medicinal & aromatic plants. Bishen Singh and Mahender Pal Singh, Dehradun, pp 158–161

Nayar MP, Sastry ARK (1987–1990) Red data book of Indian plants, vol I, II, III. Botanical Survey of India, Calcutta

Olsen CS (1998) The trade-in medicinal and aromatic plants from central Nepal to northern India. Econ Bot 279–292

Pandey A, Belwal T, Sekar KC, Bhatt ID, Rawal RS (2018) Optimization of ultrasonic-assisted extraction (UAE) of phenolics and antioxidant compounds from rhizomes of Rheum moorcroftianum using response surface methodology (RSM). Ind Crops Prod 119:218–225

Pandith SA, Dar RA, Lattoo SK et al (2018) *Rheum australe*, an endangered high-value medicinal herb of North Western Himalayas: a review of its botany, ethnomedical uses, phytochemistry, and pharmacology. Phytochem Rev 17:573–609

Prasad P, Purohit MC (2001) Altitude acclimatization and concentration of active constituents and calorific value of two medicinal plant species *Rheum emodi* and R. nobile (Rhubarb) in Sikkim Himalaya. Curr Sci 80(6):734–736

Rajasekharan PE, Ganeshan S (2002) Conservation of medicinal plant biodiversity—an Indian perspective. J Trop Med Plants 3:125–140

Rana SK, Rana HK, Stöcklin J et al (2022) Global warming pushes the distribution range of the two alpines 'glasshouse' Rheum species north and upwards in the Eastern Himalayas and the Hengduan Mountains. Front Plant Sci 13

Rana MS, Samant SS (2011) Diversity, indigenous uses, and conservation status of medicinal plants in Manali wildlife sanctuary, Northwestern Himalaya. Indian J Tradit Knowl 10:439–459

Rao SK, Page N, Kumar D (2020) Pan India bouquets. http://flora-peninsula-indica.ces.iisc.ac.in/pan/plants.php?name=Rheumtibeticum. Downloaded on 27 May 2023

Rashid S, Kaloo ZA, Singh S, Bashir I (2014) Callus induction and shoot regeneration from rhizome explants of *Rheum webbianum* Royle—a threatened medicinal plant growing in Kashmir Himalaya. J Sci Innov Res 3(5):515–518

Rokaya MB, Maršík P, Münzbergová Z (2012) Active constituents in *Rheum acuminatum* and *Rheum australe* (Polygonaceae) roots: a variation between cultivated and naturally growing plants. Biochem Syst Ecol 41:83–90

Samant SS, Dhar U, Palni LMS (1998) Medicinal plants of Indian Himalaya. Gyanodaya Prakashan, pp 35–46

Samant SS, Pal M (2003) Diversity and conservation status of medicinal plants in Uttaranchal State. Indian For 129:1090–1108

Santapau H, Henry AN (1973) Dictionary of the flowering plants in India

Schippmann U, Leaman DJ, Cunningham AB (2002) Impact of cultivation and gathering of medicinal plants on biodiversity: global trends and issues. In: Biodiversity and the ecosystem approach in agriculture, forestry, and fisheries

Shah R (2006) Nature's medicinal plants of Uttaranchal: herbs. Grasses and Ferns, Gyanodaya Prakashan, Nainital, Uttarakhand, India

Shankar R, Tripathi AK, Neyaz S et al (2015) Distribution of medicinal plants in Kurung Kumey, East Kameng and West Kameng districts of Arunachal Pradesh: their systemic conservation. World J Pharm Res 4:977–993

Shi T, Liao Z, Ye R, Xu L, Sun H, Ji L (2014) Study on chemical constituents of *Rheum nobile* Hook. f. et Thoms. Northwest Pharmaceut J 29(6):571–573

Singh PB (1999) Commercially important medicinal and aromatic plants of Himachal Pradesh. Society for Herbal Medicine and Himalayan Biodiversity, Ram Nager Mandi, Himachal Pradesh, India

Singh R, Tiwari T, Chaturvedi P (2017) *Rheum emodi* Wall ex. meissn (Indian Rhubarb): highly endangered medicinal herb. J Med Plants Stud 5:13–16

Singh P, Dash SS, Sinha BK (2019) Plant of Indian Himalayan region (an annotated checklist and pictorial guide). Botanical Survey of India, Kolkata, p 650

Singh JS (2006) Sustainable development of the Indian Himalayan region: linking ecological and economic concerns. Curr Sci 784–788

Smith Olsen C, Overgaard Larsen H (2003) Alpine medicinal plant trade and Himalayan mountain livelihood strategies. Geogr J 169:243–254

Song B, Gao Y, Stöcklin J et al (2020) Ultraviolet screening increases with elevation in translucent bracts of *Rheum nobile* (Polygonaceae), an alpine 'glasshouse' plant from the high Himalayas. Bot J Linn 193:276–286

Srivastava RC (2014) Family polygonaceae in India. Indian J Plant Sci 3:112–150

Srivastava SK, Shukla AN (2015) Flora of Cold Desert Western Himalaya, India. Government of India, Botanical Survey of India 2:4–5

Stewart RR (1916–1917) The flora of Ladhak, Western Tibet. The New Era Printing Company, Lancaster, p 650

Sun Y, Wang A, Wan D et al (2012) Rapid radiation of *Rheum* (Polygonaceae) and parallel evolution of morphological traits. Mol Phylogenet Evol 63:150–158

Tabin S, Kamili AN, Ganie SA et al (2016) Genetic diversity and population structure of *Rheum* species in Kashmir Himalaya based on ISSR markers. Flora 223:121–128

Tabin S, Kamili A, Gupta RC (2018) Micropropagation and conservation of *Rheum webbianum* collected from Zanaskar valley via tissue culture. Ann Plant Sci 7:2187–2203

Takhtajan A (1986) Floristic regions of the world. Univ California press, London

Tamang M, Pal K, Kumar Rai S et al (2017) Ethnobotanical survey of threatened medicinal plants of West Sikkim. Int J Bot Stud 2:116–125

Tayade A, Dhar P, Ballabh B et al (2012) *Rheum webbianum* royle: a potential medicinal plant from trans-Himalayan cold deserts of Ladakh, India. Plant Arch 12:603–606

Thakur RS, Puri HS, Hussain A (1989) Major medicinal plants of India. Central Institute of medicinal and aromatic plants, Lucknow, India, pp 1–100

Tsering J, Tag H (2015) High altitude ethnomedicinal plants of Western Arunachal Himalayan landscape. Pleione 9:18–25

Uniyal RC, Uniyal MR, Jain P (2000) Cultivation of medicinal plants in India: a reference book. TRAFFIC-India

Uniyal BP, Sharma JR, Choudhery U et al (2007) Flowering plants of Uttarakhand (A checklist). Bishen Singh Mahendra Pal Singh, pp 404

Usui T, Ikeda Y, Tagami T et al (2002) The phytochemical lindleyin, isolated from Rhei rhizoma, mediates hormonal effects through estrogen receptors. J Endocrinol 175:289–296

Ved DK, Kinhal GA, Ravikumar K et al (2003) Conservation assessment and management prioritization for medicinal plants of Himachal Pradesh, Jammu, and Kashmir, and Uttarakhand. In Workshop: Foundation for Revitalization of Local Health Traditions, Bangalore, India

Ved DK, Kinhal GA, Ravikumar K et al (2005) Conservation Assessment and Management Prioritisation (CAMP) for wild medicinal plants of North-East India. Med Plant Cons 11:40–44

Ved DK, Tandon V (1998) CAMP report: conservation assessment and management plan workshop, Kullu, Himachal Pradesh. FRLHT, Bangalore, India, pp 75un

Wang A, Yang M, Liu J (2005) Molecular phylogeny, recent radiation and evolution of gross morphology of the rhubarb genus *Rheum* (Polygonaceae) inferred from chloroplast DNA trnLF sequences. Ann Bot 96:489–498

Zheng QX, Wu HF, Jian GUO et al (2013) Review of rhubarbs: chemistry and pharmacology. Chin Herb Med 5:9–32

Zhu L, Wei W, Zheng YQ et al (2005) Effects and mechanisms of total glucosides of paeony on joint damage in rat collagen-induced arthritis. J Inflamm Res 54:211–220

Extent, Abundance, Threats and Information Gap on Alpine Medicinal and Aromatic Plants of Western Himalaya, India: A Review

Gajendra Singh, Naveen Chandra, Dhani Arya, and Lalit M. Tewari

Abstract Alpine region of western Himalaya arbours rich array of medicinal and aromatic plants (MAPs) owing to diverse topography and climatic conditions. These MAPs are significant for traditional system of medicine. However, our understanding on their extent, abundance and management strategies are meagre, leading their population declined at an unprecedented rate. Hence, the present study assesses the directionality, quality and sufficiency of past and ongoing research for conservation and management of MAPs wealth in the region. We analysed over 300 research articles on MAPs of alpine region to consolidate the knowledge on population status, ethnobotany, biotechnological studies, anthropogenic pressure and information gaps and future scope for research. The results reveal that the landscape has long history of MAPs studies dates back to 1849 and progressed remarkably after 1990. Initially (until 2000) most of the studies were on surveys (diversity) and ethnobotanical aspects, extended to biochemistry and biotechnology after 2000. Studies are very site specific (of the total reported alpine meadows (82) of the state, about 45% of the studies conducted only in 10 meadows) and species specific, where of the total MAPs reported from the region, only 8–10 species are studies across meadows viz., *Angelica glauca* in 25 meadows, *Picrorhiza kurroa* in 16 meadows and *Dactylorhiza hatagirea* in 14 meadows. Majority of the studies are confined to biochemistry (23%), followed by biotechnology (12.2%) and ethnobotany (9.4%). Till date, the abundance of only 31 MAPs is studied in few meadows, having average 1–2 individuals/m^2. Only 16 species have been studies for multiplication in laboratories using biotechnological tools, however none of them succeed in producing at large scale. Inadequate information on geo-spatial extant of species, habitat suitability, lack of information on demand and supply ratio, climate change impact, no clear-cut management

G. Singh (✉)
Forestry and Climate Change Division, Uttarakhand Space Application Centre, Dehradun, Uttarakhand, India
e-mail: gajendrawat@yahoo.com

N. Chandra · D. Arya
Department of Botany, S.S.J. University Campus, Almora, Uttarakhand, India

L. M. Tewari
Department of Botany, Kumaun University, Nainital, India

A. P. Mishra et al. (eds.), *Threatened Medicinal Plants in the Indian Himalayan Region*, World Sustainability Series, https://doi.org/10.1007/978-3-031-73687-2_5

strategies for MAPs were identified as area of immediate concern. Therefore, despite some national, state and scattered local initiatives on management, conservation and sustainable utilization of MAPs, there still exist gaps in our knowledge and understanding in extent and availability of MAPs resource base in the alpine region of western Himalaya.

Keywords Medicinal plants · Alpine · Population status · Ethnobotany · Himalaya

1 Introduction

Medicinal and Aromatic plants gained enough popularity globally as a source of raw material for pharmaceuticals and traditional system as principal health care resource (Kandari et al. 2012; Kumar et al. 2023). In India, plants are being used as the main therapeutic agent since ancient times, as described in Charak and Susrut Samhita (Prajapati et al. 2003). India, one of the twelve mega diversity in the world has about 15,000 medicinal plant species that include 7000 plants used in Ayueveda, 700 in Unani, 600 in Siddha, 450 in Homeopathy and 30 in modern medicine (Das 2008). Indian Himalayan Region (IHR) cover 18% of the India's geographical area have reported 1748 species of medicinal plants (Samant et al. 1998), of which about 118 species yield essential oils (Samant and Palni 2000). Of the total medicinal plant species, 62 species of medicinal plants are endemic to the Himalayan region and 208 extend their distribution to the adjacent areas, and are therefore, classified as near endemic (Chatterjee 1939). The maximum number of species (964) are reported from Uttarakhand (Kala et al. 2004) followed by North Bengal (707) and Sikkim (707) (Samant et al. 1998). Today, more than 90% of plant species used in the pharmaceutical industry are harvested from the wild, many of these come from the sub-alpine and alpine regions of the Himalaya (Singh and Rawat 2011). Presently, the pharmaceutical sector in India is using about 280 medicinal plant species, of which 175 are come from the Indian Himalayan Region (Dhar et al. 2000). These plant species, having slow growth rate, low population density and narrow geographic ranges (Kala 1998; Mishra et al. 2023a), are more prone to extinction due to over utilization (Jablonski 2004). Documentation of knowledge systems will help in conserving the biodiversity and domestication of medicinally important plants. Most of MAPs found in different terrain in high altitude region of Himalaya and thus very little information is available on variability and multiplication of these species (Mishra et al. 2023b).

Western and North-West Himalaya, including Uttarakhand, Himanchal Pradesh, Jammu-Kashmir and Ladakh a distinct botanical sub-region arbour maximum alpine region in the IHR (Lal et al. 1991). Jammu-Kashmir and Ladakh (33,000 km^2), Himachal Pradesh (14,000 km^2) and Uttarakhand have 8000 km^2 area under alpine vegetation. More than 964 species of MAPs have been reported from Uttarakhand (Kala 2010), 360 from Himachal Pradesh (Badola 2001) and about 650 species from Kashmir Himalaya. The physiognomic unit of alpine vegetation in Western Himalaya include the stunted forest, alpine scrub, alpine meadows and pioneer communities on scree slopes and moraines. There are 1350–1400 angiospermic flora reported from alpine region of Uttarakhand, 400 having medicinal properties (Rawat et al. 2005). Nearly thirty species from the Garhwal Himalaya have been listed as threatened in the India Red Data Books, of which 24 species are from high alpine region (Nautiyal 1996). There are approximately 15–20 species are endemic to the Himalayan high-altitude region of Uttarakhand. Six MAPs from alpine region of Uttarakhand are under the list of Convention on International Trade in Endangered Species of Wild Flora and Fauna (CITES).

Medicinal plants are subject to various level of threats; some of them are anthropogenic in nature (viz., over exploitation, grazing, trampling, construction of roads and unregulated tourism) and some are non-anthropogenic viz., low population growth owing to habitat specificity, low population size, narrow distribution range, invasion by exotic species, climate change and genetic drifts (Kala 1998, 2000; Rawat et al. 2021). These plant species, having slow growth rate, low population density and narrow geographic ranges (Kala 1998; Dhyani and Kala 2005), are prone to extinction (Jablonski 2004). Most of the MAPs found in difficult terrain in high altitude region of Himalaya and thus very little information is available on variability and multiplication of these species.

Documentation of existing knowledge helps in planning and re-orientation of the strategies for conservation and management of any area in particular to biodiversity. Lack of data restricts resource managers and decision makers from making accurate evaluations for interventions for the biodiversity of a targeted region or taking scientifically based conservation decision. Such information has been used to study the taxonomy and biogeography of species for hundreds of years. Species listing which contains primary and secondary species data are critical to obtain information about the baseline conditions, understand distribution of species and analyse the pattern of area coverage. Secondary data from the literature facilitates comparison among and between different studies to understand landscape-level changes in distribution and population trends. Keeping all these points into the consideration, the major objectives of the present review are: (i) to assess the current state of knowledge on MAPs of alpine region covering various disciplines in Western Himalaya, (ii) to find out the major research and information gaps along with suggested measures for management.

2 Material and Methods

2.1 Study Area

Western Himalaya comprises rich array of high value MAPs since ancient times, where alpine landscapes contributes significantly having about 24.11% of the geographical area above 3000 m asl. The physiognomic unit of alpine vegetation in western Himalaya include the stunted forest, alpine scrub, alpine meadows and pioneer communities on scree slopes and moraines. Based on altitude, aspect, moisture availability and duration of growing season, six types of alpine meadows are recognised in WH viz., tall forbs, mixed herbaceous formations or short forbs, matted shrubs/shrubberies, *Danthonia* grasslands, *Kobresia* sedge meadows and cushionoid vegetation (Rawat 2005). Rich habitat diversity of the alpine region of western Himalaya provides habitat to many threatened floral and faunal species.

In all more than 80 alpine meadows have been listed (Rawat 2005) in Uttarakhand having most extensive and prominent meadows around basins of Nanda Devi, Panchachuli, Kedarnath, Gangotri and Bandarpoonch. These meadows range in size from a few to 400 sq. km. The life of the local inhabitants in high altitudes is intimately associated with the alpine meadows as evident from a large number of folklores. Several religious sites are located in and around meadows where, local people visit for worshiping the deities. A large number of migratory and local pastoral communities uses alpine pastures as summer grazing grounds. Increase in number of livestock and overuse of certain pastures has led to degradation of high-altitude grasslands and habitats of several MAPs, RET species.

2.2 Data Collection and Analysis

We carried out a systematic review of accessible literature related to the medicinal aromatic plants of alpine region of western Himalaya. We used 'Science Direct', 'Google Scholar', and Research Gate web-based searches, using specific key words 'Medicinal plant' independently and also combined with other terms viz., alpine landscape, Himalaya, Western Himalaya, population, ethnobotany, biotechnology, biochemistry, diversity, threats etc. Information those did not appear with these search terms or were in the form of hard copies (books) were collected from various libraries and personnel interaction with the authors. A total of 285 peer reviewed articles were used for analysis. Once collected, articles were chronologically listed in a Microsoft Excel sheet and analysed for the following sub-categories: study sites, publication year and area focus of the publication. We have included the published paper until 2018. To understand the past distribution (occurrence) of MAPs in the alpine region of WH various herbarium viz., Wildlife Institute of India (WII), Botanical Survey of India (BSD), Forest Research Institute (DD), Kumaun University (KU), Regional Ayurvedic Research Institute (RARI), Almora were consulted.

3 Results and Discussions

Current status of MAPs and diversity in alpine region of Western Himalaya: The availability of diverse medicinal plants in the Himalaya has been recorded in ancient Indian scripts as far back as 1000 BC. The earliest studies on MAPs of the region were started with Sir Richard Strachey and J. E. Winterbottom, two army officers who passed through hills of Kumaun and Garhwal to Tibet in 1846–1849. These two surveyors have made best collections in the Himalayan region of the state as well as in Tibet and adjoining parts of Garhwal. Profundity literature survey of High-altitude medicinal plants reveal that most of studies were conducted after the year 2000. The majority of publication were found in Population and distribution (11.2%), Biochemistry (15%), Biotechnology (14%), Ethnobotany (12.9), Anthropogenic pressure (~3%), Climate change (~2.4%) (Figs. 1 and 2).

There are about 82 alpine meadows are reported from Uttarakhand state (Rawat 2005), however, maximum studies are confined to only 10 alpine meadows covering about 45% of the total studies published (Fig. 3). Most of the studied are confined to Tungnath and Kedarnath area, both are part of Kedarnath Wildlife Sanctuary followed by Valley of Flowers (Chamoli), Dayara (Uttarkashi) and Pindari (Bageshwar).

Fig. 1 Decadal studies conducted in alpine region of Uttarakhand

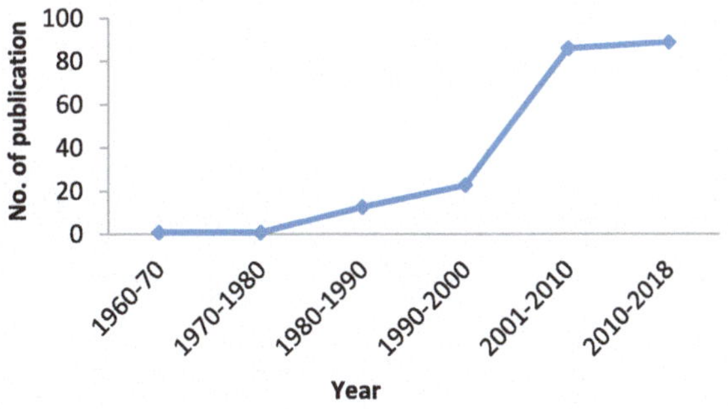

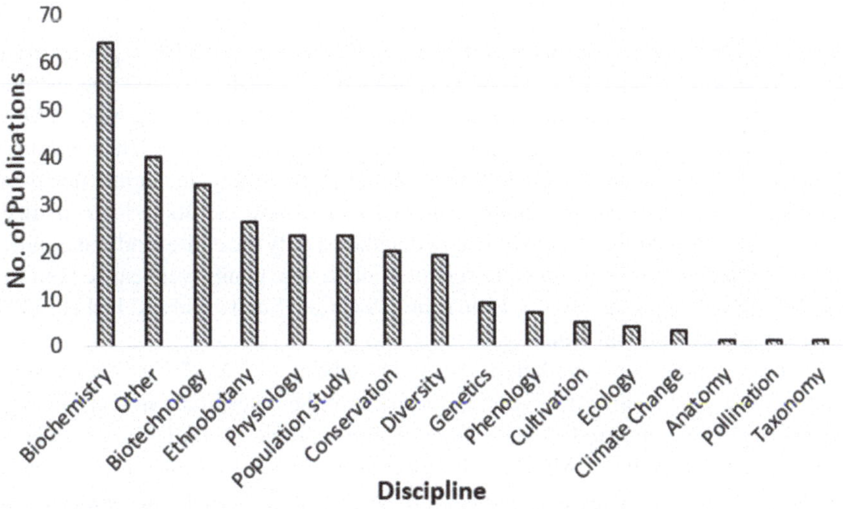

Fig. 2 Subject focus of publications in the alpine region of Uttarakhand, WH

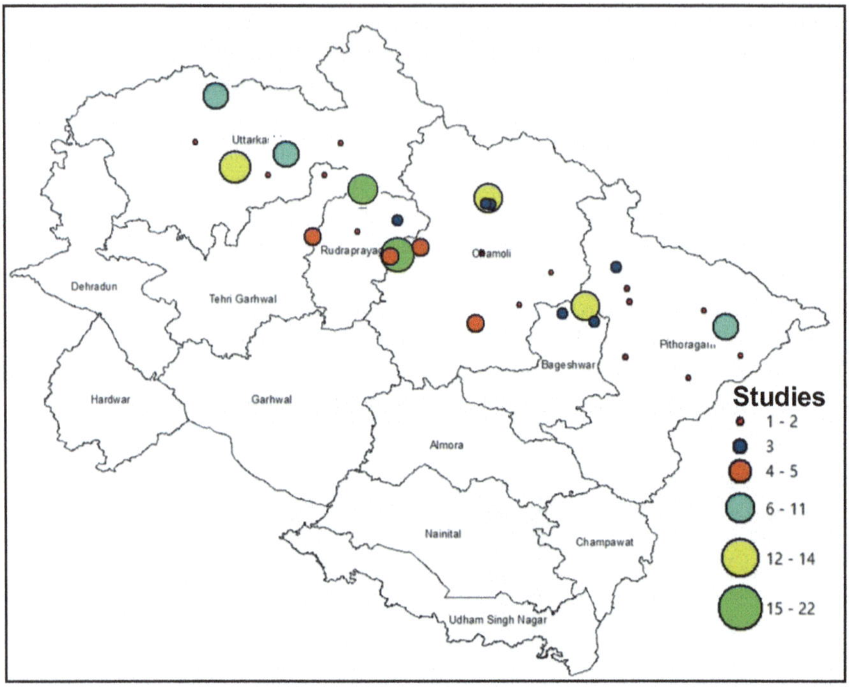

Fig. 3 Distribution of studies in the alpine region of Uttarakhnad, WH

3.1 Diversity and Population Status of Various MAPs

There are 1350–1400 plant species (80% of total alpine flora of WH) reported from the alpine region (3000–5300 m) of Uttarakhand, of which 350–400 species are expected to have medicinal properties (Rawat 2005). Of the total reported MAPs, 90% are herbaceous followed by shrub (5%) and tree (5%) species. There are few studies carried out on the population discipline of High-altitude medicinal plants in Uttarakhand. Rawat (2005) estimated density of 14 woody and 36 herbs medicinal plants species from alpine region of Uttarakhand. It is only the entire landscape study for high altitude medicinal plants. Diversity of medicinal plants were carried out by Badoni (1989), Negi et al. (1985), Joshi et al. (1982) and Sundariyal and Joshi (1987) in some parts of Garhwal Himalaya.

Intensive studies on population status (availability) of MAPs is started very recently after year 2000, where species-specific population estimation of one or two species or site-specific studies were more common (Table 1). Initially, researchers have studied the status of MAPs in forest communities e.g., Rawal et al. (1994) estimated status of *Hipphophae salicifolia* and found 703 individuals/hectare in *Quercus* dominated forest in Pindari valley. Airi (2000) estimated population density of *Podophyllum hexandrum* in *Quercus-Abies* forest of Kumaun Himalaya. Singh and

Rawat (2011) have discussed about the status of 32 MAPs between 2700–3300 m elevation in the part of Kedarnath Wildlife Sanctuary. There are limited studies on the population status of MAPs at landscape level. Rawat (2005) on his survey of 100 MAPs in the WH, estimated the population status (density) of 14 woody and 36 herbaceous species over 80 alpine meadows of WH. In the similar line, Tiwari et al. (2010) have assessed the population status and geo-spatial extent of 50 high altitude MAPs in 15 forest divisions of Garhwal Himalaya.

3.2 *Ethnobotanical Studies on High Altitude Medicinal Plants*

Ethnobotanical knowledge plays a significant role in plant diversity conservation and curing of various ailments in remote rural areas of the Himalaya. In last few decades, ethnobotany has become an important thrust area of research for documentation and preservation of historical traditional knowledge of communities. Dollard first published the comprehensive work on 'Medicinal topography of Kumaon and Shore valley' in 1840. However, 'Glimpses of Indian Ethnobotany' (Jain 1981) is the first book dealing with Indian Ethnobotany. Among the notable ethnobotanical studies in the alpine region of WH, Rawat and Pangtey (1987) had enumerated 148 plant species from Kumaun, used by various ethnic groups viz., Bhotiyas, Joharees, Darmess, Chaudansees and Byansees to cure various ailments. Similarly, Singh and Rawat (2011) enumerated 126 species while they were doing ethnobotany of Kedarnath WS. Detailed ethnobotanical studies conducted in the alpine of Uttarakhand in (Table 2).

Historically, the geographic barriers played a major role to weaken the linkages and localized distribution of ethnobotanical knowledge developed by various ethnic communities on the use and management of surrounding biological resources. The extreme ecological conditions of high altitude have shaped a life-style characterized by periodic movement between two settlements, one of which is in the valley where people reside in winter due to heavy snowfall on higher reaches, where they stay during summer. Studies on ethnobotany are not only to tap our traditional knowledge but also to add ample understanding on culture, traditions and advent of civilization with the nature (Badoni and Badoni 2001). A consolidated account and analysis of the total use pattern with regard to ailment and traditional treatments by different communities is not available. In course of time, the traditional knowledge systems have started to decline due to several reasons. The erosion of traditional knowledge is mainly due to lack of sustainable development policy towards promoting traditional knowledge-based programmes (Negi et al. 2002).

Traditionally information on the use of plant species for therapeutic purpose passed from one generation to the next through oral tradition, this knowledge of therapeutic plants has started to decline and become outmoded through the lack of recognition by younger generations because of a shift in attitude and ongoing socio-economic changes. Many remote areas in the Himalaya contain a rich traditional

Table 1 The population density of MAPs across the various Bugyals in Western Himalaya

MAPs name	Study area/meadows (density-individuals/km^2)	Key references
Aconitum balfourii	Dyara (2.4), Har Ki Doon (0.8), Kedarnath (1.4), Madhmaheshwar (1.6), Panwali Kantha (1.5), Tungnath (1.3), Valley of Flowers (1.8)	Nautiyal et al. (2002)
A. ferox	Har Ki Doon (1), Madhmaheshwar (1.3), Panwali Kantha (1.2), Milam (0.9), Martoli (0.8), Laspa (0.6), Phurkia (1.1), Dwali (0.9)	Bhatt et al. (2014a, b, c)
A. heterophyllum	Har Ki Doon (1.0), Kedarnath (1.8), Madhmaheshwar (1.2), Panwali Kantha (1.5), Tungnath (2.57), Milam (1.1), Martoli (1), Laspa (1.1), Phurkia (0.7), Dwali (1)	Bhatt et al. (2014a, b, c), Nautiyal et al. (2002)
A. violaceum	Kedarnath (4.5), Panwali Kantha (7.8), Hemkund (6.7)	Nautiyal et al. (2002)
	Kyarki (0.27)	Rawat (2005)
	Devikund (0.2)	Chandra Sekar and Rawat (2011)
Aconogonum tortuosum	Byans (0.09)	Rawat (2005)
Allium wallichii	Ralam (0.30), Johar (0.04), Kyarki (0.01)	Rawat (2005)
Angelica glauca	Chipkon (0.6), Garpak (0.6), Phagati forest (0.3), Rishikund (0.5), Lata Kharak (0.2),	Kandari et al. (2011)
	Sunderdunga (0.95), Phurkia (0.95)	Bisht et al. (2006)
	Devikund (1.2)	Chandra Sekar and Rawat (2011)
	Milam (0.90), Martoli (0.70), Laspa (0.77), Phurkia (0.97), Dwali (0.83), Sumdum (0.80), Tejum (0.87)	Arya et al. (2015)
	Bharnala (0.7), Tungnath (1.2), Kunwari pass (0.9), Panwali Kantha (0.8), Rudranath (0.8), Kedarnath (0.7), Valley of Flowers (1.5), Dayara (0.8)	Vashistha et al. (2006)
Arnebia benthami	Chipkon (0.9), Garpak (1), Phagati forest (0.4), Rishikund (0.3), Lata Kharak (0.2)	Kandari et al. (2011)
Bergenia strecheyi	Pindari (1.63)	Chandra Sekar and Rawat (2011)
	Gori valley (1.1)	Uniyal et al. (2000)
	Kedarnath Wildlife Sanctuary (13.2)	Semwal et al. (2007)
Carum carvi	Byans (0.33), Johar (0.09)	Rawat (2005)
Chaerophyllum villosum	Johar (0.43)	Rawat (2005)
Corydalis govaniana	Johar (0.01), Kyarki (0.03), Khatling (0.05)	Rawat (2005)

(continued)

Table 1 (continued)

MAPs name	Study area/meadows (density-individuals/km^2)	Key references
Dactylorhiza hatagirea	Valley of Flowers (2.02), Nagtal (2.19), Pindari (1.27), Lata (1.13), Donidhar (1.89), Kedarnath (1.64)	Bhatt et al. (2005)
	Darma (0.02), Johar (0.03), Gidara (0.08)	Rawat (2005)
	Kedarnath Wildlife Sanctuary (3.5)	Semwal et al. (2007)
	Devikund (0.25)	Chandra Sekar and Rawat (2011)
	Gori valley (4.7)	Uniyal et al. (2002)
Fritillaria roylei	Dayara (3.8), Dronagiri (2.4), Kedranath (4.20), Kunwari pass (1.1), Rudranath (3.8), Tungnath (0.40), Valley of Flowers (1.8)	Chauhan et al. (2008)
	Byans (0.1), Kyarki (0.1)	Rawat (2005)
Habernaria intermedia	Madhmaheshwar (1.5), Chopta (1.2), Ghimtoli (3.4), Mussoorie (2.1), Rambara (1.4), Rudranath (1.5)	Chauhan et al. (2007)
Iris kumaonensis	Byans (0.47), Ralam (3.70), Girthi (0.23), Kyarki (0.78), Khatling (0.47)	Rawat (2005)
Malaxis acuminata	Dayara (7.8), Madhmaheshwar (2.8), Panwali Kantha (2.5), Tungnath (0.6), Baniyakund (10), Chopta (0.6), Kunwari pass (7.2), Rudranath (4), Maggu (2), Dronagiri (8)	Chauhan et al. (2008)
Meconopsis aculeata	Devikund (0.3)	Chandra Sekar and Rawat (2011)
Nardostachys jatamansi	Pindari (17.49), Sunderdhunga (15.12), Kaphani (10.59)	Airi et al. (2000)
Paeonia emodi	Maliyadhaura (1.60), Khati (2.40)	Rawal and Rawat (2012)
Parnassia nubicola	Byans (0.3), Darma (0.2), Johar (0.5), Gidara (0.23), Kyarki (0.78)	Rawat (2005)
Picrorhiza kurrooa	Laspa (3.7), Burfu (1.7), Bilju (1.8), Martoli (1.6), Ganghar (1.3), Milam (1.2), Sumdum (1.8)	Arya et al. (2015)
	Gori valley (3.9)	Uniyal et al. (2002)
	Ralam (0.33), Johar (0.24), Gidara (1.61), Kyarki (0.11)	Rawat (2005)
	Kedarnath Wildlife Sanctuary (2.2)	Semwal et al. (2007)
	Devikund (1.1)	Chandra Sekar and Rawat (2011)
Pleurospermum angelicoides	Chipkon (0.5), Garpak (0.6), Phagati forest (0.3), Rishikund (0.4), Lata Kharak (0.3)	Kandari et al. (2011)
Polygonatum verticillatum	Dayara (2.6), Tungnath (2.8)	Bhatt et al. (2014a, b, c)
	Johar (0.6). Gidara (0.33), Khatling (0.10)	Rawat (2005)
Polygonatum cirrhifolium	Dayara (2.3), Tungnath (1.7)	Bhatt et al. (2014a, b, c)

(continued)

Table 1 (continued)

MAPs name	Study area/meadows (density-individuals/km^2)	Key references
Podophyllum hexandrum	Darma (0.05), Gidara (0.03)	Rawat (2005)
	Kedarnath Wildlife Sanctuary (4.8)	Semwal et al. (2007)
	Gori valley (1.93)	Uniyal et al. (2002)
Rheum australe	Chipkon (0.8), Garpak (0.6), Phagati forest (0.6), Rishikund (0.7), Lata Kharak (0.4)	Kandari et al. (2011)
	Gori valley (1.14)	Uniyal et al. (2002)
	Darma (0.1)	Rawat (2005)
R. moorcroftianum	Byans (0.01), Johar (0.05), Girthi (0.4)	Rawat (2005)
	Gori valley (8.98)	Uniyal et al. (2002)
R. webbianum	Devikund (1.2)	Chandra Sekar and Rawat (2011)
Swertia chirayita	Byans (0.04), Darma (0.23), Ralam (1.05), Johar (0.10), Gidara (0.17), Kyarki (0.72), Khatling (2.42)	Rawat (2005)
	Kedarnath Wildlife Sanctuary (2.3)	Semwal et al. (2007)
Trillidium govanianum	Khaliya top (1.85), Taati (0.70), Martoli (1.45), Tungnath (3.95)	Chauhan et al. (2017)

knowledge on the use of medicinal plants, which is still not documented. Documentation of traditional knowledge and use of plants for a multiplicity of purposes assumes greater significance towards making it available for future generations (Kala 2005). Because of the fast acceleration of market demand for herbal medicines, and recent controversies related to access, benefit sharing and bio-piracy, the documentation of indigenous knowledge is of urgent priority (Mukherjee 2005). The importance of ethno-biological knowledge on species-ecology can provide new paths in resource management and conservation. To maintain long-term sustainability of MAPs, it is important to explore indigenous traditional knowledge (ITK), document and exploit it to develop conservation strategies.

3.3 Tissue Culture-Based Studies on High Altitude Medicinal Plants

Advancement in plant tissue culture provides new means for conservation and propagating the valuable, rare and endangered medicinal plants. Plant tissue culture thus promises efficient and reliable production system for phyto chemicals as well as plants as a whole (Davies and Deroles 2014). Several MAPs successfully propagated by tissue culture and used as the source of active constituents across the globe (Nalawade and Tsay 2004). Unfortunately, most of the research remains up to the laboratory and commercial propagation of these plants is meagre. Literature survey

Table 2 Ethnobotanical studies conducted in the alpine region of Uttarakhand, WH

Study area	Purpose of work	Species	References
Alpine region of Kumaun	Ethnobotany	55	Rawat and Pangtey (1987)
Byans Valley	Traditional knowledge	53	Negi et al. (2011)
Dhauli Ganga	Ethnobotany	50	Kandari et al. (2012)
Madhmaheshwar (KWLS)	Traditional knowledge	152	Bhat et al. (2013)
Gori Valley	Ethnobotany	28	Jalal and Kumar (2006)
Devikund	Ethnobotany	62	Sekhar and Rawat
Sunderdunga	Ethnomedicine	76	Rawat et al. (2013a, b, c)
Johar Valley	Ethnomedicinal	22	Pandey et al. (2017)
Tungnath Chopta	Ethnomedicinal	71	Ratha et al. (2017),
Niti Valley	Ethnomedicine	86	Phondani et al. (2010)
Pindari area	Ethnobotany	242	Rawal and Rawat (2012)
Uttarakhand	Ethnomedicinal orchid	12	Jalal et al. (2008)
Nanda Devi Biosphere Reserve (NDBR) (17 village)	Ethnomedicine	21	Dangwal et al. (2011)
17 village (NDBR buffer zone) and 7 village in Pithoragarh and Bageshwar	Ethnobotany	173	Nautiyal et al. (2003)
Rishi Ganga Catchment (NDBR)	Ethnomedicine	100	Nautiyal et al. (2001)
Buffor zone, Chamoli	Ethnomedicine	90	Rana et al. (2013b)
Higher Himalayan region (47 village)	Ethnobotany	80	Rana et al. (2013a)
Nanda Devi Biosphere Reserve	Ethnobotany	41	Tiwari et al. (2010)
Niti valley 7 village	Ethnomedicine	38	Kumar et al. (2015)
Buffer Zone of Nanda Devi Biosphere Reserve	Ethnomedicine	101	Silori and Badola (1999)
5 villages in Pithoragarh District of the buffer zone	Socioeconomic	12	Silori and Badola (1999)
Rawain Valley	Traditional healthcare practices	63	Negi et al. (2011)
Bagori and Berpur village in Uttarkashi District	Ethnomedicinal use	39	Balodi et al. (2018)
Kedarnath WLS	Ethnobotany	126	Singh and Rawat (2011)
Entire alpine and sub-alpine region	Ethnobotany	139	Singh and Dixit
Tons Valley	Ethnobotany	32	Jain and Saklani (1991)

reveals that, most of the biotechnological studies in the alpine MAPs are conducted after year 2000. Several protocols are developed for propagation of alpine MAPs through tissue culture, however, the propagation of MAPs at commercial level is meagre.

Rawat et al. (1992) was perhaps the pioneer worker in using the biotechnological approach for propagation of alpine MAPs in the WH. He propagated *Aconitum balfourii* through tuber. Mathur (1992) did somatic embryogenesis from callus in *Nardostachysjatamansi.* Giri et al. (1993) also conducted somatic embryogenesis and regeneration of *Aconitum heterophyllum* from callus culture. Pandey (2002), Manjkhola and Dhar (2002), Joshi and Dhar (2003), Bist et al. (2011), carried out in vitro propagation of various alpine MAPs (Table 3). Out of the 350–400 MAPs occur in the alpine region, less than 20 species are used for tissue culture of which, *Picrorhiza kurroa* is most widely studies by various workers. There are limited studies exist on the vegetative propagation of Himalayan alpine MAPs. Only species like *Picrorhiza kurroa* (Nautiyal et al. 2001), *Aconitum balfourii* (Rawat et al. 1992; Pandey 2002), *Heracleum candicans, Angelica archangelica* (Joshi 2002) are vegetatively propagated for commercial means.

3.4 Anthropogenic Pressure Studies on Alpine MAPs

Understanding effect of anthropogenic pressure on high altitude MAPs is a mammoth task. In alpine meadows of WH, grazing by livestock (nomadic and local people) and developmental activities (e.g., road construction) are an important biotic factor which alter the vegetation composition and phyto-mass of the region. Grazing, allow little time for herbage replacement, reduced leaf area, affects tiller development and leaf root growth. It also affected the phyto-sociological parameter of plants. So far, limited studies are conducted on impact of grazing on alpine MAPs in WH. These studies are insufficient to draw any conclusive conclusion for restoration of these MAPs in the region. Major studies on impact of grazing on alpine plants are conducted by Sundariyal and Joshi (1990, 1987), Silori (2001), Kala et al. (2002), Nautiyal et al. (2004) (Table 4).

The degree of anthropogenic pressure varies within region, man-made disturbance has become a widespread feature in the Himalaya. Two recent phenomena pose conservation and management issues in the alpine region of WH. One is the great increase in livestock density spatially heavy body animals (buffaloes and mules); the other is a great increase in demand for MAPs in international market. In the WH, excessive anthropogenic pressures are the main cause of population decline and availability of high value MAPs (Kala et al. 1998). There is scanty information exists on the anthropogenic pressure pertaining to MAPs in alpine region, and no systematic study has been conducted on MAP species diversity and abundance in relation to environmental factors, including anthropogenic activities.

Table 3 The various biotechnological study on alpine medicinal plants

MAPs	Disciplines	References
Aconitum balfouri	In-vitro propagation	Rawat et al. (1992), Pandey et al. (2004), Bist et al. (2011), Sharma et al. (2012)
Aconitum heterophyllum	Somatic embryogenesis	Giri et al. (1993)
Aconitum violaceum	In-vitro propagation	Rawat et al. (2013a, b, c)
Angelica glauca	Vegetative propagation	Bisht et al. (2006), Vashista et al. (2009)
Arnebia benthami	In-vitro propagation	Manjkhola and Dhar (2002)
Arnebia euchroma	Organogenesis, embryogenesis	Manjkhola et al. (2005)
Dactylorhizahatagirea	In vitro propagation	Giri and Tamta (2012)
	Organogenesis plantlets, somatic embryo	Lata and Nautiyal (1999)
Fritillaria roylei	In-vitro bulblet regeneration	Joshi et al. (2007)
Habenaria edgeworthii	In vitro propagation	Giri et al. (2012)
Lilium polyphyllum	Propagation	Dhyani et al. (2014)
Nardostachys jatamansi	Shoot, bud regeneration	Mathur (1992)
	Callus, somatic embryos, plantlets	Mathur (1993)
Podophyllum hexandrum	Mass propagation	Nadeem et al. (1997)
	Organogenesis plantlets, somatic embryo	Lata and Nautiyal (1999)
Picrorhiza kurroa	Root culture	Mishra et al. (2011)
	Propagation	Chandra et al. (2006)
	Cryopreservation	Sharma and Sharma (2003)
	Regeneration	Nanda and Ahuja (1996)
	Tissue culture	Rawat et al. (2013a, b, c)
Saussurea obvallata	In-vitro propagation	Joshi and Dhar (2003)
Selinum wallichianum	In-vitro propagation	Pandey et al. (2010)
Swertiachirayta	Micro-propagation	Balaraju et al. (2009)

3.5 Management Strategies for Alpine MAPs

There are six protected areas (PAs) Wildlife Sanctuaries (WLS) and National Parks (NP) namely Askote WLS, Nanda Devi NP, Valley of Flower NP, Kedarnath WLS, Gangotri NP, Govind WLS and one biosphere reserve viz., Nanda Devi BR in the alpine region of the Uttarakhand, WH. These PAs cover entirely or partially the alpine region of Uttarakhand. Except a few PAs viz., Valley of Flowers most of them have a focused attention on the conservation of faunal diversity. There is not even a single PA is basically known for conservation of medicinal plants. However, there

Table 4 Various anthropogenic studies conducted on alpine MAPs of Uttarakhand, WH

Study area	Area of work	Species enumerated	References
Tungnath	Livestock grazing	29 (3grass, 2 sedge & 24 forbs)	Nautiyal et al. (2004)
Khiron valley	Livestock grazing	21	Kala and Rawat (1999)
Western Himalaya	Livestock grazing	198(Sub-alpine), 175 (Lower alpine) 140 Higher alpine GHNP. 209(Sub-alpine), 194 (Lower alpine) 153 Higher alpine VOFNP	Kala et al. (2002)
NDBR, Buffer Zone	Anthropogenic pressure	35	Silori (2001)
Tungnath	Grazing	25 (4 Grass, 21 forbs)	Sundariyal and Joshi (1990)
Rudranath	Anthropogenic pressure	41 (3 grass, 3 sedge, 35 forbs)	Sundariyal and Joshi (1990)
Tungnath	Anthropogenic pressure	10 MAPs	Chauhan (2014)

are seven Medicinal Plants Conservation Area (MPCA) of maximum 250-hectare area, designated across the state. Of which only two, namely Kandara and Khalia top are designated to management of high-altitude MAPs. The management plan of these MPCAs is lacking. Therefore, there is an urgent need for identification and notification of areas for conservation of high-altitude MAPs on a priority basis.

In the state, the Government has a policy to promote cultivation of MAPs, with an aim that increased cultivation will reduce the demand for the material collected from wild. However, there is hardly any correlation between collection and cultivation. Even if cultivation is adopted by a large number of farmers, collection from wild is likely to continue. Therefore, it is advised to promote more in-situ conservation and sustainable utilization of MAPs from the alpine region.

A comprehensive database of MAPs needs to be established at State Medicinal Plant Board (SMPB) and Forest Headquarters for further research, monitoring and identification of future MPCAs. In-situ conservation will be the best approach for conservation of medicinal plants in their natural habitats by setting up medicinal plants Conservation and Development Areas (MPCDAs), as well as strengthening of existing Medicinal Plants Conservation Areas (MPCAs) through survey inventory, documentation, protection and main streaming medicinal plants in habitat management approaches.

3.6 Information Gap and Scope for Future Research

Much of research on high altitude MAPs has so far focused on listing, status assessment and distribution patterns of a few species, analyse economic implications of elite identification (in terms of producing biomass and availability of secondary metabolites), in vitro propagation protocols, mass propagation through conventional (vegetative and seeds) methods, development of agro-techniques and understanding reproductive biology. However, there are limited landscape level studies on status of MAPs to draw a conclusive strategy for management and conservation of MAPs in the alpine region of WH. Quantitative information on the micro-habitat preferences and population status of high-altitude medicinal plants is virtually lacking from the Himalayan region. Studies on these aspects of commercially useful plants would be very vital in understanding the ecology of such species and formulating the conservation plans for the areas.

With changing various socio-economic dynamics and developmental initiatives in the alpine region there is a growing opportunity to initiate studies for addressing effects of climate change on phenology, species range shifts, habitat alternation and secondary metabolite production of high-altitude medicinal plants. Most of the work under elevated CO_2 conditions in India is being carried out on crop plants, whereas the alpine plants remain untouched for such studies.

There are many potential causes of rarity in MAPs. Despite these, studies are lacking on impact of wild herbivores on MAPs populations including impact of domestic animals. Therefore, it is imperative to carry out detailed investigations on the habitat utilization patterns, feeding ecology, geographical distribution patterns and impact of herbivores on medicinal plant population. Similarly, most of the documentation and research on indigenous uses of medicinal plants is focused on the human aspect. Animal husbandry is a backbone of economy in the Himalayan region. Since there are many medicinal plant species used in curing various animal-related disorders and diseases, research should also be carried out and emphasized on the uses of medicinal plants for curing livestock diseases.

In the high-altitude areas, the traditional knowledge needs to be enhanced mainly to younger generation. Establishment of high-altitude nursery or cultivation practices through local body would help in management of MAPs and also preserved ethnobotanical knowledge. The indigenous traditional knowledge of MAPs and therapies of various local communities has been transmitted orally for centuries is becoming extinct, due to changes in traditional culture and introduction of modern technologies. Hence, these traditional practices need proper documentation.

4 Conclusion

The available information on the MAPs of the alpine region of the WH, broadened our knowledge on species distribution, composition, abundance and traditional knowledge and utilization pattern by the local communities in general. Most of the studies are site specific and primary objective of several publication is listing of MAPs occurring in the area. Except for scattered information often severely limited to certain parts of the landscape, no attempts have been made to provide cumulative data on MAPs of the alpine region of WH. For instance, the checklist of species for some areas, including protected areas (e.g., Govind Wildlife Sanctuary) are scanty, whereas some areas (e.g., Kedarnath Wildlife Sanctuary and Valley of Flowers) are intensively studied. Furthermore, often this information is not easily accessible or translated into the action at ground because it is either unpublished, archived at various sources such as libraries or private collection. Alpine meadows of WH is not a single gregarious habitat, there are six major habitat types reported namely, tall forbs, mixed herbaceous meadows (short forbs), *Danthonia* grassy slopes, *Kobresia* sedge meadows, cushionoid and matted shrub (Rawat 2005), however, extent of these habitats, their status in terms of suitability and availability of MAPs is not known. About 350–400 species have been reported as medicinal importance in the alpine region, the geo-spatial extent and distribution of these MAPs is limited. The study reveals that, the most of the available information on various aspects of the alpine MAPs is collected from the easily accessible areas which are also accessed by the local people for the collection of MAPs for indigenous use. Therefore, despite the national, state and scattered local initiatives on the management, conservation and sustainable utilization of MAPs, there still exist gaps in our knowledge and understanding in availability of MAPs resource base in the alpine region of WH. It is feared that if exploitation rate and increasing habitat loss is not checked and management measure are not taken properly, important MAPs may be extinct from the Himalayan region in near future.

Acknowledgements Authors are thankful to Director, Uttarakhand Space Application Centre, Govt. of Uttarakhand for all the logistic support. Authors duly acknowledge Dr. G. S. Rawat, Former Director, Wildlife Institute of India, Dehradun for giving unceasing flows of ideas regarding the said work. We duly acknowledge GB Pant National Institute of Himalayan Environment GBPNIHE), Almora and Ministry of Environment, Forest and Climate Change (MoEFCC), Government of India for financial assistance under National Mission for Himalayan Studies (NMHS) program.

References

Airi S, Rawal RS, Dhar U, Purohit AN (2000) Assessment of availability and habitat preference of Jatamansi—a critically endangered medicinal plant of west Himalaya. Curr Sci 79(10):1467–1470

Arya D, Bhatt D, Joshi GC, Tewari LM (2015) Trade, population study and conservation aspects of Choraka/Choru in Kumaun Himalayas. Sci Res Essays 10(2):64–70

Badola HK (2001) Medicinal plant diversity of Himachal Pradesh. Himalayan Medicinal plants: potential and prospects. Himavikas Occasional pusl No. 14, Gyanodaya Prakashan, Nainital, pp 407–430

Badoni AK (1989) Ramarks on high altitude medicinal plants of Garhwal Himalaya

Badoni AK, Badoni K (2001) Ethnobotanical heritage. In: Kandari OP, Gusain OP (eds) Garhwal Himalaya: nature, culture and society. Transmedia, Media House, Srinagar, Garhwal, pp 127–147

Balaraju K, Agasthan P, Iguacinauthu S (2009) Micropropagation of SwertiaChirataBuch.-Hames. ex Wall: a critically endangered medicinal herb. Acta Physiol Plant 31:487–494

Balodi KN, Purohit MV, Shridhar V, Arunachalam K (2018) Ethno-medicinal uses of various plants species among the Jaad Bhotiya Community of Uttarakhand, Western Himalaya. Ethno Med 12(3):189–197

Bhat JA, Kumar M, Bussmann RW (2013) Ecological status and traditional knowledge of medicinal plants in KWL Sanctuary of Garhwal Himalaya, India. J Ethnobiol Ethnomed 9:1

Bhatt A, Joshi SK, Gairola S (2005) *Dactylorhiza hatagirea* (D. Don) Soo—a West Himalayan orchid in peril. Curr Sci 89(4)

Bhatt D, Joshi GC, Kumar R, Tiwari LM (2014a) Phytosociological feature and threat categorization of*Aconitum heterophyllum* and A. Ferox in Kumaun Himalaya. J Ecol Nat Environ 6(3):11–118

Bhatt D, Kumar R, Joshi GC, Tiwari LM (2014b) Successive variation in phytosociological aspect and threat categorization of *Picrorhiza kurroa* Royle ex Benth. In: Kumaun Himalaya of Uttarakhand. J Med Plants Res 8(23):829–833

Bhatt D, Kumar R, Tiwari LM, Joshi GC (2014c) *Polygonatum cirrhifolium* Royle and *Polygonatum verticillatum* (L.) Allioni: status assessment and medicinal uses in Uttarakhand, India. J Med Plant Res 8(5):253–259

Bisht AK, Bhatt A, Rawal RS, Dhar U (2006) Assessment of extraction pattern and restoration through propagation of Angelica glauca, a critically endangered Himalayan medicinal herb. Int J Biodivers Sci Manag 2:105–113

Bist R, Pathak K, Hatwal D, Punetha H, Gaur AK (2011) In vitro propagation of *Aconitum balfourii* Stapf.: a rare medicinal herb of the alpine Himalayas. Indian J Hortic 68(3):394–398

Chandra B, Palni LMS, Nandi SK (2006) Propagation and conservation of Picrorhiza kurrooa Royle ex Benth: an endangered Himalayan medicinal herb of high commercial value. Biodivers Conserv 15:2325–2338

Chandra Sekar K, Rawat B (2011) Diversity, utilization and conservation of ethnomedicinal plants in Devikund—a high altitude, sacred wetland of Indian Himalaya. Med Plants 3(2)

Chatterjee D (1939) Studies on the endemic flora of India and Burma. J R Asiatic Soc Bengal 5:19–67

Chauhan RS (2014) Anthropogenic pressure mediated reduction in population size of medicinal and aromatic plant species in western Himalaya, India. Medicinal Plants - Int J Phytomedicines Related Industries 6(2):94–99

Chauhan RS, Nautiyal MC, Prasad P (2007) *Habernaria intermedia*—an endangered medicinal orchid. MacAllen Int Orchid Soc Gen 8(10):15–20

Chauhan RS, Nautiyal MC, Prasad P, Purohit H (2008) Ecological feature of an endangered medicinal orchid *Malaxis muscifera* in west Himalaya. McAllen Int Soc J 9(6):8–12

Chauhana HK, Bisht AK, Bhatt ID, Bhatt A, Gallacherd D, Santoe A (2017) Population change of *Trillium govanianum* (Melanthiaceae) amid altered indigenous harvesting practices in the Indian Himalayas

Dangwal LR, Rana CS, Sharma A (2011) Ethno-medicinal plants from transitional zone of Nanda Devi Biosphere Reserve, district Chamoli, Uttarakhand (India). Indian J Nat Prod Resour 2(1):116–120

Das JS (2008) The largest genetic paradise of India lacks biotechnological implementation. Curr Sci 94(5):558–559

Davies KM, Deroles SC (2014) Prospects for the use of plant cell cultures in food biotechnology. Curr Opin Biotech 26:133–140

Dhar U, Rawal RS, Upreti J (2000) Setting priorities for conservation of medicinal plants—a case study in the Indian Himalaya. Biol Conserv 95:57–65

Dhyani A, Sharma G, Nautiya BP, Nautiyal MC (2014) Propagation and conservation of Lilium polyphyllum D. Don ex Royle. J Appl Res Med Aromat Plants 1:144–147

Giri D, Tamta S (2012) Propagation and conservation of *Dactylorhiza hatagirea* (D. Don) Soo, an endangered alpine orchid. Afr J Biotechnol 11(62):12586–12594

Giri A, Ahuja PS, Kumar PVA (1993) Somatic embryogenesis and plant regeneration from callus culture of *Aconitum heterophyllum* Wall. Plant Cell Tissue Organ Cult 32:213–218

Giri L, Jugran A, Rawat S, Dhyani P, Andola H, Bhatt ID, Rawal RS, Dhar U (2012) In vitro propagation, genetic and phytochemical assessment of *Habenariaedge worthii*: an important Astavargaplant. Acta Physiol Plant. https://doi.org/10.1007/s11738-011-0884-8

Jablonski D (2004) Extinction: past and present. Nature 427:589

Jain SK, Saklani A (1991) Observations on the ethnobotany of the tons valley region in the Uttarkashi district of the northwest Himalaya, India. Mount Res Dev 11(2):157–161

Jalal JS, Kumar P (2006) Some ethno-medicinal plants of Gori River basin, Kumaon Himalaya. Flora and Fauna Jhansi 12(2): 292–294

Jalal JS, Kumar P, Pangtey YPS (2008) Ethnomedicinal orchids of Uttarakhand, Western Himalaya. Ethnobot Leaflets 12(1227–30):2008

Joshi M (2002) Developing propagation protocols of selected high altitude endemic medicinal umbellifers of the Himalaya with a focus on *Angelica archangelica* L. var *himalaica* Cl. Ph.D thesis, Kumaun University, Nainital India

Joshi M, Dhar U (2003) In vitro propagation of *Saussurea obvallata* (DC.) Edgew—an endagered ethnoreligious medicinal herb of Himalaya

Joshi DN, Sah BCL, Suri RK (1982) Some medicinal plants of Rudranath Bugyal (dist. Chamoli) UP Bull. Medicoethnobot Res 3:27–43

Joshi SK, Dhar U, Andola HC (2007) In vitro bulblet regeneration and evaluation of *Fritilaria roylei* Hook, a high value medicinal herb of Himalaya. Acta Hort 756:75–83

Kala CP (1998) Ethnobotanical survey and propagation of rare medicinal herbs in the buffer zone of the Valley of Flowers National Park, Garhwal Himalaya. International Centre for Integrated Mountain Development, Kathmandu, Nepal

Kala CP (2005) Indigenous uses, population density and conservation of threatened medicinal plants in protected areas of the Indian Himalayas. Conserv Biol 19:368–378

Kala CP (2006) Problems and prospects in the conservation and development of the Himalayan medicinal plant sector. Int J Sustain Dev 9:370–388

Kala CP (2010) Medicinal plants of Uttarakhand: diversity livelihood and conservation. Biotech Books, Delhi, India

Kala CP, Rawat GS, Uniyal VK (1998) Ecology and Conservation of the Valley of Flowers National Park. Himalaya, WII, Dehradun 1998:99

Kala CP, Singh SK, Rawat GS (2002) Effect of sheep and goat grazing on the species diversity in the alpine meadows of Western Himalaya. Environmentalist 22(183–189):2002

Kala CP, Farooquee N, Dhar U (2004) Prioritization of medicinal plants on the basis of available knowledge, existing practices and use value status in Uttaranchal, India. Biodivers Conserv 13:453–469

Kandari LS, Phondani PC, Payal KC, Rao KS, Maikhuri RK (2012) Ethnobotanical study towards conservation of medicinaland aromatic plants in upper catchments of Dhauli Ganga in the Central Himalaya. J Mt Sci 9:286–296

Kandari LS, Rao KS, Maikhuri RK, Kharkwal G, Chauhan K, Kala CP (2011) Distribution pattern and conservation of threatened medicinal and aromatic plants of Central Himalaya, India. J For Res 22(3):403−408

Kumar A, Mitra M, Adhikari BS, Rawat GS (2015) Depleting indigenous knowledge of medicinal plants in cold-arid region of Nanda Devi Biosphere Reserve, Western Himalaya. Med Aromat Plants 4(3):1–6

Kumar A, Adhikari BS, Mishra AP (2023) Status of threatened medicinal and Biosphere Reserve, Uttarakhand. A report (TR No./2022/35) submitted to International Union for Conservation of Nature—India, p 103

Lal JB, Gulati AK, Bist MS (1991). Satellite mapping of alpine pastures in the Himalaya. Int J Remote Sensing 12(3):435–443

Lata H, Nautiyal SC (1999) Micropropagation of different populations of *Podophyllum hexandrum* Royle from Garhwal Himalaya. In: Micropropagation of some Himalayan alpine herbs 321 National symposium on role of plant tissue culture in biodiversity conservation and economic development. Kosi-Katarmal, Almora, 7–9 June 1999, 40 (Abstract)

Manjkhola S, Dhar U, Joshi M (2005) Organogenesis, embryogenesis and synthetic seed production in *Arnebia euchroma*—a critically endangered medicinal plant of the Himalaya. In Vitro Cell Dev Biol Plants 41(3):244–248

Manjkhola S, Dhar U (2002) Conservation and utilization of *Arnebia benthamii* (Wall. ex G. Don) Jonston—a high value Himalayan medicinal plant. Curr Sci 83(4):484–488

Mathur J (1992) In vitro morphogenesis in *Nardostachys jatamansi* DC: shoot regeneration from callus derived roots. Ann Bot 70:419–422

Mathur J (1993) Somatic embryogenesis from callus cultures of *Nardostachys jatamansi*. Plant Cell Tissue Organ Cult 33:163–169

Mishra J, Bhandari H, Singh M, Sandeep Rawat S, Agnihotri RK, Mishra S, Purohit S (2011) Hairy root culture of *Picrorhiza kurroa* Royle ex Benth.: a promising approach for the roduction of picrotin and picrotoxinin. Acta Physiol Plant 33:1841–1846

Mishra AP, Kumar A, Yadav SN (2023a) Ecology and conservation of threatened medicinal plants in the Trans-Himalayan region of Nanda Devi Biosphere Reserve, Western Himalaya. Trees For People 14:100451. https://doi.org/10.1016/j.tfp.2023.100451

Mishra AP, Chandra N, Mandy JJ, Dwivedi SK, Alruzuq A, Pande CB (2023b) Analytical Hierarchy Process (AHP) based on the spatial assessment of an endangered alpine medicinal herb Aconitum heterophyllum in the Western Himalayan environment. In: Climate change impacts on natural resources, ecosystems and agricultural systems. Springer International Publishing, Cham, pp 579–594. https://doi.org/10.1007/978-3-031-19059-9_25

Mukherjee PK (2005) Exploring green resources for drug development through ethnobotany. In: Chemistry for green environment. Narosa Publishing House, New Delhi, p 98

Nadeem M, Palni LMS, Nandi SK, Purohit AN (1997) Mass propagation of *Podophyllum hexandrum* Royle. Hima Paryavaran 9:17–19

Nanda L, Ahuja PS (1996) Plantlet regeneration from callus in Picrorhiza kurroa royle ex Benth—an endangered medicinal plant. Plant Tissue Cult 6(2):127–134

Nautiyal S, Maikhuri RK, Rao KS, Saxena KG (2001) Medicinal plant resources in Nanda Devi Biosphere Reserve in the Central Himalaya. J Herbs Spices Med Plants 8(4):47–64. https://doi.org/10.1300/J044v08n04_06

Nautiyal BP, Prakash V, Bahuguna R, Maithani U, Bisht H, Nautiyal MC (2002) Population study for monitoring the status of rarity of three Aconitespecies in Garhwal Himalaya. Trop Ecol 43(2):297–303

Nautiyal S, Maikhuri RK, Rao KS, Saxena KG (2003) Ethnobotany of the tolchha Bhotiya tribe of the buffer zone village in NDBR. J Econ Taxon Bot 27:119–142

Nautiyal MC, Nautiyal BP, Prakash V (2004) Effect of grazing and climatic changes on alpine vegetation of Tungnath, Garhwal Himalaya, India. Environmentalist 24(125–134):2004

Nautiyal MC (1996) Cultivation of medicinal plants and Biosphere reserve management in alpine zone. In: Ramakrishan PS, Purohit AN, Saxena KG, Rao KS, Maikhuri RK (eds) Conservation and management of biological resources in Himalaya. Oxford and IBH, New Delhi, pp 569–583

Negi KS, Tiwari JK, Gaur RD (1985) A contribution to the flora of Dodital—a high altitude lake in the Garhwal Himalaya (Uttarkashi), U. P. J Bombay Nat Hist Soc Bombay 82(2):258–270

Negi CS, Nautiyal S, Dasila L, Rao KS, Maihkuri RK (2002) Ethnomedicinal plant uses in a small tribal community in a part of Central Himalaya, India. J Hum Ecol 14(1):23–31

Negi VS, Maikhuri RK, Vashishtha DP (2011) Traditional healthcare practices among the village of Rawain valley, Uttarkashi, Uttarakhand. Indian J Tradit Knowl 10(3):533–537

Pandey H (2002) Studies in relation to cotyledonary senescence and tissue culture of medicinally important alpine herbs. Ph.D thesis, Kumaun University, Nainital, India

Pandey H, Nandi SK, Kumar A, Palni UT, Chandra B, Palni LMS (2004) In vitro propagation of *Aconitum balfourii* Stapf; an important aconite of Himalayan alpine. J Hortic Sci Biotechnol 21:69–84

Pandey M, Dhar U, Bhatt ID, Thengane SR (2010) In vitro propagation of *Selinum wallichianum* (DC), a high value medicinal plant of Himalaya. Asian J Exp Sci 24(1):179–183

Phondani PC, Maikhuri RK, Rawat LS et al (2010) Ethnobotanical uses of plants among the Bhotiya tribal communities of Niti valley in Central Himalaya, India. Ethnobot Res Appl 8:233–244

Prajapati ND, Purohit SS, Sharma AK, Kumar T (2003) A handbook of medicinal plants. Agrobios. Jodhpur

Rana CS, Ballabha R, Tiwari JK, Dangwal LR (2013a) An ethnobotanical study of plant resources in the Nanda Devi Biosphere Reserve, Uttarakhand, India. J Ethnobiol Tradit Med 120:591–601

Rana CS, Tiwari JK, Dangwal LR, Gairola S (2013b) Faith herbal healer knowledge document of Nanda Devi Biosphere Reserve, Uttarakhand, India. Indian J Tradit Knowl 12(2):308–314

Ratha KK1, Arya JC, Joshi GC, Rao MM (2017) Contribution of regional ethno medicinal flora of Tungnath-Chopta Region, Uttarakhand to Ayurveda. Inventi Rapid Planta Activa Issue 1

Rawal RS, Rawat B (2012) Nanda Devi Biosphere Reserve, Western Himalaya, India. In: Palni LMS, Rawal RS (eds) Compendium on Indian Biosphere Reserves. Progression during two decades of conservation. G.B. Pant National Institute of Himalayan Environment and Development, Govt. of India, pp 42–55

Rawat GS (2005) Alpine Meadows of Uttaranchal: ecology, landuse and status of medicinal and aromatic plants. Bishen Singh Mahendrapal Singh, Dehra Dun

Rawat GS, Pangtey YPS (1987) A contribution to the ethnobotany of alpine regions of Kumaun. J Econ Taxon Bot 11(1):139–148

Rawat AS, Pharswan AS, Nautiyal MC (1992) Propagation of Aconitum atrox, a regionally threatened medicinal herb. Econ Bot 46(3):337–338

Rawat B, Rawat JM, Mishra S, Mishra SN (2013a) *Picrorhiza kurrooa*: current status and tissue culture mediated biotechnological interventions. Acta Physiol Plant 2013(35):1–12

Rawat B, Sekar KC, Gairola S (2013b) Ethnomedicinal plants of Sunderdhunga Valley, Western Himalaya, India—traditional use, current status and future scenario. Indian For 139(1):61–68

Rawat JM, Rawat B, Agnihotri RK, Chandra A, Nautiyal S (2013c) In vitro propagation, genetic and secondary metabolite analysis of *Aconitum violaceum* Jacq.: a threatened medicinal herb. Acta Physiol Plant. https://doi.org/10.1007/s11738-013-1294-x

Rawat N, Mishra AP, Chandra N, Painuly V, Negi GS, Bisht MPS, Purohit S (2021) Maxent modeling for predicting the potential distribution of *Valeriana jatamansi* jones. in chakrata forest division of Garhwal Himalaya. Indian J Ecol 48(2):367–372

Samant SS, Dhar U, Palni LMS (1998) Medicinal plants of Indian Himalaya: diversity distribution and potential value. Gyanodaya Prakashan, Nainital

Samant SS, Palni LMS (2000) Diversity, distribution and indigenous use of essential oil yielding plants of Indian Himalayan region. J Med Aromat Plant Sci 671–684

Semwal DP, Saradhi PP, Nautiyal BP, Bhatt AB (2007) Current status, distribution and conservation of rare and endangered medicinal plants in KWLS, Central Himalaya, India. Curr Sci 92(12):25

Sharma N, Sharma B (2003) Cryopreservation of shoot tips of Picrorhiza kurrooa Royle ex Benth., an indigenous endangered medicinal plant through vitrification. Cryo Lett 24:181–190

Sharma E, Gaur K, Punetha H, Gaur AK (2012) In vitro regeneration of *Aconitum balfourii* stapf: a rare medicinal herb from Himalayan alpine through root explant. Res J Med Plant 6:318–325

Silori CS (2001) Status and distribution of anthropogenic pressure in the buffer zone of Nanda Devi Biosphere Reserve in Western Himalaya, India. Biodivers Conserv 10(1113–1130):2001

Silori CS, Badola R (1999) Nanda Devi Biosphere Reserve: a study on socio-economic aspects for the sustainable development of dependent population. Wildlife Institute of India, Dehradun

Singh G, Rawat GS (2011) Ethnomedicinal survey of Kedarnath Wildlife Sanctuary in Western Himalaya, India. Indian J Fundam Appl Life Sci 1:35–46

Sundariyal RC, Joshi NAP (1987) Some important medicinal plants of Alpine pasture of Garhwal Himalaya

Sundariyal RC, Joshi AP (1990) Effect of Grazing on standing crop, productivity and efficiency of energy capture in a grassland ecosystem at Tungnath (Garahwal Himalaya), India. Trop Ecol

Tiwari JK, Dangwal LR, Rana CS et al (2010) Indigenous uses of plant species in Nanda Devi Biosphere Reserve, Uttarakhand, India. Rep Opin 2(2):67–70

Uniyal, S, Awasthi A, Rawat GS (2002) Current status and distribution of commercially exploited medicinal and aromatic plants in upper Gori Valley, Kumaon Himalaya, Uttaranchal. Curr Sci 82(10):1246–1252

Vashista RK, Chaturvedi AK, Nautiyal BP, Nautiyal MC (2009) Vegetative propagation of *Angelica glauca* Edgew. and *Angelica arch angelica* Linn.: two high value medicinal and aromatic herbs of the Himalaya. Nat Sci 7(8)

Vashistha R, Nautiyal BP, Nautiyal MC (2006) Conservation status and morphological variations between populations of *Angelica glauca* Edgew and *A. archangelica* Linn. in Garhwal Himalaya. Curr Sci 91(11):1537–1542

Status, Threats, Diversity and Distribution of Threatened Medicinal Plants of Kumaun Himalaya, Uttarakhand, India

Geetanjali Upadhyay, Naveen Ch. Pandey, Lalit M. Tewari, Disha Upreti, Vasundhra Lodhiyal, and Sheetal Koranga

Abstract The Himalaya is characterized by its bio-physical diversity and socio-cultural heritage, traditional systems and an ample quantity of indigenous knowledge. The present study attempts to comprehensively document the diversity of threatened plants of Uttarakhand. A total of 82 species belonging to 45 families, 69 genera of different habits such as herbs (58 species), shrubs (15 species) and trees (9 species) were recorded. Out of 45 families, Asteraceae, Apiaceae and Ranunculaceae (5 species each), followed by Lamiaceae and Orchidaceae (4 species each), Apocynaceae, Asparagaceae, Ericaceae, Liliaceae and Polygonaceae (3 species each) were the dominant families in the region. The present study compilation is aimed to serve as a comprehensive database especially for researchers, conservationists, foresters, pharmaceutical professionals as well as policy makers.

Keywords Conservation · Diversity · Kumaun Himalaya · Medicinal plants · Threatened species

1 Introduction

The Indian Himalayan Region (IHR) is well known throughout the globe due to its representative, unique, natural and socio-economically important flora and fauna (Samant et al. 1998). The IHR ranging from Arunachal Pradesh to Jammu and Kashmir and rising up to an elevation of >8000 m asl is one of the major repositories of biodiversity (Barman et al. 2021; Kaur et al. 2022). IHR represents tropical, subtropical, sub-alpine, alpine, and xerophytic vegetation and supports about 10,000 species of vascular plants, out of which 4000 species are endemic (Mittermeier et al. 2004; Manish and Pandit 2018). Himalaya is the heritage depository of biodiversity having total 1748 species of medicinal plants, 279 species of fodder, 118 species of

G. Upadhyay · N. Ch. Pandey · L. M. Tewari (✉) · D. Upreti · V. Lodhiyal · S. Koranga
Department of Botany, D.S.B. Campus, Nainital, Uttarakhand, India
e-mail: l_tewari@rediffmail.com

A. P. Mishra et al. (eds.), *Threatened Medicinal Plants in the Indian Himalayan Region*, World Sustainability Series, https://doi.org/10.1007/978-3-031-73687-2_6

essential oil yielding, 675 species of wild edibles and 155 species of sacred plants (Samant et al. 1998; Samant and Palni, 2000; Samant and Shreekar 2003). Owing to ever increasing human population and anthropogenic activities, species extinction rate has increased to hundred or thousand-fold compared to background extinctions, bringing to forefront the 'sixth mass extinction' crisis (Shivanna 2020; Kumar et al. 2023; Mishra et al. 2023a, b).

Over-exploitation of important plants in drugs and pharmaceutical industries led these floras toward extinction as 121 plant species of India have been logged in the Red Data Book (Nayar and Sastry 1987, 1988, 1990) and more than 70 species are under other threat categories and many species are extinct or on the verge of extinction before they are known for their uses (Singh and Anand 1996). Countries around the world are adapting criteria to prioritize species for conservation, to determine the conservation approaches and to frame regulatory and policy decisions for conservation and multilateral agreements (Maxted et al. 1997; Rodrigues et al. 2006).

The problems in assessing the plant species diversity increased in the Himalayan region, in respect of high-altitude areas because of inhospitable climatic conditions, rough and detachment of the terrain, and short life cycle of plants (Kala 2006). The indigenous communities and commercial herb gatherers also raid these same landscapes to collect valuable medicinal plants (Kala 2004). For short- and long-term management planning, area-specific threat categorization of medicinal plant species is important (Singh et al. 2009). In literature there is staggered information regarding list of threatened plants, including medicinal plants of India, compiled at either state level or national or international level. The use of data-based decision in identifying site-specific conservation actions is increasing day by day, which in turn is yielding significant impacts with limited resources (Wilson et al. 2007; Chandra et al. 2020). The present study has enabled the compilation of data on threatened plants of Kumaun Himalaya, Uttarakhand.

2 Materials and Methods

2.1 Study Area

Kumaun Himalaya comprises six north-eastern Himalayan districts of Uttarakhand i.e. Almora, Bageshwar, Nainital, Pithoragarh, Champawat and Udham Singh Nagar. The region extending over an area of 21,033 km^2 comprises of 52 tehsils and 41 development blocks lies between the latitudes of 28°44′ and 30°49′ N and Longitudes of 78°45′ and 81°05′ E along the eastern and south-eastern part of the Central Himalaya. In the north, the natural water parting separates it from Tibet, while towards the east the river Kali marks its border with Nepal all along. Towards west lie the Himalayan districts of Chamoli and Pauri Garhwal within Garhwal Division. In the south-west and south, the region touches the boundary of the district of Bijnor,

Moradabad, Rampur, Bareilly and Pilibhit—the southern extent of the region, more or less coincide with the characteristic artesian condition of the Terai.

Kumaun Himalaya also exhibits great ethnic and cultural diversity of socio-cultural aspect. The region also comprises habitat for some ethnic-tribal communities, i.e. *Bhotiyas, Tharus, Bukshas, Rajis or Banrajis, Kiratas, Khasas* (Srivastava and Singh 2005; Joshi et al. 2021). The inhabitants who are living in the Himalayan region are entirely dependent on the natural resources for their livelihood. This region harbours about 4,000 plant species of different groups (Samant 1998), 20 species of gymnosperms and more than 350 species of pteridophytic flora (Khullar 1994 and 2000). Uttarakhand supports more than 720 medicinal plant species, 279 fodder yielding species and 344 wild edible plant species, 250 crude drugs plant species, 50 threatened medicinal plants and 522 species of legume and thistle (Samant and Dhar 1997; Samant 1998; Samant et al. 1998; Samant and Pal 2003; Arya et al. 2018; Bhatt et al. 2020; Joshi et al. 2021).

2.2 *Data Collection*

A systematic literature review was carried out (during the year 2022–2023) to assess the threatened plants of Uttarakhand. For the identification of threatened species, inventory of plants was prepared and their conservation status was evaluated from IUCN Red list, CAMP (CAMP 1998; Ved et al. 2003) workshops, other existing literatures (Red Data Book of Indian Plants) (Nayar and Sastry 1987, 1988, 1990; Srivastava and Singh 2005; Kumari et al. 2012; Bhatt et al. 2020) and field observations of the authors. To understand the distribution of threatened plant in the region, various herbarium viz., Kumaun University, Nainital and Regional Ayurvedic Research Institute (RARI), Thapla, Ranikhet were also consulted. Genera and species are arranged alphabetically under each family. In addition, different websites like APG-IV system of classification (APG 2016) and International Plant Name Index (IPNI 2012), Plants of the World Online (POWO 2019) have also been consulted to update the nomenclature.

3 Results and Discussion

The present study recorded 82 species of threatened plants belonging to 69 genera and 45 families. Out of which 9 species were trees, 15 shrubs and 58 herbs (Table 1; Photo Plate 1, 2, 3, 4). Out of 45 families, Asteraceae, Apiaceae and Ranunculaceae (5 species each) Lamiaceae, and Orchidaceae (4 species each), Apocynaceae, Asparagaceae, Ericaceae, Liliaceae and Polygonaceae (3 species each) were the dominant families (Fig. 1). As per IUCN, 24 species are recognized under different threat categories. Whereas, 11 species are least concern, 5 species each are endangered and vulnerable and 3 species are critically endangered. According to (CAMP 1998),

43 species are under different threat categories. Most of these (18) are critically endangered, 11 species are vulnerable, 8 species are endangered and 6 species are near threatened. As per Ved et al. (2003), 60 species are under different threat categories. Whereas, 24 species are vulnerable, 17 species are endangered, 11 species are near threatened and 8 species are critically endangered. Considering other existing literatures such as Shah (1983), Nayar and Sastry (1987, 1988, 1990), Joshi et al. (2010), Srivastava and Singh (2005), Kumari et al. (2012) and Bhatt et al. (2020), 51 species are threatened (22 vulnerable, 20 endangered, 5 critically endangered, 3 near threatened and 1 least concern) (Fig. 2).

The geographical peculiarities make the Himalayan region a very diverse system subtending a wide range of vegetation types. Natural and anthropogenic disturbances severely threaten the biodiversity of this region. Local inhabitants utilize these plants to cure the various ailments by traditional knowledge and are used as food and other purposes. While comparing with threatened plants reported by others, Nayar and Sastry (1987, 1988, 1990) reported 121 species of threatened plants from India, Srivastava and Singh (2005) reported 60 threatened species from Uttarakhand, Joshi et al. (2010) reported 33 threatened species from Alpine region of Uttarakhand, Sharma et al. (2011) reported 30 threatened species from Shiwalik Himalaya of Uttarakhand, Kumari et al. (2012) reported 21 species of threatened plants from Almora district of Kumaun Himalaya, Bhat and Kumar (2013) reported 18 threatened species from Kedarnath Wildlife Sanctuary of Garhwal Himalaya, Verma (2014) reported 62 species of threatened plants from Hango Valley of Kinnaur District, Himachal Pradesh, Upreti et al. (2019) reported 34 species from Pithoragarh district of Kumaun Himalaya, Uttarakhand and 50 species of threatened plants have been reported by Bhatt et al. (2020) from the Kumaun Himalaya.

The number of threatened species is an important measure or indicator of the immediate need for conservation in an area. At least 153 plant species are now extinct, and of the 268,000 flowering plants, only 22,566 species have been evaluated by IUCN that constitutes only 8% of the total. This calls for a concerted effort by the biologists to complete the threat evaluation of all the plant species following the IUCN protocol. The Himalayan region has rich biological diversity that is now under threat from rapidly expanding human populations, habitat destruction, and concomitant environmental degradation occurring at a fast pace. At the regional, national and global levels various attempts have been made to identify threatened species, including medicinal plants, using different attributes such as biogeographical range, habitat preference, population size and utilization pattern (Kala 2005, 2006; Srivastava and Singh 2005; Ved et al. 2003; Joshi et al. 2010; Tiwari et al. 2010; Jyotsana et al. 2011; Kumari et al. 2012; Saha et al. 2015; Rai et al. 2017; Rawat et al. 2021; Pandey et al. 2022; Chandra et al. 2022, 2023; Mishra et al. 2023a, b; Kumar et al. 2023).

Table 1 List of threatened plants reported from Kumaon, Uttarakhand

Taxa	Vernacular Name	Altitudinal range (m)	Habit	Fl. & Fr	Threat category			
					CAMP		IUCN	Existing literature
					1998	2003		
Family-Acoraceae								
Acorus calamus L.	Vacha	700–2300	Herb	Jun–Sep	VU	–	LC	–
Family-Amaryllidaceae								
Allium stracheyi Baker	Jambu	3500–4500	Herb	Jul–Sep	–	VU	–	VU
Family-Apiaceae								
Angelica glauca Edgew	Chhipi	3500–4000	Herb	Jul–Sep	CR	EN	EN	EN
Bunium persicum B. Fedtsch	Kala Jeera	2500–3200	Herb	May–Sep	EN	–	–	EN
Ferula jaeschkeana Vatke	Jangli Heeng	2400–3500	Herb	Jun–Sep	–	VU	–	–
Heracleum lanatum Michx	Kakariya	1700–3800	Herb	May–Oct	EN	VU	–	–
Pleurospermum angelicoides (DC.) Benth. ex C.B. Clarke	Choru	2500–4000	Herb	Jun–Aug	–	VU	–	VU
Family- Apocynaceae								
Ceropegia bulbosa Roxb	Ceropegia	300–1000	Herb	Jun–Jul	–	EN	–	–
Gymnema sylvestre (Retz.) R. Br	Gurmar	300–1000	Shrub	Apr–Jul	VU	–	–	–
Rauvolfia serpentina (L.) Benth. ex Kurz	Sarpgandha	Up to 1000	Shrub	May–Aug	EN	VU	–	EN
Family-Asparagaceae								
Drimia indica (Roxb.) Jessop	Ban Pyaj	300–1600	Herb	May–Jun	VU	–	–	EN
Polygonatum cirrhifolium (Wall.) Royle	Meda	1500–4600	Herb	Jun–Jul	–	VU	–	VU
Polygonatum verticillatum (L.) All	Mahameda	2000–4700	Herb	Jul–Aug	EN	VU	–	VU

(continued)

Table 1 (continued)

Taxa	Vernacular Name	Altitudinal range (m)	Habit	Fl. & Fr	Threat category			
					CAMP		IUCN	Existing literature
					1998	2003		
Family-Asteraceae								
Artemisia maritima L.	Pati	3000–4000	Herb	Jun–Oct	–	NT	–	–
Jurinea dolomiaea Boiss	Dhoop	3400–4500	Herb	Aug–Oct	NT	EN	–	EN
Saussurea costus (Falc.) Lipsch	Kuth	3000–4000	Herb	Jun–Sep	CR	–	–	EN
Saussurea gossypiphora D. Don	Fini Kamal	4000–5700	Herb	Aug–Oct	CR	VU	–	CR
Saussurea obvallata (DC.) Edgew	Braham Kamal	3700–4600	Herb	Jul–Sep	EN	EN	-	EN
Family-Berberidaceae								
Berberis aristata DC	Daru Haridra	1700–3000	Shrub	May–Aug	EN	–	LC	EN
Podophyllum hexandrum Royle	Ban Kakri	3300–4000	Herb	Apr–Sep	CR	EN	–	EN
Family-Betulaceae								
Betula utilis D. Don	Bhojpatra	2700–4300	Tree	Apr–Oct	–	NT	LC	–
Family-Bignoniaceae								
Oroxylum indicum (L.) Kurz	Syonake	400–1200	Tree	Jun–Dec	–	CR	–	VU
Family-Boraginaceae								
Arnebia benthamii (Wall. ex G. Don) I.M. Johnst	Bal-Char	2800–3800	Herb	Jun–Aug	CR	CR	–	CR
Arnebia euchroma (Royle ex Benth.) I.M. Johnst	Lal Jhari	3200–4400	Herb	Jun–Sep	–	EN	–	–

(continued)

Table 1 (continued)

Taxa	Vernacular Name	Altitudinal range (m)	Habit	Fl. & Fr	Threat category			
					CAMP		IUCN	Existing literature
					1998	2003		
Family-Caprifoliaceae								
Nardostachys jatamansi (D. Don) DC	Jatamansi	3400–5000	Herb	Jun–Oct	CR	CR	CR	CR
Valeriana jatamansi Jones	Sameo	1200–3500	Herb	Mar–Jun	CR	VU	–	VU
Family-Celastraceae								
Celastrus paniculatus Willd	Malkangani	300–1500	Shrub	Apr–Nov	NT	–	–	VU
Family-Colchicaceae								
Gloriosa superba L.	Kalihari	400–2200	Herb	Aug–Sep	EN	VU	LC	VU
Family-Costaceae								
Costus speciosus (J. Koenig) Sm	Keva	300–1100	Herb	Aug–Oct	NT	–	–	NT
Family-Crassulaceae								
Rhodiola heterodonta (Hook. f. & Thomson) Boriss	Rhodiola	3500–5000	Herb	Jun–Aug	–	VU	–	–
Family-Cupressaceae								
Juniperus polycarpos K. Koch	Jhora	3000–3800	Shrub	Mar–Nov	–	EN	–	–
Family-Datiscaceae								
Datisca cannabina L.	Bajarbhang	900–1500	Herb	Jul–Sep	–	EN	–	–
Family-Dioscoreaceae								
Dioscorea deltoidea Wall. ex Griseb	Tarur	1500–3000	Herb	Jul–Sep	CR	EN	–	EN

(continued)

Table 1 (continued)

Taxa	Vernacular Name	Altitudinal range (m)	Habit	Fl. & Fr	Threat category			
					CAMP		IUCN	Existing literature
					1998	2003		
Family-Elaeagnaceae								
Hippophae rhamnoides L.	Chuk	2000–3800	Shrub	Apr–Mar	–	NT	LC	–
Hippophae salicifolia D. Don	Chuk	2000–3500	Shrub	Apr–May	–	NT	–	–
Family-Ephedraceae								
Ephedra gerardiana Wall. ex C.A. Mey	Somlata	2500–4500	Shrub	Apr–Sep	–	EN	VU	EN
Family-Ericaceae								
Rhododendron anthopogon D. Don	Bhotia Chai	3000–4000	Shrub	Apr–Oct	VU	NT	–	VU
Rhododendron campanulatum D. Don	Ratpa	2800–4400	Shrub	May–Dec	–	NT	–	NT
Rhododendron lepidotum Wall. ex G. Don	Tajees	2900–4500	Tree	Jun–Nov	–	NT	–	–
Family-Euphorbiaceae								
Baliospermum montanum (Willd.) Muell.-Arg	Danti	300–1200	Shrub	Nov–May	NT	–	–	NT
Family-Gentianaceae								
Gentiana kurroo Royle	Karru	3200–4200	Herb	Aug–Oct	CR	CR	CR	CR
Swertia chirayita (Roxb.) H. Karst	Chireta	2500–3000	Herb	Jul–Nov	CR	EN	–	EN
Family-Gesneriaceae								
Didymocarpus pedicellatus R. Br	Ekpatta	500–2500	Herb	Aug–Oct	–	VU	–	VU

(continued)

Table 1 (continued)

Taxa	Vernacular Name	Altitudinal range (m)	Habit	Fl. & Fr	Threat category			
					CAMP		IUCN	Existing literature
					1998	2003		
Family-Hypericaceae								
Hypericum perforatum L.	Basant	1200–2600	Herb	May–Sep	–	VU	–	–
Family-Hypoxidaceae								
Curculigo orchioides Gaertn	Kali Musli	300–1800	Herb	Jul–Nov	VU	–	–	VU
Family-Lamiaceae								
Clerodendrum serratum (L.) Monn	Bharangi	300–1600	Shrub	Aug–Jun	VU	–	–	VU
Eremostachys superba Royle ex Benth	Gajar moola	400–1000	Herb	Mar–May	–	VU	–	–
Hyssopus officinalis L.	Lamg	2000–3500	Herb	Jun–Sep	–	NT	–	–
Roylea cinerea (D. Don) Baill	Titpati	1100–3700	Shrub	Aug–Oct	–	VU	–	–
Family-Lauraceae								
Cinnamomum tamala (Buch.-Ham.) T. Nees & Nees	Tejpat	400–2500	Tree	Feb–Oct	NT	VU	LC	–
Litsea glutinosa (Lour.) C.B. Rob	Meda	500–1600	Tree	Mar–Sep	–	NT	LC	–
Family-Liliaceae								
Fritillaria roylei Hook	Ksheer- Kakoli	3000–4000	Herb	Jun–Oct	CR	EN	–	EN
Lilium polyphyllum D. Don	Kakoli	1500–3400	Herb	Jun–Jul	–	CR	CR	–
Trillidium govanianum (Wall. ex Royle) Kunth	Satwar	2500–4000	Herb	May–Jun	–	–	EN	–

(continued)

Table 1 (continued)

Taxa	Vernacular Name	Altitudinal range (m)	Habit	Fl. & Fr	Threat category			
					CAMP		IUCN	Existing literature
					1998	2003		
Family-Melanthiaceae								
Paris polyphylla Sm	Satua	1500–2600	Herb	May–Jun	–	VU	VU	–
Family-Orchidaceae								
Crepidium acuminatum (D. Don) Szlach	Jivak	1200–2400	Herb	Jun–Aug	–	–	–	EN
Dactylorhiza hatagirea (D. Don) Soó	Hatajari	2800–4000	Herb	Jun–Oct	CR	CR	–	EN
Habenaria intermedia D.Don	Ridhi	1500–2500	Herb	Jul–Sep	–	EN	–	EN
Malaxis muscifera (Lindl.) Kuntze	Rsabhak	1600–3800	Herb	Jun–Aug	–	EN	VU	–
Family-Paeoniaceae								
Paeonia emodi Royle	Chandrayan	2000–2700	Herb	May–Jun	VU	–	LC	VU
Family-Papaveraceae								
Meconopsis aculeata Royle	Kanda	3200–4000	Herb	Jun–Sep	CR	EN	–	EN
Family-Pittosporaceae								
Pittosporum eriocarpum Royle	Agni	900–2000	Tree	Mar–Nov	–	–	EN	VU
Family-Plantaginaceae								
Bacopa monnieri (L.) Pennell	Brahmi	600–1500	Herb	Jun–Sep	–	–	LC	VU
Picrorhiza kurrooa Royle ex Benth	Kutki	3300–4800	Herb	Jun–Sep	EN	CR	–	EN
Family-Polygonaceae								

(continued)

Table 1 (continued)

Taxa	Vernacular Name	Altitudinal range (m)	Habit	Fl. & Fr	Threat category			
					CAMP		IUCN	Existing literature
					1998	2003		
Rheum australe D. Don	Dolu	3000–4500	Herb	Jul–Aug	VU	EN	–	VU
Rheum moorcroftianum Royle	Archa	3500–4600	Herb	Jun–Aug	–	NT	–	–
Rheum webbianum Royle	Dolu	3200–4600	Herb	Jun–Sep	–	VU	–	VU
Family-Primulaceae								
Embelia tsjeriam-cottam A. DC	Bayabirang	400–1600	Shrub		–	VU	–	–
Family-Ranunculaceae								
Aconitum lethale Griff	Mitha	3500–4100	Herb	Jul–Sep	CR	VU	–	–
Aconitum heterophyllum Wall. ex Royle	Atis	3000–4500	Herb	Jul–Sep	CR	CR	EN	EN
Aconitum violaceum Jacq. ex Stapf	Dudh Atis	3500–4600	Herb	Jul–Sep	CR	VU	VU	VU
Delphinium denudatum Wall. ex Hook. f. & Thoms	Nirbishi	600–2000	Herb	Apr–Jun	CR	–	–	CR
Thalictrum foliolosum DC	Mamiri	1200–2600	Herb	Jul–Sep	VU	–	–	LC
Family-Rutaceae								
Zanthoxylum armatum DC	Timur	900–2500	Shrub	Mar–Oct	–	VU	LC	VU
Family-Saxifragaceae								
Bergenia ciliata (Haw.) Sternb	Silphora	1500–3200	Herb	Apr–Jun	VU	–	LC	VU
Bergenia stracheyi (Hook. f. & Thomson) Engl	Silphora	3000–4100	Herb	Jun–Sep	–	NT	–	VU

(continued)

Table 1 (continued)

Taxa	Vernacular Name	Altitudinal range (m)	Habit	Fl. & Fr	Threat category			
					CAMP		IUCN	Existing literature
					1998	2003		
Family-Solanaceae								
Hyoscyamus niger L	Langtang	2000–3500	Herb	Jun–Sep	–	VU	–	–
Family-Symplocaceae								
Symplocos paniculata (Thunb.) Miq	Lodh	1600–3000	Tree	May–Oct	–	VU	–	–
Family-Taxaceae								
Taxus wallichiana Zucc	Thuner	1800–3400	Tree	Mar–Nov	CR	EN	EN	EN
Family-Ulmaceae								
Ulmus wallichiana Planch	Chamarmau	1600–3000	Tree	Mar–Jun	–	–	VU	–
Family-Zingiberaceae								
Curcuma angustifolia Roxb	Amahaldi	800–1300	Herb	Jun-Aug	NT	–	–	VU
Hedychium spicatum Buch.-Ham. ex Sm	Ban Haldi	1000–2500	Herb	Jul-Oct	VU	–	–	–

Source IUCN, CAMP (1998, 2003); Existing literature: Shah (1983), Nayar and Sastry (1987, 1988, 1990), Srivastava and Singh (2005), Joshi et al. (2010), Kumari et al. (2012), Bhatt et al. (2020), Chauhan (2020) and Chauhan and Bisht (2020). Abbreviations used—CR: Critically Endangered; EN: Endangered; VU: Vulnerable; NT: Near Threatened; LC: Least Concern

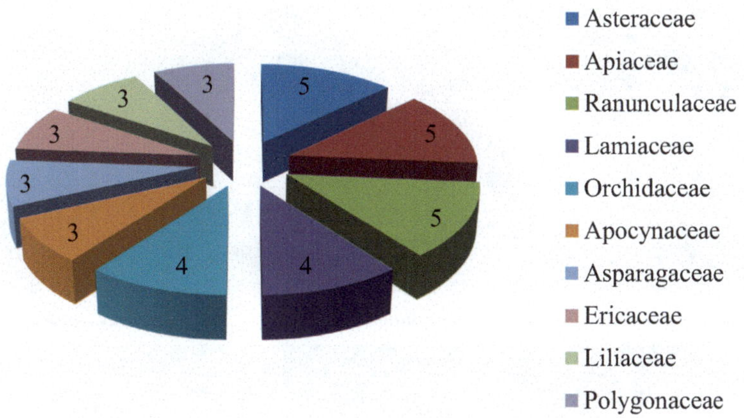

Fig. 1 Dominant families with number of threatened plants in Uttarakhand

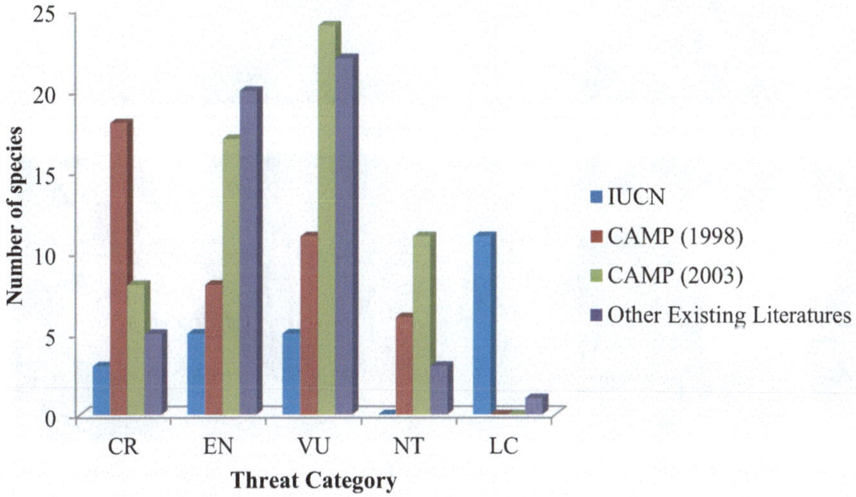

Fig. 2 Threatened plants reported in Kumaon, Uttarakhand

4 Conclusion

This study conducted in the Kumaun Himalaya region provides a comprehensive assessment of the threatened plant species, revealing a concerning status of biodiversity. A total of 82 threatened plant species, spanning 69 genera and 45 families, were identified, with diverse representation among trees, shrubs, and herbs. Dominant families such as Asteraceae, Apiaceae, and Ranunculaceae emerged as particularly vulnerable. The evaluation, considering IUCN, CAMP (1998), and other

Photo Plate 1 Threatened medicinal plants of Kumaun Himalaya

literature, demonstrated that a significant number of species face various threat categories, including critically endangered, endangered, and vulnerable. The comparison with previous studies highlights the persistent challenges to plant biodiversity in the Himalayan region. The utilization of these plants by local communities for medicinal and other purposes highlights the socio-economic importance of conserving this flora. Furthermore, the study emphasizes the limited coverage of threat evaluations, with only 8% of flowering plants evaluated by IUCN. The identification of extinct plant species and the escalating anthropogenic pressures on the Himalayan ecosystem call for urgent and concerted conservation efforts. As human activities,

Photo Plate 2 Threatened Medicinal Plants of Kumaun Himalaya

habitat destruction, and environmental degradation intensify, collaborative initiatives at local, national, and global levels become imperative to mitigate the threats faced by these plant species and sustain the rich biodiversity of the Kumaun Himalaya.

Photo Plate 3 Threatened Medicinal Plants of Kumaun Himalaya

Photo Plate 4 Threatened Medicinal Plants of Kumaun Himalaya

Acknowledgements The authors are thankful to local people, informants for their immense help during the tenure of the study. We are also thankful to IERP, GBPNIHE, Dr. Deepshikha Arya and Dr. Deepika Bhatt, Regional Ayurvedic Research Institute (RARI), CCRAS, Thapla, Ranikhet for providing the plants photographs and Head, Department of Botany, Kumaun Universtiy, Nainital for providing infrastructure, facilities and encouragement.

References

Arya JC, Joshi GC, Tewari LM (2018) Medicinal flora of Tarai region of Uttarakhand. Shree Publishers & Distributors New Delhi

Barman T, Samant SS, Singh A (2021) Structural diversity and regeneration pattern of forest communities in Parbati Valley, North Western Himalaya, India: implications for conservation, Indian. J Ecol 48:332–348

Bhat JA, Kumar M (2013) Ecological status and traditional knowledge of medicinal plants in Kedarnath Wildlife Sanctuary of Garhwal Himalaya, India. J Ethnobiol Ethnomed 9(1):1–18

Bhatt D, Joshi GC, Tewari LM, Pandey NC (2020) Traditional medicinal systems and threatened medicinal plants of Kumaun, Western Himalaya, India. Indu Book Services (Publishers & Distributors), New Delhi

Chandra N, Joshi VC, Mishra AP (2020) Ecological status of fodder and fuelwood species in Banari Devi sacred grove of Kumaun Himalaya, Uttarakhand. Indian J Ecol 47(4):1044–1048

Chandra N, Singh G, Rai ID, Mishra AP, Kazmi MY, Pandey A, Abdo HG (2023) Predicting distribution and range dynamics of three threatened cypripedium species under climate change scenario in Western Himalaya. Forests 14(3):633. https://doi.org/10.3390/f14030633

Chandra N, Rai ID, Mishra AP, Dwivedi SK, Kotiya A, Tiwari UK, Singh G (2022) Assessing potential habitats and populations of selected medicinal herbs in Alpine areas of Uttarakhand, Western Himalaya. Indian J For 45(3):144–155. https://doi.org/10.54207/bsmps1000-2023-ITRFNL

Chauhan HK, Bisht AK (2020) *Trillium govanianum*. The IUCN red list of threatened species. e.T175804005A176257695. https://doi.org/10.2305/IUCN.UK.2020-3

Chauhan HK (2020) Paris polyphylla. The IUCN red list of threatened species. https://doi.org/10.2305/IUCN.UK.2020

Conservation Assessment and Management Plan (C.A.M.P.) (1998) Workshops report

International Plant Name Index (2012) Published on the Internet. http://www.ipni.org. Accessed 28 Jan 2022

IUCN Red List for conservation. Trends Ecol Evol 21:71–76. https://doi.org/10.1016/j.tree.2005.10.010

Iv APG (2016) An update of the Angiosperm Phylogeny Group classification for the orders and families of flowering plants: APG III. Bot J Linn Soc 161:105–121

Joshi GC, Tewari LM, Lohani N, Kumar S, Kumari P, Bhatt D (2010). Studies on status, threat and strategies of endangered medicinal plants of the alpine regions of India. In: Tewari LM, Pangtey YPS, Tewari G (eds) Biodversity potential of Himalaya, Gyanodaya Prakashan, Naintal, pp 427–436

Joshi GC, Pangtey YPS, Pandey NC, Singh G, Tewari LM (2021) Legume and Thistle flora of Kumaun, Western Himalaya, India. Indu Book Services (Publishers & Distributors) New Delhi

Jyotsana S, Gaur RD, Painuli RM (2011) Conservation status and diversity of some important plants in the Shiwalik Himalaya of Uttarakhand, India. Int J Med Aromatic Plants 1(2):75–82

Kala CP (2005) Indigenous uses, population density, and conservation of threatened medicinal plants in protected areas of the Indian Himalayas. Conserv Biol 19(2):368–378

Kala CP (2006) Medicinal plants of the high-altitude cold desert in India: diversity, distribution and traditional uses. Int J Biodiversity Sci Manage 2(1):43–56

Kala CP (2004) Revitalizing traditional herbal therapy by exploring medicinal plants: a case study of Uttaranchal State in India. In: Indigenous knowledge's: transforming the academy, proceedings of an international conference Pennsylvania: Pennsylvania State University, pp 15–21

Kaur D, Tiwana AS, Kaur S, Gupta S (2022) Climate change: concerns and influences on biodiversity of the Indian Himalayas. In: Rani S, Kumar R (eds) Climate change. Springer, pp 265–281

Khullar SP (1994 and 2000) An illustrated fern flora of Western Himalaya, vol II. International Book Distributors, Dehradun

Kumar A, Adhikari BS, Mishra AP (2023) Status of threatened medicinal and Biosphere Reserve, Uttarakhand. A report (TR No./2022/35) submitted to International Union for Conservation of Nature—India, p 103

Kumari P, Joshi GC, Tewari LM (2012) Biodiversity status, distribution and use pattern of some ethno-medicinal plants. Int J Conserv Sci 3(4):309–318

Manish K, Pandit MK (2018) Geophysical upheavals and evolutionary diversification of plant species in the Himalaya. PeerJ 6:e5919

Maxted N, Hawkes JG, Guarino L, Sawkins M (1997) Towards the selection of taxa for plant genetic conservation. Genet Resour Crop Evol 44:337–348

Mishra AP, Kumar A, Yadav SN (2023a) Ecology and conservation of threatened medicinal plants in the Trans-Himalayan region of Nanda Devi Biosphere Reserve, Western Himalaya. Trees, Forests People 14:100451. https://doi.org/10.1016/j.tfp.2023.100451

Mishra AP, Chandra N, Mandy JJ, Dwivedi SK, Alruzuq A, Pande CB (2023b) Analytical hierarchy process (AHP) based on the spatial assessment of an endangered alpine medicinal herb aconitum heterophyllum in the Western Himalayan environment. In: Climate change impacts on natural resources, ecosystems and agricultural systems. Springer International Publishing, Cham. https://doi.org/10.1007/978-3-031-19059-9_25

Mittermeier RA, Robles GP, Hoffmann M, Pilgrim J, Brooks T, Mittermeier CG, Lamoreux J, Fonseca GAB (2004) Hotspots revisited: earth's biologically richest and most endangered ecoregions, CEMEX

Nayar MP, Sastry ARK (1987, 1988, 1990) Red data book of Indian plants, eds. 1–3. Botanical Survey of India, Howrah (Calcutta), India

Pandey NC, Joshi GC, Pangtey YPS, Tewari LM, Jalal JS (2022) Flora of Betalghat (Nainital), Western Himalaya, India. Corvette Press 123, Ground Floor, Sector-3, Vasundhara, Ghaziabad, Uttar Pradesh, pp 1–748

POWO (2019) Plants of the world online. Facilitated by the Royal Botanic Gardens, Kew

Rai ID, Singh G, Rawat GS (2017) Plants of Kedarnath wildlife sanctuary, Western Himalaya: a field guide. Bishen Singh Mahendra Pal Singh, Dehradun

Rawat N, Mishra AP, Chandra N, Painuly V, Negi GS, Bisht MPS, Purohit S (2021) Maxent modeling for predicting the potential distribution of *Valeriana jatamansi* jones. Chakrata forest division of Garhwal Himalaya. Indian J Ecol 48(2):367–372

Rodrigues AS, Pilgrim JD, Lamoreux JF, Hoffmann M, Brooks TM (2006) The value of the sacred plants of Indian Himalayan Region. Indian J For 26(3):201–213

Saha D, Ved D, Ravikumar K, Haridasan K (2015) *Malaxis muscifera*. IUCN Red List Threatened Species. https://doi.org/10.2305/IUCN.UK.2015-2

Samant SS, Dhar U (1997) Diversity, endemism and economic potential of wild edible plants of Indian Himalaya. Int J Sustain Develop World Ecol 4(3):179–191

Samant SS, Pal M (2003) Diversity and conservation status of medicinal plants in Uttaranchal State. Indian Forester 129(9):1090–1108

Samant SS, Palni LMS (2000) Diversity, distribution and indigenous uses of essential oil-yielding medicinal plants of the Indian Himalayan region. J Med Aromatic Plant Sci 22(1B):671–684

Samant SS, Shreekar P (2003) Diversity, distribution pattern and traditional knowledge of sacred plants in Indian Himalayan Region. Indian J for 26(3):222–234

Samant SS, Dhar U, Palni LMS (1998) Medicinal plants of Himalaya. Diversity distribution potential values. Gyanodaya Prakashan, Nainital

Samant SS (1998) Diversity, distribution and conservation of fodder resource of west Himalaya, India. In: Misri B (ed) Proceedings of the third temperate pasture and fodder network (TAPAFON), Pokhra, Nepal, 9–13 March, 1998, sponsored by F.A.O. Rome, pp 109–128

Shah NC (1983) Endangered medicinal and aromatic taxa of U.P. Himalaya. In: Jain SK, Rao RR (eds) Threatened plants of India. Botanical Survey of India, Botanical Gardern, Howarh, pp 40–49

Sharma J, Gaur R, Tewari LM (2011) Conservation status and diversity of some important plants in the Shiwalik Himalaya of Uttarakhand, India. Int J Med Aromat Plants 1:75–82

Shivanna KR (2020) The sixth mass extinction crisis and its impact on biodiversity and human welfare. Resonance 25:93–109. https://doi.org/10.1007/s12045-019-0924-z

Singh A, Manohar L, Samant SS (2009) Diversity, indigenous uses and conservation prioritization of medicinal plants in Lahaul valley, proposed Cold Desert Biosphere Reserve, India. Int J Biodivers Sci Manag 5(3):132–154

Singh KK, Anand P (1996) Observations on ethnobotany of Kol tribe of Varanasi district Uttar Pradesh, India. J Econ Tax Bot (Add. series) 12:133–137

Srivastava SK, Singh DK (2005) Glimpses of plant wealth of Uttarakhand. Bishan Singh Mahendra Pal Singh, Dehradun, Uttarakhand, India

Tiwari JK, Dangwal LR, Rana CS, Tiwari P, Ballabha R (2010) Indigenous uses of plant species in Nanda Devi Biosphere Reserve, Uttarakhand, India. Report Opin 2(2):67–70

Upreti BM, Tewari A, Tewari LM, Pandey NC (2019) Sacred forests of Pithoragarh, Western Himalaya. Indu, Book Services, New Delhi

Ved DK, Kinhal GA, Ravikumar K, Prabhakaran V, Ghate U, Vijayashankar R, Indresha JH (2003) Conservation assessment and management prioritization (CAMP) for the medicinal plants of Jammu & Kashmir, Himachal Pradesh & Uttaranchal. Foundation for Revitalization of Local Health Traditions, Bangalore, Karnataka, India

Verma RK (2014) Floristic diversity along an altitudinal gradients in Hango Valley of Cold Desert in District Kinnaur, Himachal Pradesh. Biol Forum–Int J 6(2):115–126

Wilson KA, Underwood EC, Morrison SA, Klausmeyer KR, Murdoch WW (2007) Conserving biodiversity efficiently: what to do, where, and when. PLoS Biol 5(9):e223. https://doi.org/10.1371/journal.pbio.0050223

Ethno-Medicinal Wealth and Its Conservation in Niti Valley, Nanda Devi Biosphere, Western Himalaya

Arun Pratap Mishra and Amit Kumar

Abstract The escalating demand for Medicinal and Aromatic Plants (MAPs) poses significant threats to the biodiversity of high-altitude areas. The Niti Valley, a cold-arid and buffer zone of the Nanda Devi Biosphere Reserve, is well known due to its remoteness and medicinal and ethno-botanical diversity. In this study, a comprehensive survey using a data collection framework provided insights on the valley's medicinal plant richness, usage, and conservation status. The study revealed a total of 72 MAPs belonging to 58 genera under 31 families that are locally utilized for consumption and curing at least 24 different human ailments. Of these species, 63% are in trade and locally consumed, highlighting the pressure on these valuable resources. Ethno-botanical analysis of studied species identified utilization of plant resources by *Bhotiyas* for various ailments, showcasing the cultural significance in traditional healthcare practices. This study underscores the need to protect the invaluable plant species and preserve the traditional knowledge held by local communities in the face of increasing demand and environmental threats.

Keywords Alpine · Conservation · Ethno-medicine · Traditional knowledge · Trans-Himalaya

1 Introduction

The Indian Trans-Himalaya (ITH), recognized as the High Altitude Cold Desert Zone (Zone 1), comprises four distinct biogeographic provinces: 1A, Ladakh mountains (Kargil, Nubra, Zanskar), including Lahaul and Spiti in Jammu and Kashmir and Himachal Pradesh; 1B, Changthang Plateau in eastern Ladakh, contiguous with the Tibetan Plateau; 1C, cold arid-regions of eastern Himachal Pradesh (Kinnaur) and

A. P. Mishra · A. Kumar (✉)
Department of Habitat Ecology, Wildlife Institute of India, Dehradun, Uttarakhand 248001, India
e-mail: amit@wii.gov.in

A. P. Mishra
Department of Forestry and Remote Sensing, Earthtree Enviro Private Limited, Shillong 793012, Meghalaya, India

A. P. Mishra et al. (eds.), *Threatened Medicinal Plants in the Indian Himalayan Region*, World Sustainability Series, https://doi.org/10.1007/978-3-031-73687-2_7

Uttarakhand, covering Nilang, Niti, Mana, Johar, Darma, and Byans valleys; and 1D, Sikkim Plateau (Rodgers et al. 2000; Kumar et al. 2017). These areas exhibit a unique physical and biological setting distinct from the adjoining alpine zones of the Greater Himalaya.

The alpine zone, locally termed bugyal or payar in Uttarakhand, stands out as a captivating biome in the Indian Himalayan Region. Spanning between the tree line (3300–3600 m amsl) and perpetual snow cover (5300 m) in the high altitudes of Western Himalaya, it serves as a specialized habitat for native and high-value medicinal plants. The alpine vegetation, characterized by matted dwarf shrubs, herbaceous meadows, grassy slopes, bogs, and snow-swept grounds with cushion-shaped plants, represents various habitat types crucial for a natural alpine ecosystem's biodiversity.

In Uttarakhand, the alpine zone covers approximately 8524 km^2, constituting 24% of the geographical area and encompassing more than 82 major bugyals (Rawat 2005). These bugyals, particularly around the basins of Nanda Devi, Kedarnath, Gangotri, and Bandarpunch, are integral to the lives of local inhabitants. Revered as repositories of a diverse array of medicinal and aromatic plant species (MAPs), these bugyals play a vital role in local medicine and supply herbal industries (Nautiyal et al. 2000). The mountainous region of the Himalaya is home to diverse tribal communities, with over 53 million tribal people belonging to 550 communities of 227 ethnic groups in India (Maikhuri and Gangwar 1993; Nautiyal et al. 2000). Among these communities, the Bhotiyas predominantly inhabit the cold-arid regions of Uttarakhand, practicing transhumance pastoralism. Possessing rich knowledge of medicinal plant usage in their locality, the Bhotiya tribal communities have traditionally collected medicinal plants from the wild during livestock grazing in the forests and alpine pastures. However, the 'All India trade survey of prioritized medicinal plants (2019)' highlighted a concerning trend—a 50% increase in the demand for high-value medicinal plants coupled with a 26% decline in availability. The escalating demand for natural resources, especially Medicinal and Aromatic Plants (MAPs), raises significant concerns for biodiversity, local economies, and public health. The high-altitude areas, critical for their biodiversity, ecological, socio-cultural, and hydrological values, are under severe threat due to over-exploitation, habitat degradation, economic development pressures, targeted removal of MAPs, and unpredictable changes induced by climate. In light of these challenges, this communication aims to assess the medicinal and ethno-botanical diversity in Niti valley, Nanda Devi Biosphere Reserve, Western Himalaya to understand the medicinal usage, extent of pressure, trade, market price and existing threats.

2 Study Area

The present study was conducted in the cold-arid and buffer zone of Nanda Devi Biosphere Reserve (NDBR) covering seven villages namely, Mahargaon, Kailashpur, Gurgutti, Pharkia, Bampa, Gamsali and Niti of Upper Dhauli Ganga catchment or Niti valley (Fig. 1) in Chamoli District of Uttarakhand. The valley falls under 1C

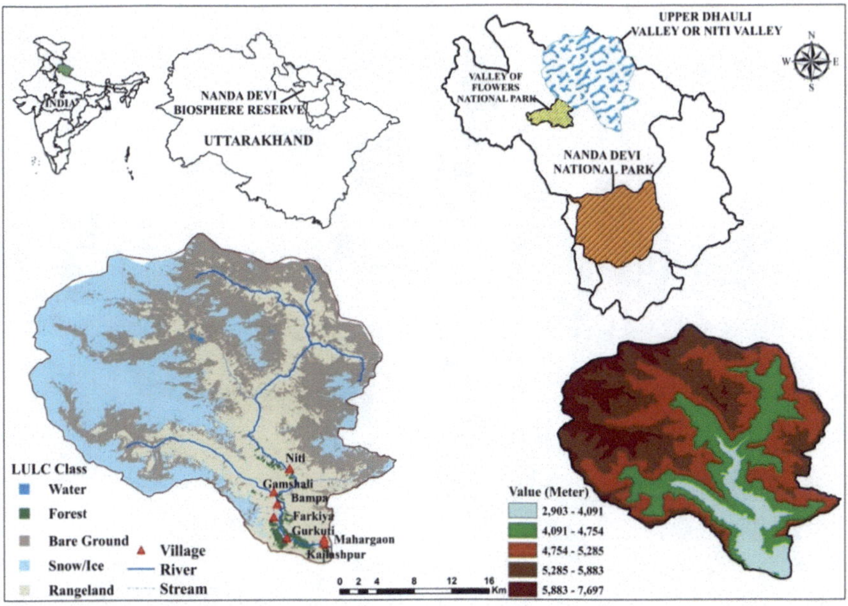

Fig. 1 Niti valley, a cold arid region of Nanda Devi Biosphere Reserve, Western Himalaya

Trans-Himalayan Biogeographic Province with an average elevation ranging from 3500 to 5000 m (Kumar et al. 2017).

2.1 *Physical Features*

Niti valley spreading over ca. 728 km^2 has three sub-watersheds namely, Amrit Ganga, Satyagad and Ganesh Ganga that comes under Trans-Himalayan region of Uttarakhand (Kumar 2017; Kumar et al. 2017). The picturesque valley is situated in the rain-shadow zone and the dryness increases towards upper reaches of the Dhauli and adjacent Girthi valley, which remain snowbound for more than six months in a year. In general, the vegetation in the valley comprises of dry temperate to dry alpine types, the flora is of steppe nature and rich at lower elevations (Kumar et al. 2016a; Kumar 2017).

2.2 *Local Inhabitants*

Niti valley is inhabited by two clans of Bhotiya community, locally known as *Tolchhas* and *Marchhas*. Of the 47 villages present within the buffer zone of NDBR,

Image 1 Bhotiya community in the Niti valley

seven migratory villages namely, Mahargaon, Kailashpur, Gurgutti, Pharkia, Bampa, Gamsali and Niti are located in the Niti valley. The Bhotiya ethnic community of Indo-Mongoloid origin has their own unique customs, folklore and religious beliefs. The inhabitants have two dwellings in cold-arid region of Niti valley between 3000 and 3600 m, where they stay during the summer and the other in the lesser Himalaya, a permanent dwelling between 1000 and 1500 m. The biotic pressure on the alpine rangelands is mainly due to heavy livestock grazing (Mitra et al. 2013). The locals depend on natural resources from the adjacent forests and alpine pastures for sustenance and livelihood, and the area continues to be used for transhumant pastoralism (Kumar and Mitra 2015) (Image 1).

2.3 *Floristic Wealth of Niti Valley*

A number of studies on floral wealth have been conducted in Nanda Devi Biosphere Reserve (NDBR). A comprehensive account on vascular plants revealed a total of 801 species of vascular plants belonging to 406 genera under 120 families from the NDBR (Hajra and Balodi 1995). The region has been extensively surveyed in terms of floral diversity by various workers, albeit highly confined to the core zones i.e., Nanda Devi National Park and Valley of Flowers National Park (Kumar et al. 2016a,

2019; Kumar 2017). Located in the buffer zone of NDBR in the Western Himalaya, Niti valley also known as upper Dhauli valley represents a cold arid region. Based on extensive surveys and systematic collections in the years 2011, 2012 and 2014, Kumar et al. (2013a), Kumar et al. (2016a) and Kumar (2017) reported 495 species of vascular plants belonging to 267 genera and 73 families adding 189 species to the existing floral wealth of the NDBR. Further, Kumar et al. (2016b) and Rai et al. (2017) recorded four noteworthy additions viz., *Dontostemon glandulosus, Potentilla pamirica, Carex sagaensis* and *Anthoxanthum flexuosum* to the flora of the Western Himalaya. Recently, Pandey et al. (2021) discovered a new species viz., *Allium negianum* from the valley.

2.4 Ecology of Habitats in Niti Valley

In ecological systems, vegetation communities are shaped by various physiographic and edaphic factors which determine their structural and functional attributes. In Trans-Himalayan region of the state of Uttarakhand, limited studies on livestock, patterns of diversity in various vegetation communities and landforms are available. Mitra et al. (2013) reported migratory route and stopover sites that face maximum anthropogenic pressure during livestock migration in Niti valley. Kumar and Mitra (2015) reported higher percentage of livestock presence, however, the blue sheep (*Pseudois nayaur*) was absent from major pastures in the region. Kumar et al. (2016c) declared *Caragana versicolor* Benth. (Fabaceae), a keystone species of high conservation concerns in the Hindu Kush Himalayan region. Kumar (2017) reported major landforms (moraine, river bed and scree) in the same region, among three major landforms, the species richness was highest in moraine (99) followed by scree (88) and river bed (44). Considering the unique floristic diversity, fauna and geology of the cold–arid regions of Kinnaur, HP and Nilang, Niti, Mana, Johar, Darma and Byans valleys in Uttarakhand along the Northern frontiers, Kumar et al. (2017) proposed inclusion of these valleys as a new Trans-Himalayan Biogeographic Province (1C) of India. Further, Mishra et al. (2023) summarized the ecology of threatened medicinal plants in the Trans-Himalayan region, emphasizing the decline in wild populations due to overgrazing, excessive harvesting, habitat destruction, and climate change. The study identified 34 high-value species, highlighted their distribution and potential threats, and recommended urgent conservation efforts, including micro-plans and sustainable management practices, to safeguard these plants from uncontrolled exploitation and ensure their survival in the face of increasing pressures.

2.5 Ethno-Botanical Wealth of Niti Valley

According to floral statistics of India (2017) hosted by Botanical Survey of India, Kolkata, a total of 2,68,600 flowering plants worldwide among them 18,386 (6.84%

of world) exited in India. In India, approximately 3000 plants species are known to have their medicinal properties. Prakasha et al. (2010) reports a total of 2500 plants that are used in traditional medicine, of which 100 plants are used regularly (Joshi and Pant 2012). In India, over 53 million tribal people belonging to 550 communities of 227 ethnic groups consume their food generally from forests for nourishment, pharmaceutical use and other prerequisites. Local communities in the Indian Himalaya fulfil their basic needs by utilizing many naturally occurring plant species including wild edibles. According to Phondani et al. (2010), 86 plant species were recognized as being utilized for the treatment of 37 common human ailments by *Tolchha* and *Marchha* sub-communities of the Bhotiya community in the Niti valley of Alaknanda catchment in the Western Himalaya. During a survey conducted in the upper catchment of Dhauli Ganga (also known as Niti valley), Kandari et al. (2012) documented 50 medicinal plants belonging to 31 families and 44 genera that are utilized by Bhotiyas. Of which, 70% were harvested from the wild, 22% were cultivated and 8% were cultivated as well as wild harvested. Kumar et al. (2013b) reported a unique archetype of conservation for safeguarding *Allium stracheyi* Baker (*Jambu-Pharan*) population in Niti valley. Kumar et al. (2015) raised a concern on depleting indigenous knowledge on medicinal plants of Bhotiya community and revealed that the locals are knowledgeable of about 38 medicinal plants used for curing 24 different human ailments. The local communities of the Himalaya have been utilizing fuelwood as one of the major sources of vitality for centuries. According to Mitra et al. (2017) Bhotiyas use 10 species for fuelwood (five species each of tree and shrub) in the same region. Singh et al. (2022) reported 27 potential wild edible fruits that are consumed by ethnic *Bhotia* community in the same region. Further, Kumar et al. (2023) emphasized the underexplored cold-arid regions of Niti valley in Uttarakhand, rich in medicinal plant diversity, facing threats from over-exploitation. The study provides a sustainable harvesting and cultivation protocol of five high value and threatened medicinal plants to address their declining populations and support the livelihoods of the Bhotiya community while preserving traditional healthcare practices.

3 Methodology

Based on the reconnaissance, the study area was delineated and a preliminary checklist of locally and commercially used plants available in the area, vegetation types, distribution pattern, trade route of the highly medicinal plants and size of the area were recorded with the help of local communities, shepherds and frontline staff of the forest department. The information on the high use value and threatened MAPs was collected through primary and secondary sources. The primary data was collected using semi-structured open-ended and closed-ended questions including individual interactions and group discussions. Surveys were carried out in selected seven villages and local/village markets in the study area. The secondary information was gathered from offline and online sources such as scientific research articles,

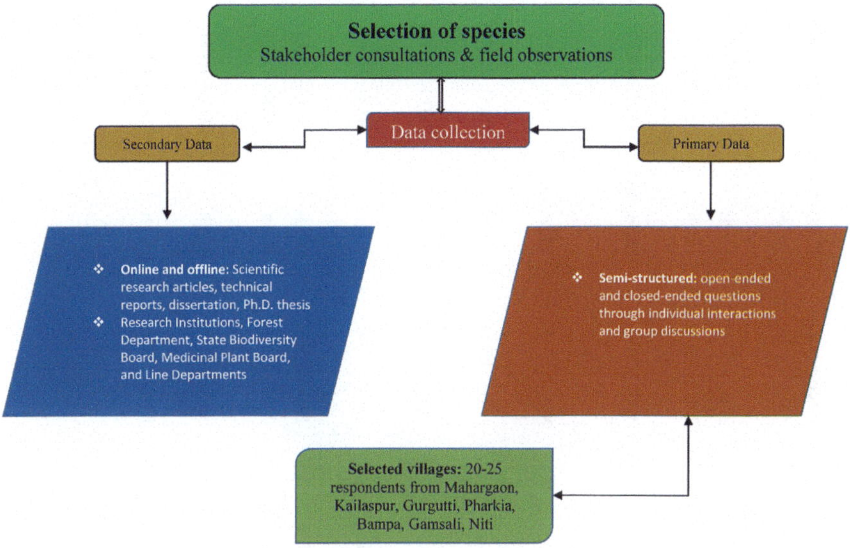

Fig. 2 Data collection framework for the current study

reports, books, dissertations, PhD thesis and using numerous databases, such as Google scholar, Research-gate, Taylor and Francis online, Springer-link, Elsevier online (Fig. 2).

4 Results and Discussion

4.1 Medicinal Plant Diversity

In the Himalayan region, the local communities depend on collection and trade of medicinal plants, handicrafts, and agriculture for their livelihood due to inhospitable conditions. The current study recorded a total of 72 medicinal and aromatic plant species belonging to 58 genera under 31 families (Table 1). Of the recorded species, 74% were wild in nature whereas, 26% were wild as well as cultivated in the valley (Fig. 3). Of the recorded medicinal plant species, 63% were in trade and locally consumed however, 37% species were only consumed by the Bhotiyas. In terms of medicinal usage, 51% of the species have high pressure due to overuse and unmanaged harvesting followed by 29% with moderate pressure and 20% have low pressure (Image 2 and Fig. 4).

Table 1 List of medicinal and aromatic plants reported in Niti valley, Nanda Devi Biosphere Reserve, Uttarakhand

S. no	Botanical name	Family	Trade/local consumption	Cultivated/ wild	IUCN*/ CITES**/ WPA***	Price per kg	Pressure/ usage	Distribution (m)
1	*Aconitum lethale* Griff.	Ranunculaceae	Trade and Lc	Wild	–	2000–2500	High	3000–4000
2	*Aconitum violaceum* Jacquem. ex Stapf.	Ranunculaceae	Trade and Lc	Wild	VU*	2000–2500	High	3500–4500
3	*Aconogonon tortuosum* (D.Don) H.Hara.	Polygonaceae	Trade and Lc	Wild	–	400–500	Moderate	3500–4500
4	*Allium carolinianum* Redouté	Amaryllidaceae	Lc	Wild	–	NA	Moderate	3000–3500
5	*Allium stracheyi* Baker	Amaryllidaceae	Trade and Lc	Cultivated and Wild	–	1000–1500	High	2800–4000
6	*Allium wallichii* Kunth	Amaryllidaceae	Trade and Lc	Cultivated and Wild	–	800–1200	High	2800–3000
7	*Angelica archangelica* L.	Apiaceae	Trade and Lc	Wild	LC*	1800–2500	High	3000–3600
8	*Angelica glauca* Edgew	Apiaceae	Trade and Lc	Wild	EN*	1800–2500	High	3000–3600
9	*Arctium lappa* L.	Asteraceae	Trade and Lc	Cultivated and Wild	LC*	2000–3000	High	3000–3800
10	*Arisaema tortuosum* (Wall.) Schott	Araceae	Trade and Lc	Wild	–	300–500	High	2800–4000
11	*Arnebia benthamii* (Wall. ex G.Don) I.M.Johnst	Boraginaceae	Trade and Lc	Wild	–	1000–1500	High	3000–4800

(continued)

Table 1 (continued)

S. no	Botanical name	Family	Trade/local consumption	Cultivated/wild	IUCN*/CITES**/WPA***	Price per kg	Pressure/usage	Distribution (m)
12	*Artemisia gmelinii* Weber ex Stechm	Asteraceae	Lc	Wild	–	NA	Low	3000–4800
13	*Artemisia maritima* L.	Asteraceae	Lc	Wild	–	NA	Low	2800–4000
14	*Astragalus candolleanus* Boiss.	Fabaceae	Trade and Lc	Wild	–	300–500	High	2800–4000
15	*Berberis pseudumbellata* R. Parker	Berberidaceae	Trade and Lc	Wild	–	1000–1500	High	2800–4000
16	*Bergenia ciliata* (Haw.) Sternb	Saxifragaceae	Trade and Lc	Cultivated and Wild	–	2000–2500	High	3000–4700
17	*Betula utilis* D.Don	Betulaceae	Trade and Lc	Wild	LC*	800–1200	High	2800–4000
18	*Cannabis sativa* L.	Cannabaceae	Lc	Cultivated and Wild	–	NA	Low	2500–3500
19	*Carum carvi* L.	Apiaceae	Trade and Lc	Cultivated and Wild	LC*	3000–5000	High	2800–3600
20	*Cedrus deodara* (Roxb. ex D.Don) G.Don	Pinaceae	Trade and Lc	Wild	LC*	5000–7000	High	2800–3800
21	*Chenopodium foliosum* Asch	Amaranthaceae	Trade and Lc	Wild	–	300–500	Moderate	3000–4800
22	*Cicerbita macrorhiza* (Royle) Beauverd.	Asteraceae	Lc	Wild	–	NA	Low	3000–3500

(continued)

Table 1 (continued)

S. no	Botanical name	Family	Trade/local consumption	Cultivated/ wild	IUCN*/ CITES**/ WPA***	Price per kg	Pressure/ usage	Distribution (m)
23	*Corydalis cornuta* Royle	Papaveraceae	Lc	Wild	–	NA	Moderate	3400–3700
24	*Cotoneaster microphyllus* Wall. ex Lindl.	Rosaceae	Lc	Wild	–	NA	Low	3200–3800
25	*Dactylorhiza hatagirea* (D.Don) Soó	Orchidaceae	Trade and Lc	Wild	EN*, Appendix II** & Appendix II***	4000–8000	High	2800–4200
26	*Delphinium denudatum* Wall. ex Hook.f. & Thomson	Ranunculaceae	Trade and Lc	Wild	–	300–500	Moderate	3000–4200
27	*Ephedra gerardiana* Wall. ex Klotzsch & Garcke	Ephedraceae	Trade and Lc	Wild	VU*	400–600	Moderate	3000–5000
28	*Fagopyrum dibotrys* (D.Don) Hara	Polygonaceae	Trade and Lc	Cultivated	–	500–700	Moderate	2800–3500
29	*Fagopyrum esculentum* Moench	Polygonaceae	Trade and Lc	Cultivated	–	500–700	Moderate	2800–3500
30	*Fragaria nubicola* (Lindl. ex Hook.f.) Lacaita	Rosaceae	Trade and Lc	Wild	–	300–500	Moderate	3200–3800
31	*Gaultheria trichophylla* Royle	Ericaceae	Trade and Lc	Wild	–	200–400	High	3000–4000

(continued)

Table 1 (continued)

S. no	Botanical name	Family	Trade/local consumption	Cultivated/ wild	IUCN*/ CITES**/ WPA***	Price per kg	Pressure/ usage	Distribution (m)
32	*Geranium wallichianum* D.Don ex Sweet	Geraniaceae	Lc	Wild	LC*	NA	Low	3200–4500
33	*Gymnadenia orchidis* Lindl.	Orchidaceae	Trade and Lc	Wild	–	1000–1200	High	3400–4000
34	*Hippophae salicifolia* D.Don	Elaeagnaceae	Trade and Lc	Cultivated and wild	–	300–500	High	2800–3800
35	*Hyoscyamus niger* L.	Solanaceae	Lc	Wild	–	NA	Low	2500–3500
36	*Hyssopus officinalis* L.	Lamiaceae	Lc	Wild	DD*	NA	Low	2500–3500
37	*Juglans regia* L.	Juglandaceae	Lc	Cultivated and wild	LC*	NA	Low	2500–3500
38	*Juniperus communis* L.	Cupressaceae	Lc	Wild	LC*	NA	High	3000–4500
39	*Juniperus indica* Bertol	Cupressaceae	Trade and Lc	Wild	LC*	500–800	High	3000–4500
40	*Malus domestica* Borkh	Rosaceae	Trade and Lc	Cultivated		100–200	High	3000–3800
41	*Mentha longifolia* (L.) Huds	Lamiaceae	Lc	Wild	LC*	NA	Moderate	2500–3500
42	*Morchella esculenta*	Morchellaceae	Trade and Lc	Wild	–	10,000–20,000	High	2800–4000
43	*Ophiocordyceps sinensis*	Ophiocordycipitaceae	Trade and Lc	Wild	VU*	10–15 lakh	High	3500–5500
44	*Origanum vulgare* L.	Lamiaceae	Trade and Lc	Wild	LC*	200–400	Moderate	2500–3500
45	*Phytolacca acinosa* Roxb.	Phytolaccaceae	Lc	Cultivated and Wild	–	NA	Moderate	2500–3500

(continued)

Table 1 (continued)

S. no	Botanical name	Family	Trade/local consumption	Cultivated/ wild	IUCN*/ CITES**/ WPA***	Price per kg	Pressure/ usage	Distribution (m)
46	*Picrorhiza kurroa* Royle ex Benth.	Plantaginaceae	Trade and Lc	Wild	EN*, Appendix II** & Appendix II***	800–1500	High	3500–4500
47	*Pinus wallichiana* A.B. Jacks	Pinaceae	Lc	Wild	LC*	NA	Moderate	3000–3500
48	*Pleurospermum brunonis* (DC.) Benth. ex C.B.Clarke	Apiaceae	Trade and Lc	Wild	–	1200–1500	High	3000–4000
49	*Polygonatum verticillatum* (L.) All	Asparagaceae	Trade and Lc	Wild	–	600–800	High	3200–3800
50	*Polygonum plebejum* R. Br.	Polygonaceae	Lc	Wild	–	NA	Moderate	2800–4000
51	*Potentilla fulgens* Wall. ex Sims	Rosaceae	Lc	Wild	–	NA	Moderate	2500–3500
52	*Prunus armeniaca* L.	Rosaceae	Lc	Cultivated and Wild	DD*	100–300	High	2500–3500
53	*Prunus cornuta* (Wall. ex Royle) Steud.	Rosaceae	Lc	Cultivated and Wild	–	NA	Low	2500–3500
54	*Prunus jacquemontii* Hook.f.	Rosaceae	Lc	Cultivated and Wild	–	NA	Moderate	2500–3500
55	*Prunus mira* Koehne	Rosaceae	Lc	Cultivated and Wild	–	NA	Moderate	2500–3500

(continued)

Table 1 (continued)

S. no	Botanical name	Family	Trade/local consumption	Cultivated/ wild	IUCN*/ CITES**/ WPA***	Price per kg	Pressure/ usage	Distribution (m)
56	*Rheum moorcroftianum* Royle	Polygonaceae	Trade and Lc	Wild	–	1000–1200	High	3000–4500
57	*Rheum webbianum* Royle	Polygonaceae	Trade and Lc	Wild	–	1000–1200	High	3000–4500
58	*Rhododendron anthopogon* D.Don	Ericaceae	Lc	Wild	–	NA	Low	3000–4800
59	*Ribes alpestre* Wall. ex Decne.	Grossulariaceae	Trade and Lc	Wild	–	1000–1500	High	3000–4000
60	*Ribes orientale* Desf.	Grossulariaceae	Lc	Wild	–	NA	Moderate	2500–3500
61	*Rosa macrophylla* Lindl.	Rosaceae	Lc	Wild	–	NA	Low	3000–3600
62	*Rosa sericea* Lindl.	Rosaceae	Trade and Lc	Wild	–	600–800	High	3000–3750
63	*Rosa webbiana* Wall. ex Royle	Rosaceae	Trade and Lc	Wild	–	200–400	Moderate	3000–4500
64	*Rumex nepalensis* Spreng.	Polygonaceae	Lc	Wild	–	NA	Low	900–4150
65	*Saussurea costus* (Falc.) Lipsch.	Asteraceae	Trade and Lc	Cultivated and Wild	CR*, Appendix I**, Appendix II***	300–500	High	3000–3200
66	*Saussurea obvallata* (DC.) Sch.Bip.	Asteraceae	Trade and Lc	Wild	–	500–1000	High	3800–4800

(continued)

Table 1 (continued)

S. no	Botanical name	Family	Trade/local consumption	Cultivated/ wild	IUCN*/ CITES**/ WPA***	Price per kg	Pressure/ usage	Distribution (m)
67	*Selinum wallichianum* (DC.) Raizada & H.O.Saxena	Apiaceae	Trade and Lc	Wild	–	1800–3000	High	3000–3500
68	*Sinopodophyllum hexandrum* (Royle) T.S.Ying.	Berberidaceae	Trade and Lc	Cultivated and Wild	Appendix II** & Appendix II***	800–1000	High	3000–4000
69	*Taraxacum officinale* F.H.Wigg.	Asteraceae	Lc	Wild	–	NA	Low	3000–3800
70	*Thymus linearis* Benth.	Lamiaceae	Trade and Lc	Wild	–	300–500	Moderate	3000–5000
71	*Urtica dioica* L.	Urticaceae	Lc	Wild	LC*	NA	Moderate	2500–3800
72	*Valeriana hardwickii* Wall.	Caprifoliaceae	Trade and Lc	Wild	–	2000–3000	High	2500–3200

Image 2 Chippi, Dhoop, Keedajadi and Gucchi in high demand and trade in Niti valley

Fig. 3 Distribution pattern of MAPs in Niti valley

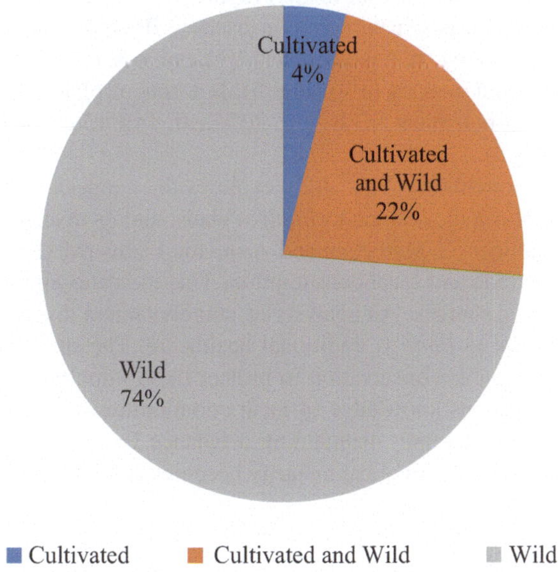

Fig. 4 Extent of pressure or usage on MAPs in Niti valley

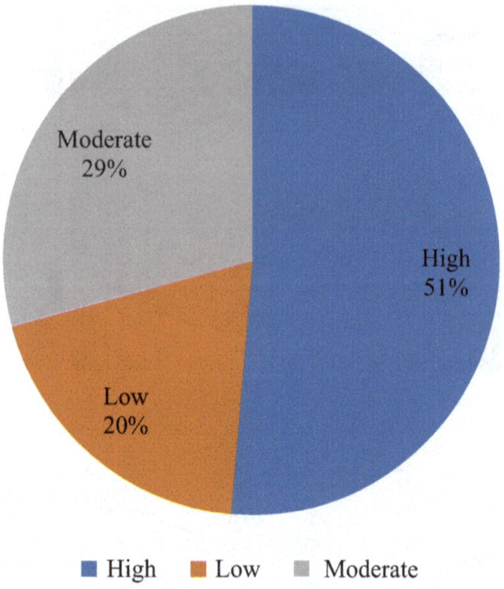

4.2 Ethno-Botanical Diversity

The Bhotiyas depend on the wild medicinal and aromatic plant species for their traditional health care system. Of total MAPs recorded in the valley, 72 species were locally utilized by the locals (Table 2). Of the recorded species, 47% of the medicinal plants are used in the common diseases. Bhotiyas use 21% medicinal species for fever followed by cold and cough (14%), nerve & joint pain and headache (7% each) and cuts and wounds (4%) (Fig. 5). In terms of plant part use, 27% species were used as a whole plant, 23% fruit, 20% leaf, 14% root, 7% bark, 5% seed and 4% tubers (Fig. 6).

Conclusion: This study explores the connection between the physical features of Niti Valley, home to *Bhotiya* clans, and its diverse plant life. A thorough survey identifies 72 Medicinal and Aromatic Plants (MAP) species, with 63% being used for trade and local consumption. This indicates significant pressure on these valuable resources. An analysis of ethnobotanical diversity pinpoints 72 plant species crucial to *Bhotiya* traditional healthcare. The study highlights the urgent need for sustainable conservation to protect these important plant species and preserve the indigenous knowledge of local communities. As demand rises and environmental threats increase, maintaining a balance between uplifting local communities and preserving ecological integrity becomes crucial for the continued existence of these botanical riches.

Way forward: It is essential to implement *in-situ* conservation strategies for the medicinal plants of Niti Valley, involving local communities in sustainable

Table 2 List of ethno-medicinal plant species utilized by Bhotiyas in Niti valley

S. no	Botanical name	Family	Vernacular name	Parts used	Medicinal uses
1	*Aconitum lethale* Griff.	Ranunculaceae	Meetha Jari	Root	Fever, rheumatism, nerve pain and sciatica
2	*Aconitum violaceum* Jacquem. ex Stapf.	Ranunculaceae	Atish or Patish	Tuberous, Root	Nerve pain, stomachache, cough and heart disease
3	*Aconogonon tortuosum* (D.Don) H.Hara.	Polygonaceae	Naylo	Root and shoot	Leucorrhoea
4	*Allium carolinianum* Redouté	Amaryllidaceae	Doom	Whole plant	Constipation, digestion, loose motion, headache, joint pain, cold and cough, diarrhea
5	*Allium stracheyi* Baker	Amaryllidaceae	Jambu Faran	Whole plant	Headache, joint pain, cold, cough, diarrhea, and as condiments
6	*Allium wallichii* Kunth.	Amaryllidaceae	Laynka	Whole plant	Jaundice, cold and cough
7	*Angelica glauca* Edgew	Apiaceae	Choru	Root and fruit	Acidity, vomiting, body ache, fever, headache
8	*Angelica archangelica* L.	Apiaceae	Choru	Root and fruit	Acidity, vomiting, body ache, fever, headache
9	*Arctium lappa* L.	Asteraceae	Jangali Kunth	Root	Stomachache, rheumatism, fever
10	*Arisaema tortuosum* (Wal) Schott.	Araceae	Kavakya,, Bal Mungari	Whole plant	Stomach infection
11	*Arnebia benthamii* (Wall. ex G.Don) I.M.Johnst.	Boraginaceae	Laljari, Balchadi	Root and leaves	Baldness, fever, cuts and wounds and blood pressure
12	*Artemisia gmelinii* Weber ex Stechm.	Asteraceae	Purchu	Leaf and root	Cold and cough

(continued)

Table 2 (continued)

S. no	Botanical name	Family	Vernacular name	Parts used	Medicinal uses
13	*Artemisia maritima* L.	Asteraceae	Purchu	Leaf and root	Stomachache, rheumatism, fever
14	*Astragalus candolleanus* Boiss	Fabaceae	Rudravanti	Whole plant	Blood pressure, blood purifier, asthma and cough
15	*Berberis pseudumbellata* R. Parker	Berberidaceae	Kilmora	Fruit	Antioxidant
16	*Bergenia stracheyi* (Hook.f. & Thomson) Engl.	Saxifragaceae	Silfore	Whole plant	Kidney stone
17	*Betula utilis* D.Don	Betulaceae	Bhuj, Bhoj	Bark	Rheumatism, asthma, cold, cough and fever
18	*Cannabis sativa* L.	Cannabaceae	Bhang	Whole plant	Urine infection, lack of concentration
19	*Carum carvi* L.	Apiaceae	Jangali Jira	Seed and leaf	Fever, diarrhea and stomachache
20	*Cedrus deodara* (Roxb. ex D.Don) G.Don	Pinaceae	Devdar	Inner wood	Inflammation, wrinkles, acne or any infection
21	*Chenopodium foliosum* Asch	Amaranthaceae	Bethua	Leaf	Indigestion, cold, cough
22	*Cicerbita macrorhiza* (Royle) Beauverd	Asteraceae	Karatu	Whole plant	Treating fever
23	*Corydalis cornuta* Royle	Papaveraceae	Chitra Jar	Tubers, bulbs and roots	Headache, insomnia, and dysmenorrhea
24	*Cotoneaster microphyllus* Wall. ex Lindl.	Rosaceae	Ruins	Fruit	Dermatitis
25	*Dactylorhiza hatagirea* (D.Don) Soó	Orchidaceae	Hatthajadi	Root	Dysentery, diarrhea, chronic fever, cough, stomachache, wounds, cuts and general weakness

(continued)

Table 2 (continued)

S. no	Botanical name	Family	Vernacular name	Parts used	Medicinal uses
26	*Delphinium denudatum* Wall. ex Hook.f. & Thomson	Ranunculaceae	Nirbishi	Root	Aconite poisoning, brain diseases, fungal infection, piles
27	*Ephedra gerardiana* Wall. ex Klotzsch & Garcke	Ephedraceae	Somlata	Root and shoot	Asthma, rheumatism, headache
28	*Fagopyrum esculentum* Moench.	Polygonaceae	Phaphar	Root and leaves	Rheumatism, lung diseases and fever
29	*Fagopyrum dibotrys* (D.Don) Hara	Polygonaceae	Phaphar	Root and leaves	Toothache, lung diseases and fever
30	*Fragaria nubicola* (Lindl. ex Hook.f.) Lacaita	Rosaceae	Jangali Kafal	Fruit	Profuse menstruation
31	*Gaultheria trichophylla* Royle	Ericaceae	Jheri	Fruit and leaves	Wounds, cold and cough
32	*Geranium wallichianum* D.Don ex Sweet	Geraniaceae	Neela Phool	Root	Toothache, skin disease
33	*Gymnadenia orchidis* Lindl.	Orchidaceae	Hathjod	Root	Acidity, lose motion, fever, diarrhea, cuts and wounds
34	*Hippophae salicifolia* D.Don	Elaeagnaceae	Ames, Chuk	Fruit and bark	Cold, cough, wounds, fever, chest pain
35	*Hyoscyamus niger* L.	Solanaceae	Phagun	Whole plant	Toothache, cold and cough
36	*Hyssopus officinalis* L.	Lamiaceae	Dhoop	Leaves and flowers	Colds, cough, congestion and lung complaints
37	*Juglans regia* L.	Juglandaceae	Jangali akhrot	Cotyledon	Helminthiasis, diarrhea, sinusitis, stomachache, arthritis, asthma, eczema, scrofula

(continued)

Table 2 (continued)

S. no	Botanical name	Family	Vernacular name	Parts used	Medicinal uses
38	*Juniperus indica* Bertol	Cupressaceae	Bitaru	Leaf and Fruit	Fever, cold and cough
39	*Juniperus communis* L.	Cupressaceae	Bitaru	Leaf and Fruit	Diuretic, antiseptic, gastrointestinal disorders
40	*Malus domestica* Borkh	Rosaceae	Seb	Fruit	Digestion, cancer
41	*Mentha longifolia* (L.) Huds	Lamiaceae	Jangali pudina	Whole plant	Stomach problems and allergy
42	*Morchella esculenta*	Morchellaceae	Gucchi	Whole plant	Cold, cough and immunity booster
43	*Ophiocordyceps sinensis*	Ophiocordycipitaceae	Keerajadi	Whole plant	Immunity booster
44	*Origanum vulgare* L.	Lamiaceae	Van tulsi	Leaf	Cold, fever and blood purification
45	*Phytolacca acinosa* Roxb.	Phytolaccaceae	Jagra	Whole plant	Asthma, cold and cough
46	*Picrorhiza kurroa* Royle ex Benth.	Plantaginaceae	Kutki	Whole plant	Cold, cough, fever, diabetes, jaundice, typhoid and constipation
47	*Pinus wallichiana* A.B. Jacks	Pinaceae	Kail	Seed	Skin complaints, wounds, sores, burns
48	*Pleurospermum brunonis* (DC.) Benth. ex C.B.Clarke	Apiaceae	Chippi	Whole plant	Stomachache, fever and condiments
49	*Sinopodophyllum hexandrum* (Royle) T.S.Ying.	Berberidaceae	Ban Kakadi	Fruit and Root	Stomachache, skin disease, sugar, cancer, hair-fall, intestinal infection, old fever

(continued)

Table 2 (continued)

S. no	Botanical name	Family	Vernacular name	Parts used	Medicinal uses
50	*Polygonatum verticillatum* (L.) All	Asparagaceae	Salam-misri	Root	Stamina booster, stomachache, tuberculosis (Kshay rog)
51	*Polygonum plebejum* R. Br.	Polygonaceae	Tanka Grass	Root	Skin disease, Boils
52	*Potentilla fulgens* Wall. ex Sims	Rosaceae	Vajradanti	Root	Toothache
53	*Prunus armeniaca* L.	Rosaceae	Chuli	Fruit	Body ache, diarrhea
54	*Prunus cornuta* (Wall. ex Royle) Steud.	Rosaceae	Jangali jamun	Fruit	Respiration and improve digestion
55	*Prunus jacquemontii* Hook.f	Rosaceae	Khursang	Fruit	Improve digestion, hepatitis, hyperlipidemia, hypertension
56	*Prunus mira* Koehne	Rosaceae	Kirol	Fruit	Hair-loss
57	*Rheum moorcroftianum* Royle	Polygonaceae	Dholu	Root	Skin disease, pain, cold and cough, stamina
58	*Rheum webbianum* Royle	Polygonaceae	Tatari	Root and leaves	Ulcer, cuts and wounds, boils
59	*Rhododendron anthopogon* D.Don	Ericaceae	Awon	Stem, leaves, and buds	Headache, digestion
60	*Ribes alpestre* Wall. ex Decne.	Grossulariaceae	Lipchi	Fruit	Detoxification, glaucoma, cardiovascular disease, stomachache,
61	*Ribes orientale* Desf.	Grossulariaceae	Darbag	Fruit	Treat rheumatism and joints pain
62	*Rosa macrophylla* Lindl.	Rosaceae	Kawa sidum	Fruit	Sexual and menstrual disorders, backache, and skin boils
63	*Rosa sericea* Lindl.	Rosaceae	Sidum	Fruit	Headaches

(continued)

Table 2 (continued)

S. no	Botanical name	Family	Vernacular name	Parts used	Medicinal uses
64	*Rosa webbiana* Wall. ex Royle	Rosaceae	Sidum	Fruit	Kidney stones, diabetes, eye strain, hemorrhoids, bronchitis, ulcers, gallstones
65	*Rumex nepalensis* Spreng.	Polygonaceae	Payoom	Root and aerial parts	Antioxidant and antibacterial
66	*Saussurea costus* (Falc.) Lipsch.	Asteraceae	Kuth	Whole plant	Asthma, inflammatory diseases, ulcer and stomach problems
67	*Saussurea obvallata* (DC.) Sch.Bip	Asteraceae	Brahma Kamal	Whole plant	Rheumatism, cuts and wounds, boils
68	*Selinum wallichianum* (DC.) Raizada & H.O.Saxena	Apiaceae	Bhutkesh	Flower	Folk medicine and ritual
69	*Taraxacum officinale* F.H.Wigg.	Asteraceae	Karanphool	Root	Acidity, fever and headache
70	*Thymus linearis* Benth.	Lamiaceae	Marchaghas	Whole plant	Stomachache, cuts and wounds, acidity, skin disease, Nerve pain
71	*Urtica dioica* L.	Urticaceae	Kandali	Leaves and stems	Painful muscles and joints, eczema, arthritis, gout, and anemia
72	*Valeriana hardwickii* Wall.	Caprifoliaceae	Taggar	Whole plant	Urine infection, oil, dhoop, liver disease

practices. Ethnobotanical studies should document and preserve traditional knowledge, while biodiversity monitoring and scientific research inform conservation measures. Regulatory measures and market interventions can manage the trade of medicinal plants, promoting fair practices. Diversification and skill development are key for community development, reducing dependence on plant collection. Climate-resilient practices and educational initiatives address climate change

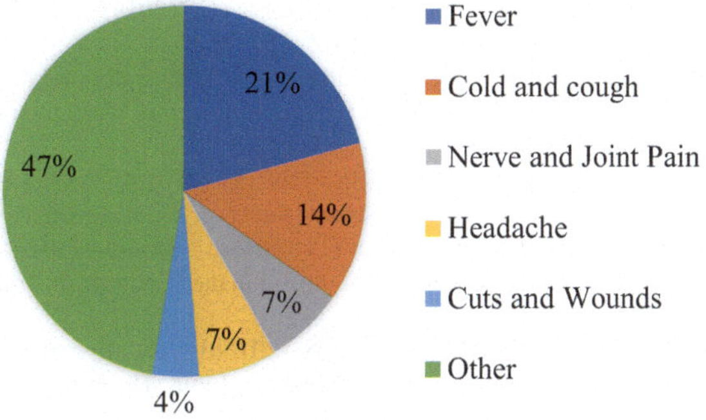

Fig. 5 Use of medicinal plants in different ailments by locals inhabitants

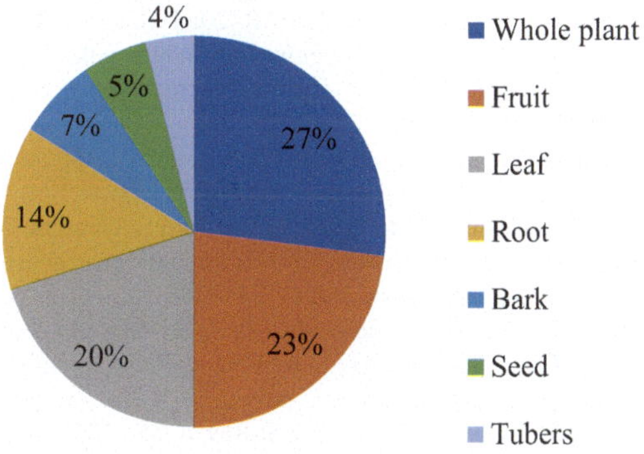

Fig. 6 Plant parts used in different ailments by locals in Niti valley

impacts and raise public awareness. Collaborative partnerships with stakeholders enhance conservation efforts for the biodiversity-rich Niti Valley.

References

Hajra PK, Balodi B (1995) Plant wealth of Nanda Devi biosphere reserve. Botanical Survey of India, Calcutta

Joshi B, Pant SC (2012) Ethnobotanical study of some common plants used among the tribal communities of Kashipur, Uttarakhand

Kandari LS, Phondani PC, Payal KC, Rao KS, Maikhuri RK (2012) Ethnobotanical study towards conservation of medicinal and aromatic plants in upper catchments of Dhauli Ganga in the central Himalaya. J Mt Sci 9(2):286–296

Kumar A, Mitra M, Adhikari BS, Rawat GS (2013a) Additions to the plant wealth of Nanda Devi Biosphere Reserve, Western Himalaya. Indian Forester 139(10):959–961

Kumar A, Mitra M, Adhikari BS, Rawat GS (2015) Depleting indigenous knowledge of medicinal plants in cold-arid region of Nanda Devi Biosphere Reserve, Western Himalaya. Med Aromatic Plants 4(195):2167–2178

Kumar A, Mitra M, Adhikari BS, Rawat GS (2016c) Flora of Niti valley: a cold arid region of Nanda Devi Biosphere reserve, Western Himalaya, India. Check List 12(1):1824–1824

Kumar A, Srivastav M, Adhikari BS, Rawat GS (2019) Plant studies in the first Himalayan biosphere reserve: a review. J Mt Sci 16(2):351–363

Kumar A, Mitra M (2015) Landforms, plant diversity and their utilization by ungulates in Upper Dhauli Valley, Nanda Devi Biosphere Reserve, Western Himalaya. Report submitted to WWF-India, New Delhi

Kumar A, Mitra M, Adhikari BS, Rawat GS (2013b) Archetype conservation in Trans-Himalayan region of Nanda Devi Biosphere Reserve, Western Himalaya. e J Appl For Ecol 1(1):84–86

Kumar A, Adhikari BS, Rawat GS (2016a) *Caragana versicolor* Benth. (Fabaceae), A keystone species of high conservation concern in the Hindu Kush Himalayan region. Curr Sci 111(6):985–987

Kumar A, Adhikari BS, Rawat GS (2016b) New phytogeographically noteworthy plant records from Uttarakhand, western Himalaya, India. Threatened Taxa 8(6):8943–8947. https://doi.org/10.11609/jott.2416.8.6.8943-8947

Kumar A, Adhikari BS, Rawat GS (2017) Biogeographic delineation of the Indian Trans-Himalaya: need for revision. Curr Sci 113(6):1032–1033

Kumar A, Adhikari BS, Mishra AP (2023) Status of threatened medicinal and aromatic plants and their use by the Bhotiya community in Niti valley, Nanda Devi Biosphere Reserve, Uttarakhand: a report (TR No./2022/35) submitted to International Union for Conservation of Nature—India, p 103

Kumar A (2017) A study on floristic diversity and plant functional types across various landforms in Upper Dhauli Valley Nanda Devi Biosphere Reserve, Western Himalaya. PhD thesis, Forest Research Institute (Deemed) University, Dehradun, 2017, p 165

Maikhuri RK, Gangwar AK (1993) Ethnobiological notes of the Khasi and Garo tribes of Meghalaya, north east India. Econ Bot 47:345–357

Mishra AP, Kumar A, Yadav SN (2023) Ecology and conservation of threatened medicinal plants in the Trans-Himalayan region of Nanda Devi Biosphere Reserve, Western Himalaya. Trees, For People 14:100451

Mitra M, Kumar A, Adhikari BS, Rawat GS (2017) Fuelwood resources and their use pattern by Bhotia community in Niti valley, Western Himalaya. Botanica Orientalis: J Plant Sci 11:1–6

Mitra M, Kumar A, Adhikari BS, Rawat GS (2013) A note on transhumant pastoralism in Niti valley, Western Himalaya, India. Pastoralism: Res Policy Pract 3(1):1–6

Nautiyal S, Rao KS, Maikhuri RK, Semwal RL, Saxena KG (2000) Traditional knowledge related to medicinal and aromatic plants in tribal societies in a part of Himalaya. J Med Aromat Plant Sci 22:528–541

Pandey A, Rai KM, Malav PK, Rajkumar S (2021) *Allium negianum* (Amaryllidaceae): a new species under subg. Rhizirideum from Uttarakhand Himalaya, India. PhytoKeys 183:77

Phondani PC, Maikhuri RK, Rawat LS et al (2010) Ethnobotanical uses of plants among the Bhotiya tribal communities of Niti Valley in Central Himalaya, India. Ethnobot Res Appl 8:233–244

Prakasha HM, Krishnappa M, Krishnamurthy YL, Poornima SV (2010) Folk medicine of NR Pura taluk in Chikmagalur district of Karnataka

Rai ID, Kumar A, Singh G, Adhikari BS, Rawat GS (2017) Three noteworthy additions to the flora of the western Himalaya, India. J Threatened Taxa 9(7):10421–10425

Rawat GS (2005) Alpine meadows of Uttaranchal: ecology, landuse and status of medicinal & aromatic plants. Bishen Singh Mahendra Pal Singh, Publication, Dehradun, India

Rodgers WA, Panwar HS, Mathur VB (2000) Wildlife protected area network in India: a review (executive summary). Wildlife Institute of India, Dehradun. pp 44

Singh B, Rai K, Bhatt KC, Chaudhury R, Dubey SK, Ahlawat SP (2022) Ethnobotanical notes on some potential wild edible fruits used by the Bhotia community of Niti Valley, Uttarakhand, India. Indian J Tradit Knowl 21(1):135–144

Synergistic Role of Plant Tissue Culture and Plant Growth-Promoting Micro-Organisms as Conservation Strategies for Himalayan Medicinal Plants Amidst Climate Change

Subhankar Das, Susmita Shukla, Samir Debnath, and Manjula Ishwara Kalyani

Abstract Over the last few decades, the world has undergone drastic climatic change due to overexploitation of natural resources and anthropogenic factors such as increased pollution, population growth and illegal over-exploitation of plants. Climate change is ambiguous and imparts various environmental stresses that include temperature rise, low temperatures, drought, heavy rainfall followed by floods and water logging, an increase in carbon dioxide concentration, mineral deficiency in soil, etc. These environmental stressors have had a negative impact and threatened various medicinal plants in the area of the Himalaya, thereby compelling them to adapt to environmental changes or go extinct under environmental influences. Nevertheless, variations in the environment and their consequences exhibit disparities among different plant species, placing several plant species at a critical risk of extinction. It is therefore important to understand and implement various innovative strategies to protect these important Himalayan medicinal plant species and, at the same time, conserve their genetic resources. Plant tissue culture has emerged as one of the most promising and preferred conservation strategies for the endeavour to preserve Himalayan medicinal plants. However, micro-propagated plants are also prone to drastic climatic impacts. Therefore, to boost the growth of

S. Das (✉)
Biotechnology Unit, Mangalore University, Mangalagangotri, Mangalore, Karnataka 574199, India
e-mail: subhankardas.sunny.92@gmail.com

S. Shukla
Applied Plant Biotechnology Lab, Centre for Plant and Environment Biotechnology, Amity University, Amity Institute of Biotechnology, Noida, Uttar Pradesh, India

S. Debnath
St. John's Research and Development Centre, Agriculture and Agri-Food Canada, St. John's, Newfoundland and Labrador, Canada

M. I. Kalyani
Department of Microbiology, Mangalore University, Jnana Kaveri Campus, Chikka Aluvara Kodagu, Mangalore, Karnataka 571232, India

A. P. Mishra et al. (eds.), *Threatened Medicinal Plants in the Indian Himalayan Region*, World Sustainability Series, https://doi.org/10.1007/978-3-031-73687-2_8

the micro-propagated plants and their survival in extreme climatic circumstances, plant growth-promoting microorganisms have been considered as a possible possibility. In this chapter, we summarise the devastating effects of climate change on the growth of Himalayan medicinal plant species, along with other additional factors that threaten their continued existence. We further emphasise the role of plant tissue culture as a conservation strategy. In addition, the beneficial impact of plant growth-promoting microorganisms on assisting micro-propagated plants to thrive in extreme environments are discussed in detail. Hence, this chapter aims to provide substantial insights into the existing research findings, with the objective of promoting more climate-resilient medicinal plants in the Himalayan region as a conservation strategy.

Keywords Climate change · Himalaya · Medicinal plants · Microorganisms tissue culture

1 Introduction

The Himalayan region is renowned for its varied climatic and environmental conditions, which provide a conducive habitat for a wide array of flora and fauna. Furthermore, the Himalayan region is well recognised as a biodiversity hotspot owing to its exceptional richness in biodiversity, accommodating a staggering number of over 9,000 species of angiosperms (Verma and Tewari 2016). The Himalayan region harbours a multitude of plant species, many of which have significant commercial and therapeutic values (Kumar et al. 2023; Mishra et al. 2023a, b). According to the World Health Organisation, approximately 80% of the global population relies on traditional medicine, as per estimates and with the considerable amount of these remedies derived from plants and their extracts serving as the primary active components. Since ancient times, plants have been contemplated by humans for applications in medicine and therapy. Plants provide important bioactive compounds, such as flavonoids, alkaloids, saponins, taxols, taxanes, etc. that are used in the manufacture of many important drugs (Mishra et al. 2023b). And as a result, plants with ethnomedicinal relevance have recently seen an upsurge of interest due to their increased use in pharmaceutical practices. Since therapeutic compounds from medicinal plants have relatively few adverse effects, they have the potential to be widely used to lower mortality rates and treat severe illnesses (Ekor 2014; Jamshidi-Kia et al. 2017; Wani et al. 2020; Banyal et al. 2021; Rawat et al. 2021).

Over the past few decades, there has been a meteoric rise in the number of species experiencing the risk of extinction due to various factors, including population growth, anthropogenic activities such as the combustion of fossil fuels, deforestation of forests for agricultural land and fuel wood etc. (Rana et al. 2017; Safdar et al. 2017). Due to all these factors, several medicinal plant species native to the Himalayan area have been classified in the International Union for Conservation of Nature and Natural Resources (IUCN) Red Data Book as well as in the Red Data Book for national status (India) as rare, endangered, and vulnerable. Furthermore, the presence of anthropogenic pressure is evident, and its magnitude is significantly amplified by the effects of climate change. However, anthropogenic influences have a significant impact on the spatiotemporal patterns of Himalayan medicinal plants, in addition to their overharvesting by local populations and worldwide trade. As further reported by the Intergovernmental Panel on Climate Change (IPCC 2014), it is anticipated that by the end of the twenty-first century, the global surface temperature will surpass 2 °C. Also, according to IUCN, 10% of vascular plants worldwide are under threat. The Himalayan area has seen degradation and fragmentation, resulting in the loss of over 70% of its original ecosystems. This has led to the inclusion of the Himalaya on the list of Global Biodiversity Hotspots, whereas only a mere 25% of the native ecosystems in the Himalaya remain unaltered (Rana and Samant 2010; IPCC 2014; Maikhuri et al. 2018; Jeon et al. 2020; Dad and Rashid 2022).

The Himalayan Mountains are well acknowledged as being very susceptible and fragile under the influence of climate change (Chandra et al. 2020). As a consequence, climate change and global warming have brought novel challenges to the sustenance of various economically important medicinal plants and also to the livelihood of indigenous populations residing in this area (Das et al. 2016; Chandra et al. 2022). With the continuous warming in the region of the Himalayas, studies have further revealed that the medicinal plants of the region have shown anomalies in their physio-biochemical characteristics. The possible influence of elevated temperatures on medicinal plants includes the induction of stress and the alteration of the synthesis of bioactive components in plants. One of the prominent causes of global warming is the elevation of carbon dioxide levels in the atmosphere. The combustion of fossil fuels, which is an inherent outcome of many human activities such as the industrial revolution, results in the oxidation of carbon into carbon dioxide, hence contributing to the escalation of this particular greenhouse gas. As a consequence of the elevation of carbon dioxide, which amplifies the retention of thermal energy emitted from the earth, leading to elevated atmospheric temperatures. In a study, it was reported that under conditions of increased carbon dioxide levels, C3 legumes have the potential to exhibit enhanced photosynthetic performance compared to C4 grasses. The presence of higher temperatures further facilitates the viability of C4 crops, such as maize, in comparison to C3 plants. Climate change might potentially impact the phenology and distribution of medicinal plants, as seen in the case of medicinal species in the Central Himalaya region of India. It has been evidenced that some of the medicinal plants have exhibited early flowering and fruiting patterns in response to the accelerated onset of the spring season, which can be attributed to alterations in weather patterns. Alterations in phenology and distribution, despite the fact that they might not appear to be substantial, might in fact provide considerable challenges to the continued existence of a species. As a result, plant phenology is seen as an indication of climate change since it is related, either directly or indirectly, to global climate change. It is necessary to understand the pattern of climate change, which is crucial to the conservation of medicinally important plants in the Himalayan regions on a global scale (Rana and Samant 2010; Rao et al. 2016; Gaira and Dhar 2020; Cahyaningsih et al. 2021; Kumari et al. 2022a, b).

Furthermore, the development of comprehensive biotechnological approaches is of paramount significance in addressing the conservation of medicinal plants in the Himalayan region. This entails the careful selection of superior genotypes and the execution of both ex-situ and in-situ conservation measures. In-situ conservation pertains to the preservation of plants within their original habitats, encompassing both wild plant groups and crop cultivars integrated within conventional agricultural systems. In contrast, ex-situ conservation encompasses several strategies that entail preserving biodiversity outside of its original environment. These methods include seed preservation or seed banks, in-vitro conservation, pollen storage, botanical gardens, DNA storage, and field gene banks. However, it is important to note that these procedures are inherently natural and require a significant investment of time. The medicinal plants conservation could be achieved through the use of tissue culture, which serves as an ex situ conservation approach largely focused on the clonal propagation of vegetative tissue or immature seeds. Moreover, the use of biotechnological interventions is crucial for expediting the propagation and genetic enhancement of medicinal plants, in conjunction with both in-situ and ex-situ approaches. Also, the biotechnological techniques encompassed categories such as micro-propagation, mycorrhization, genetic transformation, and the development of DNA banks. In terms of micro-propagation, there are notable distinctions between plants that are developed in-vitro and those that are grown in outdoor conditions. In the context of micropropagation techniques, the growth substrate utilised is devoid of microorganisms. Furthermore, the use of microbial inoculants can act as an alternative to chemical fertilisers and pesticides in the development of sustainable plant production. Thus, reducing the dependency on harmful, toxic, and non-eco-friendly chemical fertilisers in the later development of the conserved medicinal plants. Furthermore, the use of plant growth-promoting bacteria in the preservation of medicinal plants will offer further support in guaranteeing the long-term viability of these plants (Hussain et al. 2016; Verma et al. 2012; Chauhan et al. 2014; Rai et al. 2010; Chandra et al. 2010). Therefore, the main emphasis of this chapter is to discuss the potential of plant tissue culture as a tool for conservation initiatives to protect the threatened, rare and endangered medicinal plant species of the Himalaya. This chapter additionally draws attention to the need of plant growth promoting bacteria in maintaining the survival of tissue-cultured plants and thus assure its survivability in adverse environmental condition.

2 Influence of Climate Change on Himalayan Medicinal Plants

The region of the Himalayas is well recognised as a significant repository of medicinal plants and has been identified as one of the worldwide hotspots for biodiversity conservation. Due to its favourable habitat, the Himalayan region caters to a diverse array of medicinal plants, which are found throughout various altitudes ranging from lowland to highland regions. The large altitudinal range, diverse habitat types, and varying microclimatic conditions present in the Himalayas collectively render the region to create an environment that is very conducive to the cultivation and growth of medicinal and aromatic plants. High species diversity in the Himalayan region is supported by a confluence of variables, including ecological, phytogeographic, and evolutionary variables. The significance of medicinal plants may be understood by considering the reports that around 85% of medicinal products used in traditional healthcare systems are derived from these plants (Rana et al. 2017; Phondani et al. 2016a, b). Therefore, a significant majority of the global population relies heavily on plant species for addressing health issues, with more than three-fourths of individuals being dependent. Moreover, the projected size of the worldwide market for herbal pharmaceuticals is close to INR 300,000 crores, with India accounting for close to INR 3000 crores of that total. A further projection proposes that by the year 2050, the trade in medicinal and aromatic plants would attain a value of US$5 trillion, given the current annual growth rate of 15–25% for herbal items (Rathore et al. 2022; Negi et al. 2018). The number of species that are on lists of those that are in danger of becoming extinct has expanded enormously over the past several decades owing to numerous anthropogenic factors such as overexploitation and unmonitored harvesting of plants, infrastructural expansion and climate change (Mehta et al. 2021). However, the issue of global climate change is a pressing one, as the temperature fluctuations in Himalayan mountainous regions have already exerted significant impacts on the distribution, habitat, ecology, and phenology of medicinal species in the Himalaya at different elevations. The available evidence strongly indicates that regional warming is expected to surpass 2 °C (Rana et al. 2020).

The ecosystems of the Himalayas are particularly vulnerable to global warming. Thus, the anthropogenic influence in the Himalayas has led to several consequences, one of which is the deterioration of forest cover. This deterioration has given rise to a range of issues, including soil erosion, slope failures, depletion of soil fertility, scarcity of fuelwood and fodder, increased overland flows, reduced groundwater recharge, loss of biological diversity, and accelerated siltation of river beds in lowland areas (Singh et al. 2021; Negi et al. 2017). Moreover, the primary factor contributing to the widespread overharvesting of plants for their medicinal and aromatic properties has put many species of high-value plants at risk. In light of this, it has been widely acknowledged and recommended that endangered wild species be subject to protection and management (Phondani et al. 2016a, b).

The primary driver of contemporary climatic changes is the increased release of greenhouse gases, with carbon dioxide (CO_2) being the most prominent, followed by methane and ozone. These enhanced emissions have contributed significantly to alterations in the Earth's climate system. In addition to the ongoing phenomenon of climate change, a multitude of occurrences are seen, such as increased frequency and intensity of monsoon rainfall, heightened occurrence of hotter summer days, reduced frequency and magnitude of dry season rainfall, and intensified and more frequent storms characterised by high wind speeds. Moreover, the alteration in climatic conditions, coupled with the diversity in geographical locations and altitudes, exerts pressure on plants and has the potential to influence the production, quality, and quantity of secondary metabolites and other bioactive compounds in plants, which often serve as the foundation for their therapeutic properties. In the study investigated by Adhikari et al. (2021), it was found that the accumulation of bioactive compounds in *Taxus wallichiana* needles is influenced by the edaphic and climate variables of the habitat. Furthermore, there was a substantial positive association seen between altitude and many bioactive chemicals, including total phenol, flavonoid content, gallic acid, ascorbic acid, and quercetin. Similarly, according to a study conducted by Kumari et al. (2022a, b), the habitats of *Valeriana wallichii* exhibit a high degree of susceptibility to climatic fluctuations (Mishra 2016; Adhikari et al. 2022; Kumari et al. 2022a, b). In addition to the consequences of climate change, plants frequently encounter unfavourable environmental conditions, known as abiotic variables, which give rise to abiotic stressors. Climate change has been found to have a detrimental effect on the overall performance of plants, either through direct or indirect means, by influencing the severity and frequency of various abiotic stress elements (Ramazan and John 2020).

Since various plants are being utilised in several therapeutic systems, such as Ayurveda, Siddha, and Homoeopathy, due to the presence of abundant bioactive components in them. As, reports have indicated that a significant proportion of commercially available medications, approximately one-fourth, are derived from plants, and therefore it becomes crucial to comprehend and investigate the influence of diverse and deteriorating conditions, including abiotic stresses such as light exposure, temperature fluctuations, humidity levels, water, and the availability of nutrients, as well as the presence of heavy metals. Furthermore, biotic stressors, including bacteria, viruses, fungi, parasites, and insects, also contribute to the alteration of plant secondary metabolites and their corresponding therapeutic characteristics (Joshi et al. 2022).

Environmental pressures have a significant role in defining the production of secondary metabolites in plants as well as influencing the potential for survival of particular plant species in the future. It has been evident from various literature surveys that persistent and fluctuating climatic variations exert a substantial influence on the physiological, morphological, molecular, and biochemical responses of plant species inhabiting high-altitude regions. Consequently, these medicinal plant species have developed various mechanisms to biosynthesis secondary metabolites, thereby enhancing their ability to endure and thrive in adverse environmental conditions. This factor may also serve as a significant contributing factor in the accumulation of a wide range of distinct chemical substances apart from the normal biocompounds. In a study conducted by Kaur et al. (2016), five medicinal plants native to the western Himalayan area, characterised by subtropical and temperate climate zones, were investigated. The plants were grown in two distinct locations, Jammu and Srinagar, characterised by altitudes of 305 m and 17,630 m, respectively. Significant differences in the phenolic and flavonoid contents of five medicinal plants, namely *Thymus vulgaris, Hypericum perforatum, Matricaria chamomilla, Echinacea purpurea,* and *Cynara carduncculus*, were observed based on their geographical locations. Furthermore, the study has indicated that at higher altitudes, temperate sites exhibit an increase in their content by up to 13 times. The study established that alterations in climatic circumstances have a substantial impact on the accumulation of bioactive components in medicinal plants, serving as a defensive mechanism (Tiwari et al. 2023; Kaur et al. 2016; Qaderi et al. 2023).

It has been well established that plant and microbial diversity are inextricably linked due to their shared significance in promoting a wide range of ecological processes. The soils of the Himalayan region exhibit an extensive array of microbial diversity. Further, microbes in the Himalayan region have long been recognised for their contributions to numerous ecological processes, particularly biogeochemical cycles and nutrient cycling, among other ecological roles. However, it is important to point out that the microbiome of the Himalayan region is also adversely affected by the impact of climate change. Many studies have shown that biotic and abiotic variables, such as changes in litter quantity and quality, soil temperature and moisture, atmospheric CO_2 concentrations, and precipitation patterns, all have an effect on the activities of soil microorganisms. Since the Himalayan Region is renowned for hosting a significant population of diverse microorganisms, including psychrotrophic, psychophilic, and thermophilic microorganisms. The abilities of these microorganisms to adapt and thrive in highly challenging environmental conditions might be related to their ability to express certain genes that are inherently encoded in their genetic makeup. It is also due to the genetic traits of these microorganisms that enable them, even in the midst of climate change, to carry out many functions, including nitrogen fixation, phosphorus solubilization, potassium and zinc mobilisation, producing siderophores and phytohormones, as well as the synthesis of antimicrobial chemicals including antibiotics and hydrogen cyanide. The vital importance of an array of diverse microorganisms, including archaea, bacteria, and fungi, etc., cannot be overlooked in relation to the maintenance of forest health. This microbial diversity plays a pivotal role in enabling forests to effectively acclimatise to the fluctuations in weather patterns resulting from climate change. The future of forests is mostly contingent upon the activity and coordination among fungal symbiotic guilds, saprotrophic fungi and bacteria, and fungal plant diseases. Thus, the presence of the microbiome in the Himalayan region has a significant association with the growth of plants under the harsh and challenging climatic conditions of the region. Moreover, in light of ongoing and persistent climate change, there has been a notable escalation in drought occurrences, which has significantly diminished the resilience of forests. This, in turn, has initiated detrimental processes like disease outbreaks and fires that have a negative influence on the microbial composition and overall performance of forests in terms of carbon storage and nutrient turnover. Hence, it is imperative to have a comprehensive understanding of the response of microbial communities to atmospheric and climate changes. This understanding will thus enable us to accurately estimate the possible influence of different microbial communities in the Himalayan region on the mitigation of climate change (Yadav et al. 2017; Suyal et al. 2021; Baldrian et al. 2023; Castro et al. 2010; Valencia et al. 2018; Dhakar and Pandey et al. 2020).

3 A Brief Overview on Conservation Strategies and Its Beneficial Impact:

Throughout ancient history, plants have been used as medicinal remedies for various human ailments. However, the Himalaya region is widely recognised for its abundant diversity of medicinal and aromatic plant species. The Himalayan region, due to its diverse soil composition, topographical features, and presence of several climatic conditions, creates favourable conditions for nurturing and ensuring the survivability of numerous medicinal plant species that hold therapeutic significance. Plant-derived compounds are still regarded as valuable reservoirs for key components required in medications. And thus, in recent decades, despite the significant advancement in synthetic pharmaceuticals, medicinal plants have emerged throughout the years as an important component in the process of developing contemporary therapeutics, with around 20–30% of today's modern pharmaceuticals having been produced from plant-derived compounds, with a significant number being synthetic analogues of naturally occurring chemicals found in plants. Also, phytoconstituents derived from the plant have much fewer adverse effects in comparison to manufactured synthetic drugs. The current Indian herbal economy heavily relies on plant collection from its wild habitats, with an estimated 95% of various therapeutic plants. Moreover, approximately 500,000 metric tonnes of dry material are gathered using destructive methods, while an area of 165,000 ha of forest is cleared and harvested annually, causing an enormous decline in the diversity of medicinally useful plant species, with many of them currently facing the imminent threat of extinction. It has also been reported that an estimated 60,000 plant species are projected to face extinction by the year 2050 (Safdar et al. 2017; Chandra 2016; Krishna et al. 2023; Mukherjee and Chakraborty 2019; Mukherjee et al. 2009; Siwach et al. 2013). In addition, climate change has been seen to expedite the process of the extinction of medicinally important species. Also, the climate change that leads to the escalation of temperature is accompanied by an intensification of extreme weather events, degradation and irreversible impact on the ecosystem, notable alterations in species patterns of migration, seasonal activities, phenology, fluctuation in pattern of precipitation, and geographical distribution of plant species. Moreover, several plant species have a somewhat slower regeneration rate in comparison to the pace at which they are being harvested (Guan et al. 2022; Kunwar et al. 2023; Sharma et al. 2022a, b). Therefore, it is crucial to implement comprehensive conservation strategies that incorporate both in-situ and ex-situ techniques in order to effectively safeguard and ensure the long-term existence of this diverse form of rare, endangered, and threatened plant species (Wani et al. 2022; Asgher et al 2020; Avhad et al. 2023).

In-situ conservation is often regarded as the most effective approach for safeguarding biological and genetic variety. This method entails the preservation and protection of wild species within their natural habitats. The techniques employed in the practice of in-situ conservation encompass the development of various protected areas, such as biosphere reserves, wildlife sanctuaries, national parks, and holy groves, among others. Also, the effectiveness of in situ conservation programmes further relies heavily on the essential involvement of various community members, native peoples, and various stakeholders in collaborative endeavours. In addition, the foremost important consideration is the active involvement of local people in the maintenance and surveillance of preserved medicinal plants, since this cultivates a sense of ownership in those people and encourages the sharing of traditional expertise, as well as incites them to utilise such plants in a more reasonable and sustainable manner. Although the preservation of species in their natural habitats, known as in-situ conservation, is considered the optimal approach, it may not always be feasible to guarantee their survivability or proper conservation. Additionally, relying excessively on in-situ conservation poses challenges in attaining conservation objectives due to the dynamic shifts in species distribution ranges resulting from global climatic changes. Consequently, ex-situ conservation strategies might assume a crucial and supplementary function in guaranteeing the survivability of the conserved species. Also, the increasing necessity of ex-situ conservation, which guarantees the secure replication of native plant varieties, can be attributed to the phenomenon of climate change and its consequential species migration. The ex-situ strategy involves the protection of plant species by relocating them from their natural environment in a deliberate manner for long-term preservation. The ex-situ methods further involve the conservation of seeds through seed storage, gene banks, and pollen conservation through the in-vitro conservation of plant shoots that slow down or halt development. Furthermore, botanical gardens and arboreta serve a crucial function in ex-situ conservation efforts through the preservation and cultivation of medicinal plants. However, in order to formulate efficacious conservation strategies, it is imperative to possess comprehensive knowledge pertaining to the distribution of rare, endangered, and threatened species, as well as the capacity to anticipate their habitat appropriateness in the face of a changing climate. Moreover, additional research has indicated that the use of tissue culture and micro-propagation techniques can be employed as effective methods for the quick multiplication of uncommon or slow-growing medicinal plants, with the purpose of facilitating reintroduction initiatives (Dad and Rashid 2022; Gurung and Gurung 2019; Lamo et al. 2019; Saaondo and Olalekan 2023; Shukla 2023; Ye et al. 2023).

In a study conducted by Lohani et al. (2012), ex-situ conservation was implemented for the plant species *Polygonatum verticillatum*, which holds significant importance in many therapeutic applications due to the presence of various medicinal components. The study was undertaken by implementing several experiments using different types of organic fertilisers and nursery beds. Furthermore, the objective of the research was to establish ways for optimal culture practices and harvesting methods that promote the sustainable use of this plant. The findings indicate that T3 beds, namely beds planted in furrows and supplemented with forest litter, exhibit superior growth patterns and yields compared to the other treatments. Moreover, the results have also revealed that all treatments benefited when included with organic fertilisers as compared to the control, but that the beds that were planted in furrows and supplemented with forest litter achieved the highest levels of morphological growth and yield. In another study conducted by Pandey et al. (2005) that showcased the conservation strategies for *Aconitum heterophyllum* Wall., a medicinal plant native to the Himalayan region, the researchers employed ex-situ mass multiplication strategies in both open and greenhouse conditions. Furthermore, researchers have implemented and utilised hot water treatments to enhance the germination of seeds. The study also involved an examination of the alkaloid composition of this endangered plant species. Result of the study has demonstrated that a significant increase in seed germination rates (88 and 76%) occurred after subjecting the seeds to hot water treatment at temperatures of 45 and 50 °C for a duration of 90 s (Lohani et al. 2012; Pandey et al. 2005). In a study that was conducted to investigate the physiological, biochemical, growth, and reproductive properties of *Valeriana wallichii*, a critically endangered medicinal plant native to the Himalayan region, from a medicinal perspective, it is widely acknowledged that *Valeriana wallichii*, a plant belonging to the Valerianaceae family, has roots and rhizomes abundant in diverse bioactive constituents. These include valerenic acids, valepotriates, baldrinal and homobaldrinal alkaloids, as well as phenolic acids and flavonoids. Further in this investigation, the physiological examination has revealed that ex-situ populations exhibit notable levels of stress when compared to their in-situ counterparts. The observed variation in plant performance in response to their respective habitats can be attributed to the presence of superoxide and H_2O_2 molecules inside the leaves. The findings indicated that plants in in-situ conditions had enhanced photosynthetic characteristics. Despite the finding that ex-situ plant proline oxidase levels decreased, it was found that proline levels increased, along with the presence of biosynthesis enzymes such as ornithine aminotransferase and pyrroline-5-carboxylate reductase. Additionally, ex-situ plants exhibited an increase in glucose-6-phosphate dehydrogenase, shikimic acid dehydrogenase, phenylalanine lyase, and flavonoids. Furthermore, while considering ex-situ conditions, it was observed that the growth and reproductive properties of the plants were more pronounced (Asgher et al. 2020).

Seed banking remains a very effective and efficient approach within ex-situ conservation efforts. At present, there are a total of over 1750 seed banks that are dedicated to the conservation of a variety of plant genetic resources, which are considered to be of utmost significance on a global scale. The success of the seed bank lies in the remarkable adaptability of seeds across a wide range of taxa, which has resulted in a growing number of seed banks. Numerous seed banks, which are often situated inside or affiliated with botanic gardens, are responsible for the preservation and management of seed accessions containing a wide range of wild plant varieties. However, the limitations of seed banks include the exclusion of species that are incapable of seed production or whose seeds cannot withstand freezing temperatures, rendering them unsuitable for inclusion in seed banks (Walters and Pence 2021; Wambugu et al. 2023; Breman et al. 2021). In addition, the preservation of plants within botanical gardens plays a significant role in ex-situ conservation efforts and has a long-standing history of spearheading research and conservation initiatives related to plant species. Leveraging their vast collections and specialised knowledge, botanical gardens are well-positioned to develop comprehensive programmes aimed at mitigating the impact of climate change on worldwide plant biodiversity. However, it has also been shown that the growing circumstances of wild plants undergo changes when subjected to ex-situ conservation in botanical gardens. Additionally, the presence of plants in space-constrained environments such as gardens and arboreta gives rise to several genetic and demographic challenges that are linked to the small numbers of these plants. Moreover, the tiny populations are vulnerable to selective pressure from artificial circumstances in the ex-situ environment, as well as founder effects, genetic drift, and inbreeding. In research conducted by Sun et al. (2022), it was shown that wild plants exhibit alterations in their ecophysiological features as a means of adapting to the conservation environment found within botanical gardens. Consequently, this phenomenon gives rise to discernible distinctions between wild plants and their cultivated counterparts. Thus, the study findings provided valuable insights into the ecological adaptation followed by plants during extended periods of conservation in a botanical garden (Sun et al. 2022; Volis 2023). However, numerous scientific methodologies can be encompassed under the in-situ and ex-situ approaches. This might greatly benefit conservationists involved in activities where researchers are engaged in creating cultivation technologies. It would also be of interest to progressive farmers who wish to commercially produce these technologies on their farms (Zegeye 2017). Furthermore, the inadequate availability of scientifically reliable data about the effects of different threats on the distribution of plant species poses limitations on the effective implementation of conservation efforts for these plants. Therefore, a critical aspect of successful conservation efforts entails the prioritisation of economically and medicinally significant plant species for preservation. This necessitates a robust understanding of the plants themselves, including their risk of extinction and conservation status, in order to inform conservation initiatives effectively (Shivali and Namrata 2022). Trans-situ conservation is a relatively new idea that has been developed for the purpose of protecting plants effectively and that dynamically combines a wide range of in-situ and ex-situ strategies. This approach dynamically combines a wide range of conservation, research, and education initiatives at both the local and global levels. But none of these approaches worked well enough to be adopted as a standard conservation tactic (Engels and Ebert 2021; Volis and Blecher 2010).

4 The Potential of Plant Tissue Culture Techniques to Revive Medicinal Plants in the Himalayan Region

Maintaining medicinally important plants typically necessitates a multidimensional approach, encompassing the selection of superior genotypes and the implementation of ex-situ and in-situ conservation strategies. Additionally, the propagation of these plants can be achieved through conventional and biotechnological methods, which offer potential solutions to the current challenges (Hussain et al. 2016). Ex-situ conservation techniques, acknowledged as methods of conservation conducted outside the natural environment, encompass the cultivation and preservation of plants by means of the long-term storage of plant propagules in repositories of plant tissue culture. Micro-propagation, a technique that utilises plant tissue culture, involves the in-vitro culture of plant clones. This method has several advantages, including the development of rooted plantlets that are readily available for transplanting and further growth. Gottlieb Haberlandt introduced the idea of cultivating individual plant cells on an artificial nutritional medium in 1902. Plant tissue culture, or in-vitro culture, is the controlled development of plant tissues, organs, cells, seeds, protoplasts, or embryos in a specially designed environment with a nutrient-rich media that is free from any contaminants. The principle behind in-vitro plant tissue culture primarily relies on the idea of totipotency in plant cells, which pertains to the capacity of a single cell to express the whole genome through cell division. The phenomenon known as "totipotency" refers to the inherent potential of plant cells, specifically somatic plant cells, to undergo regeneration and develop into a complete plant organism. Tissue culture techniques provide a practical means for the large-scale propagation and preservation of superior medicinal plant species that are on the verge of extinction, consequently addressing the demands for various pharmaceutical and commercial purposes. The propagation methods provide approaches that have unique advantages over other conservation tactics. Some of these advantages are outlined below: (1) Collection can take place at any time, regardless of the specific flowering season for each species, and it does not necessarily require the seed material of the plant, (2) The preservation of the genetic resources of endangered plant species, (3) With the use of micro-propagation techniques, fast multiplication and development of rooted plantlets that are readily available for transplanting and further growth, (4) Plants can be resistant to pesticides, fungicides, and herbicides, (5) further ensure the availability of plants throughout all seasons, (6) Less storage space and (7) Plant multiplication can be attained when plants are sterile and incapable of producing viable seeds, or when seeds cannot be effectively stored, etc. However, there are several critical aspects that play a significant role in the process of micro-propagation. These parameters encompass the composition of the nutritional medium, the concentration and combination of plant growth regulators, the conditions of the culture environment, as well as the genotype and physiological state of the donor plant, i.e., the explant (Shahzad et al. 2017; Abdalla et al. 2022; Rawat et al. 2013a, b; Sharma and Thokchom 2014; Chatterjee and Ghosh 2020; Dias et al. 2016; Misra and Saema 2016; Thorpe 2007). The process of plant propagation by tissue culture must be conducted in a systematic manner, following a series of five stages. These procedures have been extensively documented in numerous literature studies and may be summarised as follows:

Stage 0: Process of preparing the donor plant: This first step is designed to prepare the mother plant for in-vitro cultivation. It is crucial for the mother plants to be protected from any kind of disease and virus throughout cultivation so that they have a chance of achieving healthy and favourable outcomes. The explants obtained from plants in the field often encounter issues related to contamination by microorganisms, and further achieving complete sterilisation of these explants is often challenging. Therefore, plant tissue culture relies on using healthy explants to successfully propagate plants.

The disease-free explants from the mother plants are then used to inoculate, and clones of the plant are then produced.

Stage I: establishment of aseptic cultures: The second phase of the micro-propagation method typically involves the establishment of an aseptic culture containing the desired plant material. Achieving success at this particular phase necessitates the first transfer of explants (commonly used explants such as organs, shoot tips, and axillary buds) to a sterile culture environment devoid of any discernible microbial contaminants. Subsequently, this should be followed by the initiation of growth, such as the development of a shoot tip or the creation of a callus. The successful completion of Stage I would be determined by the satisfactory survival of a sufficient number of explants devoid of any contamination and their subsequent growth.

Stage II: Rapid Multiplication Stage: During this particular stage, a significant amount of micro-propagation activity takes place inside a well-defined medium to generate a number of clonal propagules. At this point, masses of tissues are regularly subcultured onto fresh culture medium in an effort to promote the multiplication of propagules. And during this subsequent phase, the process of induction is initiated, wherein cells are distinguished for the purpose of developing shoot, root, or embryo structures. The achievement of new plant propagules at this stage is essentially contingent upon the effective optimisation of numerous growth regulators, medium composition, and in-vitro culture conditions in a coordinated and balanced manner.

Stage III: Plantlet Establishment, Shooting, and Rooting: This stage is specifically intended to inhibit the fast proliferation of cells and encourage the growth of a mature plantlet. This involves the elongation of shoots and developing roots, and, if needed, it can also be formulated for the development of storage organs that function as autonomous means of propagation, such as bulbs, corms, and tubers. This stage entails the transfer of the shots to a specialised medium designed to facilitate expeditious shoot development. Consequently, thereafter, the shoots are subjected to inoculation in order to promote root growth. This process may involve the use of either the same or a different medium. Additionally, it has been shown that the composition of the rooting media mostly comprises a greater concentration of auxin and a lower concentration of cytokinin. This specific combination is known to promote the development of roots. At this point, efforts are made to cultivate isolated plantlets or groups of them that can perform photosynthesis and survive without an external source of carbohydrates. However, in certain cases, in order to mitigate the expenses associated with micro-propagation, a growing number of laboratories have adopted the practice of extracting unrooted shoots from the in-vitro setting and subsequently inducing root growth outside the culture vessel.

Stage IV: Acclimatisation stage: This stage refers to the process by which the plants that have been cultivated in-vitro are weaned and hardened by external conditions. Owing to the elevated humidity levels within the culture vessel and the controlled environment in which they are developed, the plantlets exhibit a delicate nature and are thus not well prepared to adapt to the natural conditions of the field, which leads to low survival rates. Consequently, the propagules are acclimated by meticulously manipulating the humidity levels and modifying the light prerequisites. They are then transplanted into pots filled with a mixture of sand, clay, peat, and other suitable materials. Subsequently, they are relocated to a greenhouse environment and ultimately taken to the exterior field (Singh and Singh 2021; Gulzar et al. 2020; Bhoite and Palshikar 2014; Monthony et al. 2021; García-Gonzáles et al. 2010).

The acclimatisation phase, which is crucial in terms of the inoculation of beneficial bacterial strains, takes place. As in the previous studies that have indicated the physiological characteristics of in-vitro propagated banana plants (*Musa* sp.) have shown enhancement when subjected to the introduction of plant growth-promoting bacteria, specifically *Buttiauxella agrestis* and *Bacillus thuringiensis*, which are known to produce auxins (de Araújo et al. 2021). In a study, two fungi species, specifically *Cladosporium ramotenellum* strain PGP02 and *Phoma* spp. strain 22 PGP03, were collected alongside the bacterium *Pseudomonas oryzihabitans* PGP01. These microorganisms were obtained from Pyrus as well as Prunus in-vitro and rescued from the immature embryos. The resultant plantlets exhibited enhanced growth compared to cultures that were not contaminated. The work has proven the use of microorganisms inside a sterile in-vitro culture environment to promote the growth and development of woody plants (Cantabella et al. 2020). Therefore, a wide variety of plants may be efficiently grown using tiny plant segments in laboratory settings, allowing for large-scale production in a shorter period of time and supporting the inoculation of microbes for survival in nature. This approach has potential for supporting conservation efforts aimed at addressing the decline of medicinal plants in the Himalayas (as indicated in Table 1). The exploitation of medicinal plants for their bioactive components poses a significant danger to the survival of essential plant species. Also, given the significance of plant secondary metabolites in the pharmaceutical sector, it is imperative to ensure the accessibility of authentic herbal raw materials derived from their natural habitats. The initial endeavour to utilise plant tissue culture for the synthesis of useful metabolites can be traced back to 1950. Interestingly, regulated in-vitro cultures of plants have demonstrated the ability to manufacture these valuable natural products with precision. Consequently, the use of this approach mitigates the potential for overuse of plant resources (Gupta and Sood 2022; Kaushal et al. 2014; Karuppusamy 2009; Dias et al. 2016). In a study performed by Bhardwaj et al. (2018), a micro-propagation procedure was demonstrated for the cultivation of *Rhodiola imbricata*, an endangered medicinal plant that is indigenous to the trans-Himalayan Leh-Ladakh area of India. The study has aimed to evaluate the secondary metabolites that are of medicinal significance, as well as the antioxidant capacity, in both in-vitro and in-vivo plant samples. In addition, high-performance liquid chromatography has been used to confirm the presence of Rosavin, total polyphenols, and free radical scavenging activity in *R. imbricata*, all of which have therapeutic applications. It is noteworthy to highlight that the Rosavin concentration exhibited an increase in plants cultivated in-vitro, but the total polyphenolic content and DPPH activity were comparatively lower in the in-vitro-developed plants in comparison to the plants grown in the field (Bhardwaj et al. 2018).

Table 1 Application of Plant tissue culture techniques on some Himalayan medicinal plants:

Sl. no	Species	Explant	Therapeutic properties	References
1	*Aconitum chasmanthum*	Leaves, stems, petioles, and nodes	Heart disease, fever, anodyne, anti-inflammatory, tonsillitis, diarrhea, anti-diabetic, neurasthenic, astringent, cold, cough	Rafiq et al. (2021)
2	*Aconitum lethale*	Leaf and petiole	Analgesic, antirheumatic, vermifuge, anti-inflammatory	Gondval et al. (2016)
3	*Aconitum lethale*	Sprouted axillary buds,	Rheumatism, fever	Pandey et al. (2004)
4	*Aconitum violaceum*	Leaves shoot tips and nodal segments	Antipyretic, analgesic, renal pain, rheumatism, high fever, allergy, boils, cuts, wounds, edema, contagious Infections, disorder of gall bladder disorders, intestinal inflammation	Rawat et al. (2013a, b)
5	*Angelica glauca*	Seedlings (divided into epicotyls, hypocotyls, and cotyledonary nodes)	Stomach ailments, rheumatism, urinary disorders, mennorhagia, infantile atrophy, gynaecological disorders, rinderpest, cardiac stimulant	Pandey et al. (2011)
6	*Angelica glauca*	Leaf	Stomach ailment, bronchitis	Bisht et al. (2015)
7	*Angelica glauca*	Rhizomes	Anti-inflammatory, stimulant, appetizer, carminative, expectorant, cardioactive, cordial, diaphoretic, and also used in dyspepsia, renal disorders, digestive disorders, lungs disorders, rheumatism, menorrhea	Rawat et al. (2018)
8	*Arnebia euchroma*	Leaf and root	Antibacterial, antifungal, anti-inflammatory, wound healing, fevers, cardiac disorders	Manjkhola et al. (2005)
9	*Atropa acuminata*	Nodal	Anti-arthritic, anti-inflammatory, hepatoprotective, anticholinergic activities and are also used to treat gastrointestinal motility and hyperactive bladder	Rajput and Agrawal (2020)

(continued)

Table 1 (continued)

Sl. no	Species	Explant	Therapeutic properties	References
10	*Berberis asiatica*	Vitro-raised plants leaf	Cholera, diarrhoea, cancer, jaundice, malarial fever, eye, and ear infection, skin diseases	Bisht et al. (2022)
11	*Curcuma caesia*	Rhizome with one bud	Leprosy, bronchitis, asthma, piles, cancer, epilepsy, fever wounds, impotency, fertility, tooth ache and vomiting	Chowdhury et al. (2020)
12	*Gentiana kurroo*	Leaves, petioles, roots	Treat diseases like gout, dyspepsia, debility, torpid, liver, spleen enlargement	Sharma et al. (2014)
13	*Heracleum candicans*	Cotyledonary leaf and cotyledonary node	Leucoderma, nerve tonic, aphrodisiac, or to treat abdominal colic, flatulence and other digestive complaints	Joshi et al. (2004)
14	*Heracleum candicans*	Hypocotyl	Used as nerve tonic, aphrodisiac, leucoderma, skin diseases like eczema and itches	Jan et al. (2018)
15	*Nardostachys jatamansi*	Excised rhizome segments bearing shoot buds	Anti-lipid peroxidative, hypolipidemic, antioxidant, hepatoprotective, sedative, tranquilizing, hypotensive, anti-inflammatory, antidepressant, anticonvulsant, anti-asthmatic, anti-estrogenic	Dhiman et al. (2021)
16	*Nardostachys jatamansi*	Petiole, young leaf and shoot tip explants	Sedative, hepatoprotective, cardioprotective, neuroprotective, antidiabetic, antioxidant, antifungal, antibacterial and anticancer properties	Bose et al. (2016)
17	*Nardostachys jatamansi*	Shoot tip and nodal segments	Used as nerve, digestive, circulatory, effective memory booster	Pant et al. (2021)
18	*Picrorhiza kurroa*	Auxiliary shoot tips	Infection of the liver as well as the upper respiratory tract, reduce fever (pyrexia), dyspepsia, chronic diarrhea, scorpion sting, stomachic, cathartic, cholagogue, laxative, cardiotonic, antipyretic, anthelmintic	Ganeshkumar et al. (2017)
19	*Picrorhiza kurroa*	Leaf and nodal segments	Treat disorders of the liver and upper respiratory tract	Sharma et al. (2010)

(continued)

Table 1 (continued)

Sl. no	Species	Explant	Therapeutic properties	References
20	*Picrorhiza kurroa*	Nodal segments	Hepato-protective, anti-inflammatory, anti-asthmatic	Patial et al. (2017)
21	*Picrorhiza kurrooa*	Immature leaves	Liver, chronic diarrhea, laxative in small dosage, cathartic in large dosage, antimicrobial agents	Rehman et al. (2014)
22	*Picrorhiza scrophulariiflora*	Shoot tips and rhizomes of the collected plants, 2 weeks old sprouted buds from rhizomes	Liver diseases, fever, antiasthamatic; immunomodulators	Bantawa et al. (2010)
23	*Pittosporum napaulensis*	Lateral node	Treat chronic bronchitis	Dhar et al. (2000)
24	*Podophyllum hexandrum*	Excised embryos	Anticancer activity, Warts treatment	Sultan et al. (2009)
25	*Podophyllum hexandrum*	Rhizome	Lung cancer, testicular cancer, leukemia rheumatoid	Chakraborty et al. (2010)
26	*Potentilla fulgens*	Apical segments of the shoots	Treat stomach problems, cough and cold, toothache, antitumor activity and hypoglycemic	Sambyal et al. (2006)
27	*Rheum australe*	Seeds, leaves, shoots and rhizomes	Gastrointestinal hemorrhage, skin diseases, antiviral, nephron-protective, antimicrobial, anti-tumorous, anti-inflammatory, anti-cholesteroleamic, antifungal, anti-Parkinsonal, hepatoprotective, anti-SARS, antidiabetic, anti-cancerous	Tabin et al. (2018)
28	*Rhodiola imbricata*	Seed	Wound healing, anticancer, immuno-modulatory, radioprotective, cryoprotective, anti-aging, anti-fatigue anti-stress high altitude sickness, depression, headache, schizophrenia, sexual dysfunction	Bhardwaj et al. (2018)
29	*Saussurea costus*	Seeds, leaf, petiole and internode	Antiseptics, bronchial asthma, diarrhoea, jaundice, stomachache, respiratory tract infections, antispasmodic agents	Khan et al. (2021)

(continued)

Table 1 (continued)

Sl. no	Species	Explant	Therapeutic properties	References
30	*Saussurea obvallata*	Epicotyle segments	Bone-ache, intestinal ailments, urinary track problems and coughs	Joshi and Dhar (2003)
31	*Sinopodophyllum hexandrum*	In vivo plants explant: Leaf and petioles in vitro plants: roots and leaves	Lung and testicular cancer; antiviral, antimicrobial, antidote, anthelminthic	Sharma et al. (2022a, b)
32	*Taxus wallichiana*	Juvenile meristem, stem and leaf	Anticancer activity like Breast, ovarian cancer	Hussain et al. (2013)
33	*Thalictrum foliolosum*	Hypocotyl, cotyledon and petiole	Diuretic, stomachic, antiseptic, jaundice, rheumatism	Mishra et al. (2020)
34	*Valeriana jatamansi*	Shoot bud	Obesity, skin diseases, epilepsy	Pandey et al. (2020)
35	*Viola canescens*	Leaf	Cold, cough and sore throat, jaundice, renal problems, cancer and respiratory tract problems	Khajuria et al. (2021)
36	*Zanthoxylum armatum*	Nodal	Stomach pain, toothache, intestinal worms, rheumatism, scabies, fever, cholera	Purohit et al. (2017)
37	*Zanthoxylum armatum*	Leaf	Stomach and tooth ache, intestinal worms, rheumatism, scabies, fever, cholera, antiseptic	Purohit et al. (2020)

The utilisation of tissue culture technology has emerged as a highly valuable approach in the propagation of several medicinal plants and serves as a valuable method for both the large-scale multiplication of existing germplasm stocks and the preservation of endangered species. Also due to the fact that numerous plant species require several months for propagation by seed, and in certain instances, seeds are generated years after pollination. Therefore, the use of in-vitro micro-propagation technology appears to be a suitable solution for enhancing the efficacy and expanding the scope of plant cultivation. As reported in a study conducted by Parray et al. (2018), which aimed to establish a tissue culture procedure for the micro-propagation of the diminished germplasm of *A. benthamii*, Furthermore, plant tissue culture techniques are now being applied in an attempt to enhance the quantity and quality of several commercially significant and endangered medicinal plant species. In addition, micro-propagated plants have the potential to be grown in bioreactors in order to meet the growing demands of the pharmaceutical sector. The fundamental requirement for achieving such a scaling-up process is the utilisation of micro-propagation techniques to cultivate the desired plants in a liquid culture system. And thus, it is possible to broaden the scope of this plant tissue culture approach to get diverse pharmacological chemicals without causing harm or excessive exploitation of important medicinal plant species. As in the realm of plant tissue culture, cell and organ culture presents a viable and practical alternative for the mass production and elicitation of specific phytochemicals, irrespective of varying environmental circumstances. As in the study conducted by Sharma et al. (2022a, b), they uncovered the podophyllotoxin synthesis in *Sinopodophyllum hexandrum*, an endangered medicinal plant, as well as an optimised single-step procedure for high-frequency in-vitro multiplication via rhizome buds. The findings of this study demonstrate that several plant parts, including leaves, petioles, roots, and in vitro leaves, were used to initiate callus formation for the production of podophyllotoxin. The results indicate that the maximum rate of callus induction from both in vivo and in-vitro explants was observed when they were subjected to dark incubation on MS and B5 medium. The findings of the investigation were further substantiated by the HPLC analysis, which demonstrated that the highest levels of podophyllotoxin synthesis were observed in the in vitro roots and calli (Gupta and Sood 2022; Sharma et al. 2014, 2022a, b; Parray et al. 2018; Dhiman et al. 2021).

5 The Significance of Plant Growth-Promoting Bacteria in Improving Climate Resilience in Himalayan Medicinal Plants Grown Under in Vitro Condition

Plants establish intricate associations with microorganisms, encompassing bacteria, fungi, etc., that can exert both advantageous and detrimental impacts on plant well-being. However, the positive benefits of rhizospheric microorganisms, which represent those microorganisms that are found in the immediate proximity of plant roots and engage in beneficial interactions with the plant, are worth highlighting. These rhizospheric microorganisms perform a vital role in aiding plants in nutrient uptake, disease resistance, and tolerance to adverse environmental conditions such as high temperatures or water scarcity. Similarly, plant-growth-promoting rhizobacteria are beneficial bacteria that colonise the rhizosphere while synthesising chemicals that promote the growth of plants and defend them against disease. Moreover, it is vital to recognise the presence of a variety of microbial populations residing within and on the surfaces of the plants, which further significantly impacts several aspects such as plant development, well-being, resistance to diseases, and the nutritional value of the produced food. During various phases of plant growth, there is significant communication between microorganisms and plants, whereby signalling molecules from both sides play a crucial role. Fungal and bacterial species have the capability to perceive the presence of the plant host and commence their colonisation tactics within the rhizosphere through the synthesis of well-established plant growth-regulating compounds, such as auxins, cytokinins, etc. (Niazi et al. 2023; Ortíz-Castro et al.

2009; Vafadar et al. 2014). Hence, the current increase in research on host-microbe interactions and the observed beneficial effects of microbes on host development provide substantial evidence for the need for "microbiome conservation." As with the ongoing climate change brought various events, including precipitation patterns, ambient carbon dioxide levels, and temperature variations, have had a profound impact on plant growth and development. These changes further impose several limitations that can lead to reduced crop yields or even plant mortality. However, the introduction of microorganisms in the implementation of conservation methods for Himalayan medicinal plants possesses the potential to augment plant nourishment, development, yield, and resilience to environmental stressors. The exploration and comprehension of diverse beneficial microorganisms, including fungi and bacteria and their symbiotic interactions, offer a promising strategy for improving plant performance and productivity, especially in the context of climate-induced stressors (Banerjee et al. 2020; Adhikari et al. 2021; Sangiorgio et al. 2020; Fadiji et al. 2022; Tripathi et al. 2022).

It is important to acknowledge that there are notable distinctions between plants developed under in-vitro condition and those grown in outdoor conditions. A significant increase in death rates is found when micro shoots are transferred to ex-vitro environments, mostly due to the presence of non-functional stomata, a weak root system, and an underdeveloped cuticle in the cultured plants. In addition, the in-vitro plants that have been shifted to the field do not have adequate resistance against harmful soil microorganisms and eventually die from pathogen assault. In order to address such conditions, the use of biopriming appears to be a viable approach. The term "biopriming" pertains to the use of microbial inoculants, specifically bacteria and mycorrhizal, as agents for priming propagules. This priming can occur both in-vitro through co-cultures and during the process of transplantation. When subjected to stress, plants that have been pre-sensitised exhibit enhanced and expedited adaptation compared to plants that have not been primed. This heightened response contributes to increased resistance to plant diseases in many commercially significant horticultural crops. Also, rhizosphere microorganisms that promote plant development might be reintroduced to protect plants from illness or boost plant establishment and ongoing growth after transfer to ex-vitro growing conditions, respectively (Chandra et al. 2010; Thomas et al. 2010; Fiodor et al. 2023; Harish et al. 2008). However, the concept of utilising bacteria to enhance the efficiency of micro-propagation was first introduced more than three decades ago. Herman initially proposed the potential for enhancing the efficiency of plant tissue cultures through the use of certain bacterial strains as "contaminators". Consequently, Digat et al. (1985) were the first to describe the bacterization of plant cultures in-vitro, as the research further revealed the beneficial effects of *Pseudomonas fluorescens* and *P. putida* strains on the rooting and acclimatisation processes of Primula microshoots.

Another method, called biotization, involves the introduction of stress tolerance in plant propagules cultivated in invitro conditions before transplantation with the use of microbial inoculants in micro-propagation. The term "biotization" refers to the metabolic response of in vitro-grown plant material to a microbial inoculant, which results in developmental and physiological alterations that improve the resultant propagules' resilience to biotic and abiotic stress as well as synergistic or artificial symbiotic associations and partnerships (Nowak 1998; Orlikowska et al. 2017; Digat et al. 1985; Herman 1990, 1987). *Picrorhiza kurroa* Royle ex Benth is a perennial medicinal plant that holds significant importance. It is mostly found in the alpine regions of the northwestern Himalayas, making it an endemic species. The rhizome-derived extracts consist of three glycosides, specifically Picroside I, Picroside II, and Kutkoside. Consequently, the constituents of the rhizome have been reported to have demonstrated hepatoprotective, antioxidant, anticancer, and immunomodulatory properties. However, unscientific and indiscriminate harvesting practices have resulted in a significant reduction of naturally occurring populations of *P. kurroa*, leading to its classification as a vulnerable species according to the Red List. Although conservation strategies have been made through the clonal propagation of *P. kurroa* by shoot tip culture (2) and the approach of axillary bud proliferation. Hence, it has become essential to devise approaches that promote the sustainable economic use of *P. kurroa*. Plant propagation by tissue culture has emerged as a valuable technology in this context. However, in order to guarantee the viability of the micropropagated plants, the use of plant growth-promoting rhizobacteria (PGPR) has proven to be an effective strategy for fortifying the resilience of micropropagated plants, contributing to increased survival rates. This approach is one of several efforts that have been taken in recent years to enhance the viability of these plants. In research conducted by Kumar et al. (2021), it was found that *Serratia quinivorans* PKL:12, a growth-promoting rhizobacteria strain isolated from the rhizospheric soil of *P. kurroa*. This particular strain has demonstrated the most efficient enhancement of vegetative growth and survival rates in micropropagated plantlets of *P. kurroa*. Among the characterised plant growth-promoting rhizobacteria (PGPRs), *S. quinivorans* PKL:12 shows the most promising potential for plant growth promotion (PGP) activities, such as the generation of siderophores, solubilization of phosphate, and synthesis of indole-3-acetic acid (IAA). The PGPR strain has demonstrated effective biopriming capabilities through outcomes in terms of vegetative growth of the micro propagated plantlets and the development of lateral root numbers. Moreover, genomic research has provided additional evidence of the existence of crucial genes linked to cold adaptation and the induction of plant growth. There are genes that are also associated with the production of indole-3-acetic acid (IAA), siderophores, and the solubilization of phosphate (Patial et al. 2017; Habib Ur Rehman et al. 2014; Ganeshkumar et al. 2017; Sharma et al 2010; Kumar et al. 2021). *Crocus sativus* L., which is widely recognised as one of the most valuable spices globally due to its high market price, possesses a range of therapeutic characteristics, including aphrodisiac, antispasmodic, antimicrobial, anticancer, antibacterial, antifungal, antiseptic, and anti-inflammatory effects. A study was conducted to investigate the effects of co-cultivating plant growth-promoting rhizobacteria (PGPR) with in-vitro raised saffron cormlets. The findings of this study provide additional evidence that the bacterial strains *Pseudomonas* sp., *Bacillus subtilis*, and *Pantoea* sp. have a significant impact on the increase in weight of cormlets. In addition, the findings demonstrated that the combination of *Pantoea* sp., *Bacillus subtilis*, and *Pseudomonas* sp. on MS liquid medium resulted in a notable enhancement in the proliferation of cormlets. Additionally, the combined inoculation of the bacterial strains *Acinetobacter haemolyticus*, *Acinetobacter lwoffii*, and *Pantoea* sp. led to a complete germination rate of 100% for cormlets. Furthermore, it was observed that the root system of the cormlets had a higher density and thickness compared to the control cormlets (Kumar et al. 2022; Parray et al. 2015).

It is also crucial to consider the significance of bacterial inoculation during the initial stage of acclimatisation in a micropropagated plant system, particularly in terms of establishing the rhizosphere soil microbiota. In this regard, a study conducted by del Carmen Jaizme-Vega et al. (2004) aimed to assess the impact of a consortium of *Bacillus* spp. rhizobacteria on the first developmental phases of two micropropagated banana plants. Two different cultivars of banana plants, namely 'Grande Naine' and a tetraploid hybrid 'ITC 1297' derived from bananas, were subjected to inoculation with a suspension of *Bacillus* spp. during the weaning phase. The findings of this study indicate that the tested consortium of *Bacillus* spp. was capable of enhancing plant development in both cultivars, whereas the foliar mineral composition of Grande Naine had significant improvements (del Carmen Jaizme-Vega et al. 2004).

Heat stress, brought on by climate change in the form of rising temperatures, leads to additional biochemical, physiological, molecular, and morphological changes, all of which can have a negative impact on plant growth, biomass, productivity, and yield, either alone or in concert with other abiotic stresses. In the midst of climate change, there are various environmental factors that have increased, which include abiotic and biotic factors that cause the medicinally important plants of the Himalayan region to thrive. Beneficial microorganisms that can synthesise or metabolise plant hormones are crucial because they may alter the concentration of particular phytohormones, hence affecting the balance between the growth of plants and their defence against stressors and pathogens. In addition, numerous studies have substantiated the efficacy of plant-growth-promoting bacteria and fungi as viable biostimulants for the purpose of augmenting plant development. These microorganisms have been found to enhance nutrient absorption and photosynthetic activity, hence improving crop quality and bolstering plant resilience against both biotic and abiotic stressors. In the realm of plant tissue culture, a pivotal aspect of the asexual reproduction of numerous woody species is the occurrence of adventitious rooting from the stem. When combined with auxins, certain *Pantoea agglomerans* strains are known to generate compounds that stimulate the growth of adventitious roots. This process entails the alteration of the fate of stem cells in close proximity to the injured area, ultimately resulting in the establishment of a new radical meristem. The inclusion of this process is crucial for several woody species, even within in-vitro culture methods. Consequently, the production of many plants on an industrial propagation scale and the preservation of the genetic integrity of elite genotypes become challenging tasks (Luziatelli et al. 2020; Khan et al. 2020; Bhutia et al. 2018).

The study by Muganu et al. (2015) aimed to assess the impact of *Burkholderia* spp. strain IF25 on the process of adventitious rooting in micropropagated grapevine explants. The findings of the experiment indicate that the bacterization of grapevine explants in a controlled environment by *Burkholderia* spp. strain IF25 had an impact on root differentiation. It was observed that the earliest occurrence of root formation took place in shoots that were inoculated, regardless of whether exogenous IBA was administered. Additionally, the average number of roots per explant was found to be higher in the inoculated samples (Muganu et al. 2015). In recent times, there has been a shift in focus towards investigating the potential beneficial impacts of microbes in the context of in-vitro plant cultures. In the same perspective, the use of arbuscular mycorrhizal fungi (AMF) in micro-propagation is currently seeing increased attention since it has been shown to have a beneficial effect on the post-transplant performance of plants cultivated in-vitro. Enhanced nutrient absorption such as Ca, Cn, Mn, Zn, etc., water regulation, aeration, soil pH equilibrium, and their prospective application as bioregulators Studies have stated that integration of mycorrhization at the acclimation phase gives an approach to mitigating the transition stress by addressing the challenges associated with temporary transplanting, thereby lowering the duration of the acclimatisation period. Nevertheless, a significant challenge in the use of AMF in micro-propagation systems pertains to the absence of dependable and readily applicable techniques for the large-scale synthesis of inoculum (Kapoor et al. 2008; Yadav et al. 2013). Vafadar et al. (2014) demonstrated in a study that regenerated tissue-cultured plantlets of stevia were transferred to pots in a greenhouse. The greenhouse plants were subsequently subjected to inoculation with plant-growth-promoting rhizobacteria, including *Bacillus polymixa*, *Pseudomonas putida*, and *Azotobacter chroococcum*, as well as AMF, specifically *Glomus intraradices*. The findings indicated that, when compared to the control group, the introduction of a single microbe resulted in a substantial enhancement in both root and shoot biomass, as well as a rise in the levels of stevioside, chlorophyll, and NPK content within the plants. Subsequently, it has been observed that the aforementioned effects are notably amplified as a result of the combination of suitable inoculants, which exhibit strong synergistic connections with other bacterial strains. The study further demonstrated the possibility that the use of such beneficial microbes might be employed at a later stage of the Himalayan medicinal plant conservation process, particularly when the plants are transplanted to the field and encounter adverse environmental conditions (Vafadar et al. 2014; Pandey et al. 2000). Therefore, it has been shown that inoculation of the in-vitro formed plantlets with plant growth-stimulating microbes has resulted in a high survival rate in challenging climatic conditions, which is crucial for the effective conservation of valuable Himalayan medicinal plants.

6 Conclusion

The Himalayas, which serve as a habitat for a wide variety of medicinal plants and exhibit a notable level of biodiversity, therefore make a substantial contribution to the welfare of human populations. The use of botanical species from the himalayan region for their therapeutic properties has long been recognised as a valuable asset for traditional healers and pharmaceutical enterprises. The medicinal plants in the

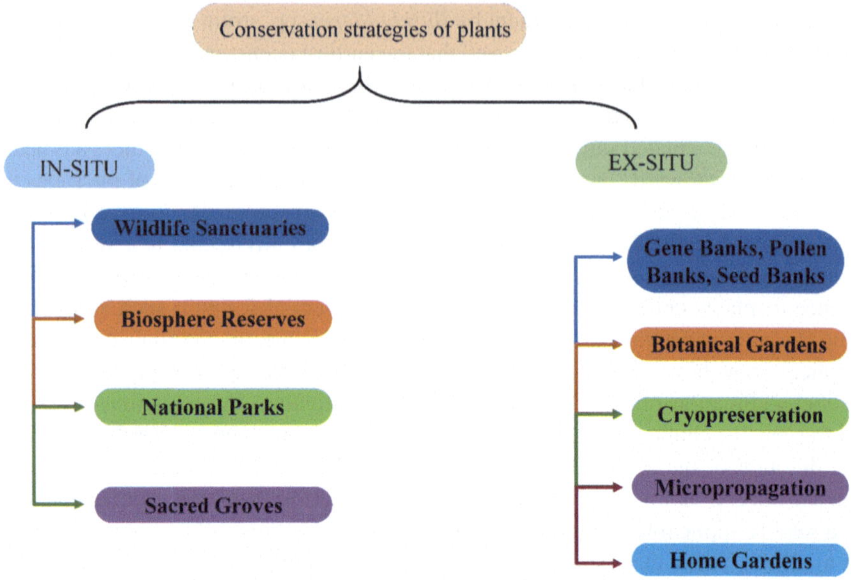

Fig. 1 Schematic illustration of several conservation strategies

Himalayan region at times encounter adverse environmental circumstances, as well as indiscriminate harvesting from their native habitats, ecological degradation, and overgrazing, resulted from the climate change. The aforementioned circumstances have presented considerable obstacles to the accessibility of these plant species, leading to a substantial number of them being at risk of becoming scarce, vulnerable, and ultimately endangered. This chapter concluded that the avenue for synergistic function of implementing plant growth-boosting microorganisms with in -vitro produced plants. To effectively promote the conservation of important Himalayan medicinal plants, it is advisable to incorporate the growing of in-vitro plantlets alongside the use of plant growth-boosting bacteria. The use of beneficial microorganisms with plant growth-promoting traits has been widely acknowledged as an effective approach for enabling the successful acclimatisation of in-vitro plants. The aforementioned methodology has exhibited a notable enhancement in the viability of these Himalayan medicinal plants when subjected to adverse weather patterns, concurrently ameliorating the adverse consequences linked to the physiological stress experienced during their relocation to indigenous environments (Figs. 1 and 2).

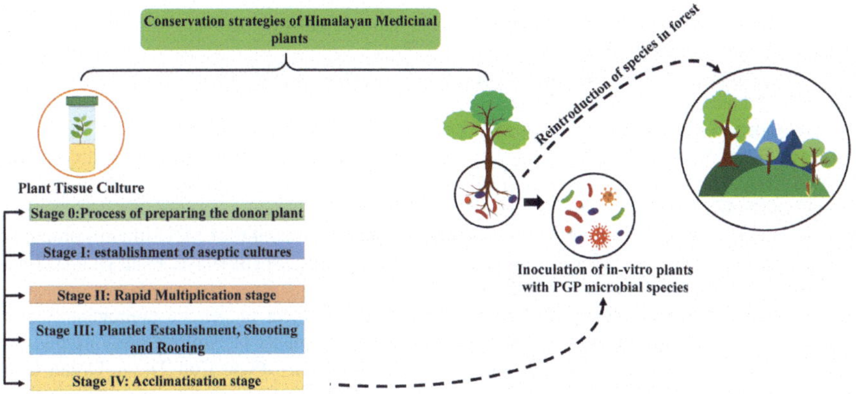

Fig. 2 Schematic representation of the conservation of Himalayan medicinal plants through the use of plant tissue culture methods and PGP microorganisms

References

Abdalla N, El-Ramady H, Seliem MK, El-Mahrouk ME, Taha N, Bayoumi Y, Shalaby TA, Dobránszki J (2022) An academic and technical overview on plant micro-propagation challenges. Horticulturae 8(8):677

Adhikari P, Joshi K, Singh M, Pandey A (2022) Influence of altitude on secondary metabolites, antioxidants, and antimicrobial activities of Himalayan yew (*Taxus wallichiana*). Plant Biosyst Int J Dealing All Aspects Plant Biol 156(1):187–195

Adhikari P, Jain R, Sharma A, Pandey A (2021) Plant growth promotion at low temperature by phosphate-solubilizing *Pseudomonas* spp. isolated from high-altitude Himalayan soil. Microbial Ecol 82(3):677–687

de Araújo RC, Rodrigues FA, Nadal MC, de Souza Ribeiro M, Antônio CAC, Rodrigues VA, de Souza AC, Pasqual M, Dória J (2021) Acclimatization of Musa spp. seedlings using endophytic Bacillus spp. and *Buttiauxella agrestis* strains. Microbiol Res 248:126750

Asgher M, Verma S, Khan NA, Vyas D, Kumari P, Rashid S, Khan S, Qadir S, Ajmal Ali M, Ahmad P (2020) Physiological, biochemical and reproductive studies on *Valeriana wallichii*, a critically endangered medicinal plant of the Himalayan Region Grown under in-situ and ex-situ conditions. Plants 9(2):131

Avhad A, Patil N, Zunjarrao R (2023) Overview of in vitro strategies to conserve rare and endangered plant species in India. Int J Sci Res Arch 9(1):650–669

Baldrian P, López-Mondéjar R, Kohout P (2023) Forest microbiome and global change. Nat Rev Microbiol 1–15

Banerjee A, Cornejo J, Bandopadhyay R (2020) Emergent climate change impact throughout the world: call for "Microbiome Conservation" before it's too late. Biodivers Conserv 29:345–348

Bantawa P, Ghosh SK, Bhandari P, Singh B, Ghosh PD, Ahuja PS, Mondal TK (2010) Micropropagation of an elite line of *Picrorhiza scrophulariiflora*, Pennell, an endangered high valued medicinal plant of the Indo-China Himalayan region. Med Aromat Plant Sci Biotechnol 4:1–7

Banyal A, Thakur V, Thakur R, Kumar P (2021) Endophytic microbial diversity: a new hope for the production of novel anti-tumor and anti-HIV agents as future therapeutics. Curr Microbiol 78:1699–1717

Bhardwaj AK, Singh B, Kaur K, Roshan P, Sharma A, Dolker D, Naryal A, Saxena S, Pati PK, Chaurasia OP (2018) In vitro propagation, clonal fidelity and phytochemical analysis of

Rhodiola imbricata Edgew: a rare trans-Himalayan medicinal plant. Plant Cell. Tissue Organ Cult (PCTOC) 135:499–513

Bhoite HA, Palshikar GS (2014) Plant tissue culture: a review. World J Pharm Sci 565–572

Bhutia K, Khanna V, Meetei T, Bhutia N (2018) Effects of climate change on growth and development of chilli. Agrotechnology 7(2):1–4

Bisht AK, Bhatt A, Dhar U, Bhatt A, Dhar U (2015) Note on somatic embryogenesis and synthetic seed production in *Angelica glauca*: a valuable medicinal plant of Himalaya. J Med Plants Res 9(12):419–425

Bisht A, Giri L, Belwal T, Pandey A, Bahukhandi A, Bhatt ID, Rawal RS (2022) In vitro propagation and antioxidant potential of *Berberis asiatica* from Western Himalaya. Plant Biosyst Int J Dealing All Aspects Plant Biol 156(2):490–496

Bose B, Kumaria S, Choudhury H, Tandon P (2016) Assessment of genetic homogeneity and analysis of phytomedicinal potential in micropropagated plants of *Nardostachys jatamansi*, a critically endangered, medicinal plant of alpine Himalayas. Plant Cell Tissue Organ Cult (PCTOC) 124:331–349

Breman E, Ballesteros D, Castillo-Lorenzo E, Cockel C, Dickie J, Faruk A, O'Donnell K, Offord CA, Pironon S, Sharrock S, Ulian T (2021) Plant diversity conservation challenges and prospects—the perspective of botanic gardens and the millennium seed bank. Plants 10(11):2371

Cahyaningsih R, Phillips J, Brehm JM, Gaisberger H, Maxted N (2021) Climate change impact on medicinal plants in Indonesia. Glob Ecol Conserv 30:e01752

Cantabella D, Dolcet-Sanjuan R, Casanovas M, Solsona C, Torres R, Teixidó N (2020) Inoculation of in vitro cultures with rhizosphere microorganisms improve plant development and acclimatization during immature embryo rescue in nectarine and pear breeding programs. Sci Hortic 273:109643

Castro HF, Classen AT, Austin EE, Norby RJ, Schadt CW (2010) Soil microbial community responses to multiple experimental climate change drivers. Appl Environ Microbiol 76(4):999–1007

Chakraborty A, Bhattacharya D, Ghanta S, Chattopadhyay S (2010) An efficient protocol for in vitro regeneration of *Podophyllum hexandrum*, a critically endangered medicinal plant

Chandra LD (2016) Bio-diversity and conservation of medicinal and aromatic plants. Adv Plants Agric Res 5(4):00186

Chandra S, Bandopadhyay R, Kumar V, Chandra R (2010) Acclimatization of tissue cultured plantlets: from laboratory to land. Biotech Lett 32:1199–1205

Chandra N, Joshi VC, Mishra AP (2020) Ecological status of fodder and fuelwood species in Banari Devi sacred grove of Kumaun Himalaya, Uttarakhand. Indian J Ecol 47(4):1044–1048

Chandra N, Rai ID, Mishra AP, Dwivedi SK, Kotiya A, Tiwari UK, Singh G (2022) Assessing potential habitats and populations of selected medicinal herbs in Alpine areas of Uttarakhand, Western Himalaya. Indian J For 45(3):144–155. https://doi.org/10.54207/bsmps1000-2023-ITRFNL

Chatterjee T, Ghosh B (2020) Micro-propagation of medicinal plants: a review. Int J Econ Plants 7(2):066–072

Chauhan RS, Tiwari D, Bisht AS, Shukla A (2014) Ex situ conservation of medicinal and aromatic plants in Bharsar, Uttarakhand, India. Med Plants-Int J Phytomed Related Indus 6(4):282–292

Chowdhury S, Pal K, Chakraborty M, Chakraborty S, Mandal S, Pandit GK, Maitra S, Sahana N (2020) Conservation and in vitro propagation of an endangered wild turmeric (*Curcuma caesia* Roxb.) species from Sub-Himalayan Terai region of West Bengal. Int J Curr Microbiol App Sci 9(2):2132–2140

Dad JM, Rashid I (2022) Differential responses of Kashmir Himalayan threatened medicinal plants to anticipated climate change. Environ Conserv 49(1):33–41

Das M, Jain V, Malhotra S (2016) Impact of climate change on medicinal and aromatic plants. Indian J Agric Sci 86(11):1375–1382

del Carmen Jaizme-Vega M, Rodríguez-Romero AS, Guerra MSP (2004) Potential use of rhizobacteria from the Bacillus genus to stimulate the plant growth of micropropagated bananas. Fruits 59(2):83–90

Dhakar K, Pandey A (2020) Microbial ecology from the Himalayan cryosphere perspective. Microorganisms 8(2):257

Dhar U, Upreti J, Bhatt ID (2000) Micro-propagation of *Pittosporum napaulensis* (DC.) Rehder & Wilson–a rare, endemic Himalayan medicinal tree. Plant Cell, Tissue Organ Cult 63:231–235

Dhiman N, Devi K, Bhattacharya A (2021) Development of low cost micro-propagation protocol for *Nardostachys jatamansi*: a critically endangered medicinal herb of Himalayas. S Afr J Bot 140:468–477

Dias MI, Sousa MJ, Alves RC, Ferreira IC (2016) Exploring plant tissue culture to improve the production of phenolic compounds: a review. Ind Crops Prod 82:9–22

Digat B, Hermelin V, Brochard P, Touzet M (1985) Interest of bacterized synthetics substrates MILCAP® for in vitro culture. In: Symposium on in vitro problems related to mass propagation of horticultural plants, vol 212, pp 375–378

Ekor M (2014) The growing use of herbal medicines: issues relating to adverse reactions and challenges in monitoring safety. Front Pharmacol 4:177

Engels JM, Ebert AW (2021) A critical review of the current global ex situ conservation system for plant agrobiodiversity. I. History of the development of the global system in the context of the political/legal framework and its major conservation components. Plants 10(8):1557

Fadiji AE, Babalola OO, Santoyo G, Perazzolli M (2022) The potential role of microbial biostimulants in the amelioration of climate change-associated abiotic stresses on crops. Front Microbiol 12:829099

Fiodor A, Ajijah N, Dziewit L, Pranaw K (2023) Biopriming of seed with plant growth-promoting bacteria for improved germination and seedling growth. Front Microbiol 14:1142966

Gaira KS, Dhar U (2020) Phenological change modelling for selected Himalayan medicinal herbs using herbarium records: a case study. Eco Inform 60:101177

Ganeshkumar Y, Ramarao A, Veeresham C (2017) Picroside I and picroside II from tissue cultures of *Picrorhiza kurroa*. Pharmacognosy Res 9(Suppl 1):S53

García-Gonzáles R, Quiroz K, Carrasco B, Caligari P (2010) Plant tissue culture: current status, opportunities and challenges. Int J Agric Nat Resourc 37(3):5–30

Gondval M, Chaturvedi P, Gaur AK (2016) Thidiazuron-induced high frequency establishment of callus cultures and plantlet regeneration in *Aconitum balfourii* Stapf.: an endangered medicinal herb of North-West Himalayas

Guan L, Yang Y, Jiang P, Mou Q, Gou Y, Zhu X, Xu YW, Wang R (2022) Potential distribution of Blumea balsamifera in China using MaxEnt and the ex situ conservation based on its effective components and fresh leaf yield. Environ Sci Pollut Res 29(29):44003–44019

Gulzar B, Mujib A, Malik MQ, Mamgain J, Syeed R, Zafar N (2020) Plant tissue culture: agriculture and industrial applications. In: Transgenic technology based value addition in plant biotechnology. Academic Press, pp 25–49

Gupta R, Sood H (2022) Optimizing nutrient media conditions for continuous production of shoot biomass enriched in major medicinal constituents, amarogentin and mangiferin of endangered medicinal herb, *Swertia chirayita*. Vegetos 1–9

Gurung C, Gurung A (2019) Ex situ conservation of three Himalayan ferns through hydroponic culture under Greenhouse conditions. Pleione 13(2):216–225

Habib Ur Rehman JA, Chester K, Tajuddin TE, Ahmad S (2014) Chemoprofiling and tissue culture studies on *Picrorhiza kurroa* Royle ex Benth. for production of picroside II. Ann Phytomed 3(2):70–77

Harish S, Kavino M, Kumar N, Saravanakumar D, Soorianathasundaram K, Samiyappan R (2008) Biohardening with plant growth promoting rhizosphere and endophytic bacteria induces systemic resistance against banana bunchy top virus. Appl Soil Ecol 39(2):187–200

Herman E (1987) Toward control of micro-propagation contamination. Agricell Rep 9:33–35

Herman E (1990) Non-axenic plant tissue culture: possibilities and opportunities. Acta Hortic 280:233–238

Hussain A, Qarshi IA, Nazir H, Ullah I, Rashid M, Shinwari ZK (2013) In vitro callogenesis and organogenesis in *Taxus wallichiana* ZUCC, The Himalayan Yew. Pak J Bot 45(5):1755–1759

Hussain M, Bibi Y, Raja NI, Iqbal M, Aslam S, Tahir N, Imran M, Iftikhar A (2016) A review of therapeutic potential of *Ajuga bracteosa*: a critically endangered plant from Himalaya. J Coastal Life Med 4(11):918–924

IPCC (2014) Summary for policymakers. In: Field CB et al (eds) Climate change: impacts, adaptation, and vulnerability. part a: global and sectoral aspects. Contribution of working group II to the fifth assessment report of the intergovernmental panel on climate change. Cambridge University Press, Cambridge, United Kingdom and New York, NY, USA

Jamshidi-Kia F, Lorigooini Z, Amini-Khoei H (2017) Medicinal plants: past history and future perspective. J Herbmed Pharmacol 7(1):1–7

Jan M, Singh S, Maqbool F, Nawchoo IA (2018) Direct shoot regeneration from hypocotyl explants of *Heracleum candicans* Wall: a vulnerable high value medicinal herb of Kashmir Himalaya. Afr J Agric Res 13(28):1419–1424

Jeon JY, Adhikari P, Seo C (2020) Impact of climate change on potential dispersal of *Paeonia obovata* (paeoniaceae), a critically endangered medicinal plant of South Korea. Environ Conserv 26:S145–S155

Joshi M, Dhar U (2003) In vitro propagation of *Saussurea obvallata* (DC.) Edgew.—an endangered ethnoreligious medicinal herb of Himalaya. Plant Cell Rep 21:933–939

Joshi M, Manjkhola S, Dhar U (2004) Developing propagation techniques for conservation of *Heracleum candicans*–an endangered medicinal plant of the Himalayan region. J Hortic Sci Biotechnol 79(6):953–959

Joshi K, Adhikari P, Bhatt ID, Pande V (2022) Source dependent variation in phenolics, antioxidant and antimicrobial activity of *Paeonia emodi* in west Himalaya, India. Physiol Mol Biol Plants 28(9):1785–1798

Kapoor R, Sharma D, Bhatnagar AK (2008) Arbuscular mycorrhizae in micro-propagation systems and their potential applications. Sci Hortic 116(3):227–239

Karuppusamy S (2009) A review on trends in production of secondary metabolites from higher plants by in vitro tissue, organ and cell cultures. J Med Plants Res 3(13):1222–1239

Kaur T, Bhat R, Vyas D (2016) Effect of contrasting climates on antioxidant and bioactive constituents in five medicinal herbs in Western Himalayas. J Mt Sci 13:484–492

Kaushal S, Sidana A, Dev K (2014) In vitro plant production through apical meristem culture of *Gentiana kurroo* Royle. J Med Plants Stud 3(1):4–9

Khajuria AK, Hano C, Bisht NS (2021) Somatic embryogenesis and plant regeneration in *Viola canescens* Wall. Ex. Roxb.: an endangered himalayan herb. Plants 10(4):761

Khan MA, Asaf S, Khan AL, Jan R, Kang SM, Kim KM, Lee IJ (2020) Thermotolerance effect of plant growth-promoting *Bacillus cereus* SA1 on soybean during heat stress. BMC Microbiol 20(1):1–14

Khan A, Shah AH, Ali N (2021) In-vitro propagation and phytochemical profiling of a highly medicinal and endemic plant species of the Himalayan region (*Saussurea costus*). Sci Rep 11(1):23575

Krishna KVVS, Chaturvedi S, Sakhare, Chand J, Sakhare DT, Mourya VK, Pathak VK, Richa, Bhaskar MN, Kola M (2023) Ajuga bracteosa: a review on endangered Indian medicinal plant. Eur Chem Bull 2(12):1380–1398

Kumar R, Borker SS, Thakur A, Thapa P, Kumar S, Mukhia S, Anu K, Bhattacharya A, Kumar S (2021) Physiological and genomic evidence supports the role of Serratia quinivorans PKL: 12 as a biopriming agent for the biohardening of micropropagated *Picrorhiza kurroa* plantlets in cold regions. Genomics 113(3):1448–1457

Kumar A, Devi M, Kumar R, Kumar S (2022) Introduction of high-value *Crocus sativus* (saffron) cultivation in non-traditional regions of India through ecological modelling. Sci Rep 12(1):11925

Kumar A, Adhikari BS, Mishra AP (2023) Status of threatened medicinal and Biosphere Reserve, Uttarakhand. A report (TR No./2022/35) submitted to International Union for Conservation of Nature—India, p 103

Kumari A, Lakshmi GA, Krishna GK, Patni B, Prakash S, Bhattacharyya M, Singh SK, Verma KK (2022a) Climate change and its impact on crops: a comprehensive investigation for sustainable agriculture. Agronomy 12(12):3008

Kumari P, Wani IA, Khan S, Verma S, Mushtaq S, Gulnaz A, Paray BA (2022b) Modeling of Valeriana wallichii habitat suitability and niche dynamics in the Himalayan Region under anticipated climate change. Biology 11(4):498

Kunwar RM, Thapa-Magar KB, Subedi SC, Kutal DH, Baral B, Joshi NR, Adhikari B, Upadhyaya KS, Thapa-Magar S, Ansari AS, Thapa GJ (2023) Distribution of important medicinal plant species in Nepal under past, present, and future climatic conditions. Ecol Ind 146:109879

Lamo T, Gurmet P, Dolkar P, Stobgais T, Dawa S, Dolma T, Angdus T, Kunphel S, Angmo T (2019) Indigenous Use of some important trans-Himalayan medicinal plants of Himachal Pradesh (North India). Int J Curr Microbiol App Sci 8(2):2682–2689

Lohani N, Kumar R, Tewari LM, Joshi GC (2012) Ex-situ conservation of *Polygonatum verticillatum* (L.) Allioni under different types of organic treatments. Int J Biodivers Conserv 4:22–31

Luziatelli F, Gatti L, Ficca AG, Medori G, Silvestri C, Melini F, Muleo R, Ruzzi M (2020) Metabolites secreted by a plant-growth-promoting *Pantoea agglomerans* strain improved rooting of *Pyrus communis* L. cv Dar Gazi cuttings. Front Microbiol 11:539359

Maikhuri RK, Phondani PC, Dhyani D, Rawat LS, Jha NK, Kandari LS (2018) Assessment of climate change impacts and its implications on medicinal plants-based traditional healthcare system in Central Himalaya, India. Iran J Sci Technol Trans A Sci 42:1827–1835

Manjkhola S, Dhar U, Joshi M (2005) Organogenesis, embryogenesis, and synthetic seed production in *Arnebia euchroma*—a critically endangered medicinal plant of the Himalaya. In Vitro Cell Develop Biol-Plant 41:244–248

Mehta P, Bisht K, Sekar KC (2021) Diversity of threatened medicinal plants of Indian Himalayan Region. Plant Biosyst-Int J Dealing All Aspects Plant Biol 155(6):1121–1132

Mishra T (2016) Climate change and production of secondary metabolites in medicinal plants: a review. Int J Herb Med 4(4):27–30

Mishra MK, Pandey S, Misra P, Niranjan A, Srivastava A (2020) An efficient protocol for clonal regeneration and excised root culture with enhanced alkaloid content in *Thalictrum foliolosum* DC.—an endemic and important medicinal plant of temperate Himalayan region. Indus Crops Prod 152:112504

Mishra AP, Kumar A, Yadav SN (2023a) Ecology and conservation of threatened medicinal plants in the Trans-Himalayan region of Nanda Devi biosphere reserve, Western Himalaya. Trees For People 14:100451. https://doi.org/10.1016/j.tfp.2023.100451

Mishra AP, Chandra N, Mandy JJ, Dwivedi SK, Alruzuq A, Pande CB (2023b) Analytical hierarchy process (AHP) based on the spatial assessment of an endangered alpine medicinal herb *Aconitum heterophyllum* in the Western Himalayan environment. In: Climate change impacts on natural resources, ecosystems and agricultural systems. Springer International Publishing, Cham, pp 579–594. https://doi.org/10.1007/978-3-031-19059-9_25

Misra P, Saema S (2016) Plant tissue culture for in vitro mutagenesis, large-scale propagation, and genetic transformation. Plant Tissue Cult: Prop Conserv Crop Improv 309–342

Monthony AS, Page SR, Hesami M, Jones AMP (2021) The past, present and future of *Cannabis sativa* tissue culture. Plants 10(1):185

Muganu M, Paolocci M, Bignami C, Di Mattia E (2015) Enhancement of adventitious root differentiation and growth of in vitro grapevine shoots inoculated with plant growth promoting rhizobacteria. Vitis 54(2):73–77

Mukherjee D, Chakraborty S (2019) Coptis Teeta: conservation and cultivation practice-a rare medicinal plant on Earth. Curr Invest Agric Curr Res 6:845–851

Mukherjee D, Chakraborty S, Roy A, Mokthan MW (2009) Differential approach of germplasm conservation of high value of medicinal plants in north eastern himalayan region. Int J Agric Environ Biotechnol 2(4):332–340

Negi VS, Maikhuri RK, Pharswan D, Thakur S, Dhyani PP (2017) Climate change impact in the Western Himalaya: people's perception and adaptive strategies. J Mt Sci 14(2):403–416

Negi VS, Kewlani P, Pathak R, Bhatt D, Bhatt ID, Rawal RS, Sundriyal RC, Nandi SK (2018) Criteria and indicators for promoting cultivation and conservation of medicinal and aromatic plants in Western Himalaya, India. Ecol Ind 93:434–446

Niazi P, Monib AW, Azizi A (2023) A review on plants and plant/microbial systems in reducing exposure. J Res Appl Sci Biotechnol 2(2):1–7

Nowak J (1998) Benefits of in vitro "biotization" of plant tissue cultures with microbial inoculants. In Vitro Cell Develop Biol-Plant 34:122–130

Orlikowska T, Nowak K, Reed B (2017) Bacteria in the plant tissue culture environment. Plant Cell Tissue Organ Cult (PCTOC) 128:487–508

Ortíz-Castro R, Contreras-Cornejo HA, Macías-Rodríguez L, López-Bucio J (2009) The role of microbial signals in plant growth and development. Plant Signal Behav 4(8):701–712

Pandey A, Palni LMS, Bag N (2000) Biological hardening of tissue culture raised tea plants through rhizosphere bacteria. Biotech Lett 22:1087–1091

Pandey H, Nandi SK, Kumar A, Palni UT, Chandra B, Palni LMS (2004) In vitro propagation of *Aconitum balfourii* Stapf.: an important aconite of the Himalayan alpines. J Horticult Sci Biotechnol 79(1):34–41

Pandey S, Kushwaha R, Prakash O, Bhattacharya A, Ahuja PS (2005) Ex situ conservation of *Aconitum heterophyllum* Wall.—an endangered medicinal plant of the Himalaya through mass propagation and its effect on growth and alkaloid content. Plant Genet Resourc 3(2):127–135

Pandey M, Dhar U, Sher Singh S, Mrudul Vijay S, Shubhada Ratnakar T (2011) Recurrent somatic embryogenesis and plant regeneration in *Angelica glauca* Edgew., a critically endangered medicinal plant of the Western Himalaya. J Horticult Sci Biotechnol 86(5):493–498

Pandey S, Sundararajan S, Ramalingam S, Baniya MK, Pant B (2020) Rapid clonal propagation and valepotriates accumulation in cultures of *Valeriana jatamansi* Jones, a high-value medicinal plant. J Appl Bot Food Qual 93

Pant HC, Pant HV, Kumar A, Tomar H, Sharma MD, Gaurav N (2021) In vitro clonal propagation of *Nardostachys jatamansi*: a traditional Himalayan medicinal plant. J Mountain Res 16(3):87–98

Parray JA, Kamili AN, Reshi ZA, Qadri RA, Jan S (2015) Interaction of rhizobacterial strains for growth improvement of *Crocus sativus* L. under tissue culture conditions. Plant Cell Tissue Organ Cult (PCTOC) 121:325–334

Parray JA, Kamili AN, Jan S, Mir MY, Shameem N, Ganai BA, Abd-Allah EF, Hashem A, Alqarawi AA (2018) Manipulation of plant growth regulators on phytochemical constituents and DNA protection potential of the medicinal plant *Arnebia benthamii*. BioMed Res Int

Patial V, Sharma M, Bhattacharya A (2017) Potential of thidiazuron in improved micro-propagation of *Picrorhiza kurroa*–an endangered medicinal herb of alpine Himalaya. Plant Biosyst Int J Dealing All Aspects Plant Biol 151(4):729–736

Phondani PC, Bhat ID, Negi VS, Kothiyari BP, Bhatt A, Maikhuri RK (2016a) Promoting medicinal plant cultivation for livelihood enhancement and biodiversity conservation in the Indian Himalayan region. J Asia-Pacific Biodivers 9:39–46

Phondani PC, Bhat ID, Negi VS, Kothiyari BP, Bhatt A, Maikhuri RK (2016b) Promoting medicinal plant cultivation for livelihood enhancement and biodiversity conservation in the Indian Himalayan region. J Asia-Pacific Biodivers 9:39–46

Purohit S, Jugran AK, Bhatt ID, Palni LMS, Bhatt A, Nandi SK (2017) In vitro approaches for conservation and reducing juvenility of *Zanthoxylum armatum* DC: an endangered medicinal plant of Himalayan region. Trees 31:1101–1108

Purohit S, Joshi K, Rawat V, Bhatt ID, Nandi SK (2020) Efficient plant regeneration through callus in *Zanthoxylum armatum* DC: an endangered medicinal plant of the Indian Himalayan region. Plant Biosyste Int J Dealing All Aspects Plant Biol 154(3):288–294

Qaderi MM, Martel AB, Strugnell CA (2023) Environmental factors regulate plant secondary metabolites. Plants 12(3):447

Rafiq S, Wagay NA, Bhat IA, Kaloo ZA, Rashid S, Lin F, El-Abedin TKZ, Wani SH, Mahmoud EA, Almutairi KF, Elansary HO (2021) In vitro propagation of *Aconitum chasmanthum* Stapf

Ex Holmes: an endemic and critically endangered plant species of the western Himalaya. Horticulturae 7(12):586
Rai MK (2010) Biotechnological strategies for conservation of rare and endangered medicinal plants. Biodiversitas 11(3):157–166
Rajput S, Agrawal V (2020) Micro-propagation of *Atropa acuminata* Royle ex Lindl. (a critically endangered medicinal herb) through root callus and evaluation of genetic fidelity, enzymatic and non-enzymatic antioxidant activity of regenerants. Acta Physiologiae Plantarum 42(11):160
Ramazan S, John R (2020) Climate change impacts and mitigation; Abiotic stress impressions and tolerance: a perspective from Kashmir Himalayas. In Proc Indian Natn Sci Acad 86(3):1139–1156
Rana MS, Samant SS (2010) Threat categorisation and conservation prioritisation of floristic diversity in the Indian Himalayan region: a state of art approach from Manali Wildlife Sanctuary. J Nat Conserv 18(3):159–168
Rana SK, Rana HK, Ghimire SK, Shrestha KK, Ranjitkar S (2017) Predicting the impact of climate change on the distribution of two threatened Himalayan medicinal plants of Liliaceae in Nepal. J Mt Sci 14:558–570
Rana SK, Rana HK, Ranjitkar S, Ghimire SK, Gurmachhan CM, O'Neill AR, Sun H (2020) Climate-change threats to distribution, habitats, sustainability and conservation of highly traded medicinal and aromatic plants in Nepal. Ecol Ind 115:106435
Rao VB, Franchito SH, Gerólamo ROP, Giarolla E, Ramakrishna SSVS, Rao BRS, Naidu CV (2016) Himalayan warming and climate change in India. Am J Clim Chang 5(4):558–574
Rathore S, Raj Y, Debnath P, Kumar M, Kumar R (2022) Ethnopharmacology, phytochemistry, agrotechnology, and conservation of *Inula racemosa* Hook f.–a critically endangered medicinal plant of the western Himalaya. J Ethnopharmacol 283:114613
Rawat B, Rawat JM, Mishra S, Mishra SN (2013a) *Picrorhiza kurrooa*: current status and tissue culture mediated biotechnological interventions. Acta Physiol Plant 35:1–12
Rawat JM, Rawat B, Agnihotri RK, Chandra A, Nautiyal S (2013b) In vitro propagation, genetic and secondary metabolite analysis of *Aconitum violaceum* Jacq.: a threatened medicinal herb. Acta Physiol Plant 35:2589–2599
Rawat JM, Bhandari A, Mishra S, Rawat B, Dhakad AK, Thakur A, Chandra A (2018) Genetic stability and phytochemical profiling of the in vitro regenerated plants of *Angelica glauca* Edgew.: an endangered medicinal plant of Himalaya. Plant Cell Tissue Organ Cult (PCTOC) 135:111–118
Rawat N, Mishra AP, Chandra N, Painuly V, Negi GS, Bisht MP S, Purohit S (2021) Maxent modeling for predicting the potential distribution of *Valeriana jatamansi* jones. in chakrata forest division of Garhwal Himalaya. Indian J Ecol 48(2):367–372
Saaondo T, Olalekan O (2023) Ex situ and in situ conservation of plant genetic resources in Nigeria. ScienceOpen Preprints
Safdar I, Bibi Y, Hussain M, Iqbal M, Saira H, Shaheen S, Tahir N, Mehboob H (2017) Review on current status of *Betula utilis*: an important medicinal plant from Himalaya. J Bot Sci 6:1–7
Sambyal M, Dogra A, Koul S, Ahuja A (2006) Rapid in vitro propagation of *Potentilla fulgens* wall—a Himalayan alpine herb of medicinal value. J Plant Biochem Biotechnol 15:143–145
Sangiorgio D, Cellini A, Donati I, Pastore C, Onofrietti C, Spinelli F (2020) Facing climate change: application of microbial biostimulants to mitigate stress in horticultural crops. Agronomy 10(6):794
Shahzad A, Sharma S, Parveen S, Saeed T, Shaheen A, Akhtar R, Yadav V, Upadhyay A, Ahmad Z (2017) Historical perspective and basic principles of plant tissue culture. Plant Biotechnol Principles Appl 1–36
Sharma S, Thokchom R (2014) A review on endangered medicinal plants of India and their conservation. J Crop Weed 10(2):205–218
Sharma S, Katoch V, Rathour R, Sharma TR (2010) In vitro propagation of endangered temperate Himalayan medicinal herb *Picrorhiza kurroa* Royle ex benth using leaf explants and nodal segments. J Plant Biochem Biotechnol 19:111–114

Sharma A, Kaur R, Sharma N (2014) In vitro morphogenic response of different explants of *Gentiana kurroo* Royle from Western Himalayas—an endangered medicinal plant. Physiol Mol Biol Plants 20:249–256

Sharma N, Thakur M, Sharma P, Sharma YP, Dutt B (2022) In vitro propagation from rhizomes, molecular evaluation and podophyllotoxin production in Himalayan May Apple (*Sinopodophyllum hexandrum* Royle TS Ying): an endangered medicinal plant. Plant Cell, Tissue Organ Cult (PCTOC) 149(1–2):159–173

Sharma T, Anmol, Sharma U, Kumar S (2022) Iridoid glycosides from Picrorhiza genus endemic to the Himalayan region: phytochemistry, biosynthesis, pharmacological potential and biotechnological intercessions to boost production. Crit Rev Biotechnol 1–16

Shivali V, Namrata S (2022) Threat status of three important medicinal Himalayan plant species and conservation implications. Nat Conserv Res 7(1):27–41

Shukla SK (2023) Conservation of medicinal plants: challenges and opportunities. J Med Bot 5–10

Singh PR, Singh LJ (2021) In vitro propagation for improvement of medicinal plants: a review. J Pharmacognosy Phytochem 10(1):1484–1489

Singh C, Chauhan N, Upadhyay SK, Singh R, Rani A (2021) The Himalayan natural resources: challenges and conservation for sustainable development. J Pharmacognosy Phytochem 10(1):1643–1648

Siwach M, Siwach P, Solanki P, Gill AR (2013) Biodiversity conservation of Himalayan medicinal plants in India: a retrospective analysis for a better vision. Int J Biodivers Conserv 5(9):529–540

Sultan P, Shawl AS, Fayaz AS, Ramteke PW (2009) In vitro embryogenesis and marker guided analysis of podophyllum hexandrum: an endangered medicinal plant. Int J Biodivers Conserv 1(7):215–222

Sun Q, Lai L, Zhou J, Yi S, Liu X, Guo J, Zheng Y (2022) Differences in ecological traits between plants grown in situ and ex situ and implications for conservation. Sustainability 14(9):5199

Suyal DC, Joshi D, Kumar S, Bhatt P, Narayan A, Giri K, Singh M, Soni R, Kumar R, Yadav A, Devi R (2021) Himalayan microbiomes for agro-environmental sustainability: current perspectives and future challenges. Microbial Ecol 1–33

Tabin S, Kamili AN, Gupta RC, Parray JA, Bansal A (2018) In vitro culture of *Rheum emodi* Wall: an endangered medicinal plant of Northwestern Himalaya. Proc Natl Acad Sci India Section B: Biol Sci 88:995–1006

Thomas J, Ajay D, Raj Kumar R, Mandal AKA (2010) Influence of beneficial microorganisms during in vivo acclimatization of in vitro-derived tea (*Camellia sinensis*) plants. Plant Cell Tissue Organ Cult (PCTOC) 101:365–370

Thorpe TA (2007) History of plant tissue culture. Mol Biotechnol 37:169–180

Tiwari D, Kewlani P, Gaira KS, Bhatt ID, Sundriyal RC, Pande V (2023) Predicting phytochemical diversity of medicinal and aromatic plants (MAPs) across eco-climatic zones and elevation in Uttarakhand using generalized additive model. Sci Rep 13(1):10888

Tripathi A, Pandey P, Tripathi SN, Kalra A (2022) Perspectives and potential applications of endophytic microorganisms in cultivation of medicinal and aromatic plants. Front Plant Sci 13:985429

Vafadar F, Amooaghaie R, Otroshy M (2014) Effects of plant-growth-promoting rhizobacteria and arbuscular mycorrhizal fungus on plant growth, stevioside, NPK, and chlorophyll content of *Stevia rebaudiana*. J Plant Interactions 9(1):128–136

Valencia E, Gross N, Quero JL, Carmona CP, Ochoa V, Gozalo B, Delgado-Baquerizo M, Dumack K, Hamonts K, Singh BK, Bonkowski M (2018) Cascading effects from plants to soil microorganisms explain how plant species richness and simulated climate change affect soil multifunctionality. Glob Change Biol 24(12):5642–5654

Verma P, Mathur AK, Jain SP, Mathur A (2012) In vitro conservation of twenty-three overexploited medicinal plants belonging to the Indian sub-continent. Sci World J 2012

Verma RK, Tewari VP (2016) Some important medicinal plants of cold desert regions of District Kinnaur of Himachal Pradesh state in India: their uses and chemical ingredients. J. Plant Chemist Ecophysiol 1:1009

Volis S (2023) Living collections of threatened plants in botanic gardens: when is ex situ cultivation less appropriate than quasi in situ cultivation? J Zool Bot Gard 4(2):462–475
Volis S, Blecher M (2010) Quasi in situ: a bridge between ex situ and in situ conservation of plants. Biodivers Conserv 19:2441–2454
Walters C, Pence VC (2021) The unique role of seed banking and cryobiotechnologies in plant conservation. Plants, People, Planet 3(1):83–91
Wambugu PW, Nyamongo DO, Kirwa EC (2023) Role of seed banks in supporting ecosystem and biodiversity conservation and restoration. Diversity 15(8):896
Wani ZA, Ridwan Q, Khan S, Pant S, Siddiqui S, Moustafa M, Ahmad AE, Yassin HM (2022) Changing climatic scenarios anticipate dwindling of suitable habitats for endemic species of Himalaya—predictions of ensemble modelling using *Aconitum heterophyllum* as a model plant. Sustainability 14(14):8491
Wani IA, Kumar V, Verma S, Tasleem Jan A, Rather IA (2020) *Dactylorhiza hatagirea* (D. Don) Soo: a critically endangered perennial orchid from the North-West Himalayas. Plants 9(12):1644
Yadav K, Aggarwal A, Singh N (2013) Arbuscular mycorrhizal fungi induced acclimatization and growth enhancement of *Glycyrrhiza glabra* L.: a potential medicinal plant. Agric Res 2:43–47
Yadav AN, Verma P, Kumar V, Sachan SG, Saxena AK (2017) Extreme cold environments: a suitable niche for selection of novel psychrotrophic microbes for biotechnological applications. Adv Biotechnol Microbiol 2(2):1–4
Ye J, Shan Z, Peng D, Sun M, Niu Y, Liu Y, Zhang Q, Yang Y, Lin Q, Chen J, Zhu R (2023) Identifying gaps in the ex situ conservation of native plant diversity in China. Biol Cons 282:110044
Zegeye H (2017) In situ and ex situ conservation: complementary approaches for maintaining biodiversity. Int J Res Environ Stud 4(1):1–12

Ecology, Phytochemistry, Threat and Conservation Status of *Polygonatum cirrhifolium* (Wall.) Royle

Renu Suyal

Abstract Medicinal plants play an important role in human life by way of not only functioning as traditional medicines, but also as trade commodities and meeting the pharmaceutical demands. The present review focuses on ecology, phytochemistry, threat and conservation status of *Polygonatum cirrhifolium* (Wall.) Royle (Asparagaceae) along with various conservation initiatives taken in the Himalayan region. *P. cirrhifolium* is one among 71 species and 4 varieties of genus of *Polygonatum* with medicinal and therapeutic properties used in ancient traditional medicine. The species has been reported as rich source of flavonoids and phenolic acids, anthraquinones, Caffeic acid, ellagic acid, 3-hydroxy cinnamic acid, and p-coumaric acid and possess antioxidant, antimicrobial, antifungal and antistress properties. The rhizomes are used as an ingredient of popular ayurvedic polyherbal formulations 'Chyavanprash'. However, the availability of species in wild is decreasing continuously due to overexploitation for different commercial purpose, habitat degradation, heavy grazing and trampling by livestock. Therefore, there is an urgent need to develop strategy for its conservation and collaborative initiatives at regional, national, and international levels are required to pool resources, expertise, and knowledge to address the conservation challenges faced by *Polygonatum cirrhifolium* and other medicinal plants in the Himalayan region.

Keywords Maha-meda · Herbal medicines · Himalayan ecosystem · Resilience · Bioactivity · Endangered flora

---

R. Suyal
G.B. Pant National Institute of Himalayan Environment (NIHE), Kosi-Katarmal, Almora, India

R. Suyal (✉)
Centre for Ecology Development and Research (CEDAR), 814, Indra Nagar, Dehradun 248006, India
e-mail: renu@cedarhimalaya.org; renusuyal04@gmail.com

A. P. Mishra et al. (eds.), *Threatened Medicinal Plants in the Indian Himalayan Region*, World Sustainability Series, https://doi.org/10.1007/978-3-031-73687-2_9

Abbreviations

ABTS	2,2′-Azino-bis (3-ethylbenzothiazoline-6-sulfonic acid) radical scavenging assay
DPPH	2,2 Diphenyl 1 picrylhydrazyl radical scavenging assay
GCMS	Gas chromatography–mass spectrometry
mM	Milli Molar
AAE	Ascorbic acid equivalent
IC_{50}	Half maximal inhibitory concentration
CR	Critically endangered
VU	Vulnerable
EN	Endangered
m asl	Meter above sea level
D	Density
IVI	Importance value index
OE	Over exploitation
HD	Habitat degradation
GR	Grazing
RD	Restricted distribution
GAs	Gibberellic acid
°C	Degree centigrade
Mg	Milli gram
L	Litre
Kg	Kilo gram
NAA	Naphthalene acetic acid
ppm	Part(s) per million
BAP	6-Benzylaminopurine

1 Introduction

The genus *Polygonatum* Mill. (Family-Asparagaceae) comprises about 71 species and 4 varieties distributed throughout the temperate Northern Hemisphere (TPL 2017; www.tropicos.org). It is a genus of perennial herbs and has a long history of traditional uses in a wide range of herbal products. Over 37 species of the genus *Polygonatum* have been used in China, India, Japan, Europe, America, etc. in traditional medicine for the treatment of various diseases such as abscess, feebleness, leprosy, aphrodisiac, appetizer, antipyretic, galactagogue, etc. (Zhao et al. 2017). Among others, *Polygonatum cirrhifolium* (Wall.) Royle is one of the species grows in Indian Himalayan Region (IHR) and known as King's Solomon's seal in English. The species has different vernacular names such as Mahameda, Basuchidra, Dharaa, Khakan, Manichhidra, Nakhchechi, Shalyaparni, Subrakand. However, it is popularly known as Maha meda (Balkrishna et al. 2012). The term 'meda' symbolize the mucilage present inside the rhizome of the species. Its different parts like rhizomes, roots, leaves, etc. being utilized for different medicinal purposes. In Ayurveda, the rhizomes of *P. cirrhifolium* are considered as important ingredients of Ashtavarga, a group of eight plant species highly prized for their rejuvenating, aphrodisiac and tonic properties (Balkrishna et al. 2012; Giri et al. 2017). Rhizomes and roots are useful in cough, leprosy, skin diseases, anorexia, fever, sexual debility and other seminal disorders. The species used in different forms as taila (oil), ghritam (medicated clarified butter), churna (powder) and formulation in traditional medicinal system including chyavanprash.

Rhizomes of *P. cirrhifolium* have also been used in traditional Chinese medicine against mastitis, cough, and respiratory disorders and as an antipyretic (Zhao et al. 2017). Other biological activities of the plant rhizome include stimulation of salivary and gastric juice secretion (Son et al. 1990). The roots of *P. cirrhifolium* are reported to be used as a strength-giving food (Watt 1892) and as a detergent (Sarin 1967). *P. cirrhifolium* (roots and rhizomes) has become an important drug in international market and an estimated annual trade of its rhizome is 100–200 metric tons per year (National Medicinal Plant Board, Government of India, 2017). However, availability of this plant in the wild is decreasing continuously due to heavy harvesting for different medicinal and commercial purposes. Restricted distribution in small pockets (Balkrishna et al. 2012; Suyal et al. 2020), loss of habitat, and excessive grazing (Bhatt et al. 2014; Pandey et al. 2019; Suyal et al. 2020) are some of the factors threatens the survival of the species in the wild and therefore there is an urgent need to develop strategy for its conservation. In India, the species is found in temperate Himalayas from Himachal Pradesh at an altitude of 1500–3000 m (Dhaliwal and Sharma 1999) eastwards to Sikkim (1670–4200 m; (Ved et al. 2003)), Manipur (1200 m), Darjeeling (1800–4000 m; (Ghosh and Mallick 2014)), and Uttarakhand (1600–4600 m; Naithani 1984)). Reviews concerning phytochemistry, ethnopharmacology, phytochemical updates, ethnobotanical review on food uses of genus *Polygonatum* (Khan et al. 2012a; Wujisguleng et al. 2012; Khan and Rauf 2015; Zhao et al. 2017) are available. From this review, it is evident that genus contain variety of chemical constituents mainly saponins, phytohormones, glycosides, flavonoids and alkaloids having different pharmacological properties. For instance, Kinganone and 3-ethoxymethyl-5,6,7,8-tetrahydro-8-indolizinone from rhizomes of *P. kingianum* showed antifungal and antibacterial activity (Wang et al. 2009). Similarly, Steroidal saponins from rhizomes of *P. odoratum* showed antimicrobial activity (Wang et al. 2009). Rhizomes of *P. verticillatum* exhibit antimicrobial, antimalarial and antinociceptive activities (Khan et al. 2010, 2012b). All the available review currently lacks information on *P. cirrhifolium*. However, it is important to collect, analyze and synthesize all the available information on the species so as to gap in research can be identified and further potential of species could be promoted. The objective of this chapter is to summarize all the research findings available on various aspects, such as botanical description and distribution, ethnopharmacology, phytochemistry, propagation, and conservation measures of *P. cirrhifolium*. Also, a systematic approach for conservation and sustainable utilization of the species has been suggested. Based on all the compiled information, research gap has also been discussed. This chapter provides the basis for further studies on conservation and development of identifying better therapeutic agents and health products from the species. Keeping in view above, present review focus on its ecology and phytochemistry and various conservation initiatives in the Himalayan region.

2 Study Area

Polygonatum cirrhifolium is found to be distributed in Himalayan region ranging from Bhutan, China, India, and Nepal to Pakistan (Figs. 1 and 2). In Indian Himalayan Region (IHR), the species is recorded from Himachal Pradesh—Rudung, Urni; Michar, Jeori, Ponda (Nair 1977); Kinnaur, Kullu (Chowdhery and Wadhwa 1984); Naggar, Manikaran (Dhaliwal and Sharma 1999); Dohra thach (Butola and Badola 2008); Parvati valley (Sharma et al. 2010); Mandi (Sharma and Samant 2014), Uttarakhand Gulabkoti, Sitapur, Sutol, Nanda devi Biosphere Reserve, Sunderdhunga, Kafni, Khati, Phurkia, Ramgarh, Mukteshwer, Gagar, kilbury, Ralam, Belju, Burfu (Naithani 1984, Gaur 1999; Rana et al. 2013; Manikandan and Srivastava 2015; Rawat et al. 2016; Suyal et al. 2020); Sikkim—Choongtam, Zemu Valley (www.sikenvis.nic.in); Tamze valley (Dahal et al. 2017), Darjeeling—Tiger hills, Neora valley (Ghosh and Mallick 2014) and Manipur (Dzukou valley (Gogoi 2010).

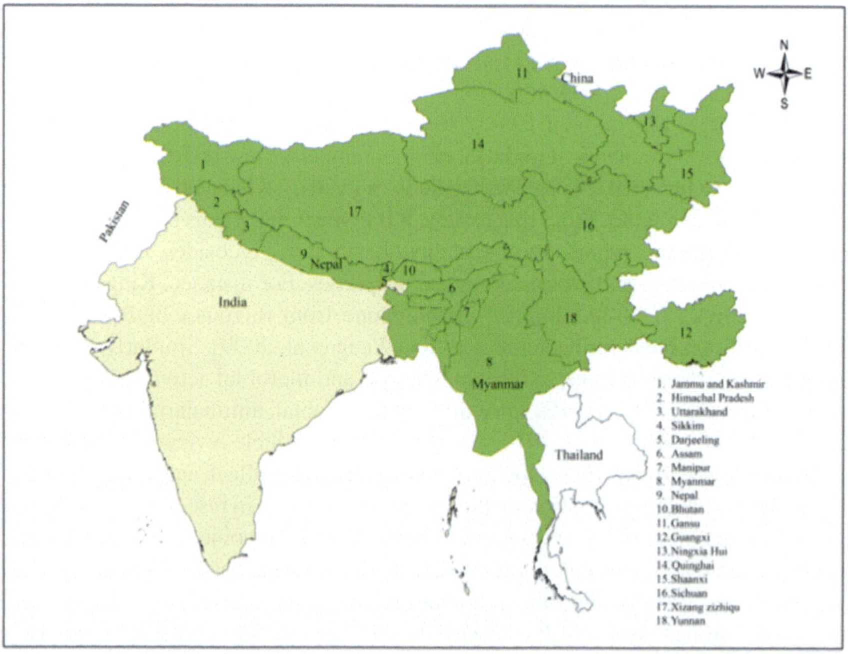

Fig. 1 Distribution range of *P. cirrhifolium* across the globe (*source* www.tropicos.org)

Fig. 2 *Polygonatum cirrhifolium*: flowers, fruits and rhizomes

3 Methodology

In a systematic approach, we conducted an extensive literature review (1967–2023) related to various aspects such as ecology, population, threat and conservation of *P. cirrhifolium*. The search was located, selected, and extracted from scientific journals, books, thesis, herbarium and reports via library and online databases (Google scholar; Wiley and Scopus). The review included the following search terms such as '*P. cirrhifolium*', 'Mahameda', 'Astavarga', 'ethnomedicinal uses', 'diversity and distribution', 'Morphology and taxonomy', 'Phytochemical and pharmacology', 'propagation and cultivation'.

4 Results

4.1 Ecology and Threat Status

Studies on quantitative assessment and ecological studies of the threatened medicinal plant along with the assessment of availability, growth preferences, distribution pattern, diversity along altitudinal gradient, species richness and habitat preference are vital steps for development of conservation measures (Rai et al. 2010; Suyal et al. 2020). However, only few studies are available on ecological aspect confined to Indian Himalayan region (Table 1), which shows low population density across the surveyed populations and indicates poor availability of the species in the wild. The reported studies on *P. cirrhifolium* revealed that the species has limited number of individuals in West Himalaya. Generally, *P. cirrhifolium* occupies an altitudinal range from 1900–4500 m asl and prefers to grow on hill slopes in shady places, oak–*rhododendron* forest, open grassy slopes, under growth of *Juniperus* species (Suyal et al. 2020). Comparative low density and localized distribution, reveals that the target species are on the edge of threat. Dominant associated species reported for *P. cirrhifolium* at most of the sites are *Thalictrum foliolosum* DC., *Arisaema speciosum* (Wall.) Mart. ex Schott & Endl., *Hedychium spicatum* Buch.-Ham. ex Sm., *Rubia cordifolia* L., *Anaphalis contorta* (D. Don) Hook. f., *Valeriana jatamansi* Jones, and *Polygonatum verticillatum* (L.) All. (Suyal et al. 2020). The intensive ecological researches on *P. cirrhifolium* have indicated that the anthropogenic disturbances like deforestation, over exploitation, habitat degradation, habitat fragmentation and grazing and trampling by livestock have been reported to adversely reduce the growth rate of the species (Bhatt et al. 2005) and also responsible for the population decline of other threatened MAPs in IHR (Chauhan et al. 2018) are the reason behind the continuous depletion of the natural populations. Its traded products are most often the roots, rhizomes or tubers. Based on the literature, it was found that the species is Endangered in Himachal Pradesh (Ved et al. 2003; Butola and Badola 2008; Rana and Samant 2011) while in Uttarakhand it is rare and vulnerable (Ved et al. 2003; Bhatt et al. 2014). It has been reported that high habitat preferential regulated by small population size often results in unrestricted exploitations of MAPs from the region, with severe implications for their availability, population decline, and species endangerment (Pandey et al. 2019; Suyal et al. 2020). Therefore, the species requires immediate conservation interventions, as the species with specific habitat requirements have a greater possibility of extinction than species with broad habitat range (Samant 1996).

Table 1 Quantitative assessment of *P. cirrhifolium* in different regions

Area	Altitude (m asl)	Density (ind/m^2)	IVI	Threat status	Threats	References
Kumaun Himalaya	1750–3900	2.20–7.20	11.28–63.38	CR	OE, HD, GR	Lohani (2012)
	1829–3680	0.93–1.90	9.63–28.85	VU	OE, HD, HF	Bhatt et al. (2014)
	2700–4500	1.6	–	VU	OE	Pandey et al. (2019)
	1900–3500	1.20–6.33	56.81–151.66		OE, HD.GR, RD	Suyal et al. (2020)
Garhwal Himalaya	3000–3800	0.12–0.46	–	VU	OE	Negi et al. (2018a)
	3520–3760	0.2–0.8			–	Negi et al. (2018b)
Himachal Pradesh	2000–3000	100/5 location	–	EN	OE, HD, RD	Rana and Samant (2011)
	2600–2800	3.55	–	VU	RD	Butola and Badola (2008)

4.2 Medicinal Uses and Traditional Methods of Application

Polygonatum cirrhifolium is reported to be used either as a single drug or in combination in the Indian systems of medicine (Ayurveda, Siddha, and Unani). Its roots are mainly used in leucorrhea, as blood purifier, aphrodisiac, and rhizomes are used in urogenital disorder, skin eruptions, bronchitis, tuberculosis and menstrual problem (Balkrishna et al. 2012; Boesi 2014). *P. cirrhifolium* has been used traditionally by different ethnic communities of the Northern India. For instance, a rhizome of *P. cirrhifolium* soaked in wine is reported to be used as a tonic for cardiac and respiratory problems. Similarly, leaves and rhizomes are used as vegetables, in cuts and wounds, cold and cough in different parts of Himalayan region (Samant et al. 2007). The available published literature (n = 24) on the species suggests that the species is largely used as aphrodisiac (17%), blood purifier (11%), in respiratory disorder (9%), bone and joint disorder (8%), energy tonic (8%), fever (8%) and skin disease, etc., (Fig. 3) and mostly the rhizomes are collected for medicinal purposes. Method of use of plants varies according to nature of disease. For example, the decoction is generally made by crushing and boiling the plant parts in water and used after cooling (Table 2). Roasted powder of roots with butter; powder of species and combination with other species are taken with milk is used as suboleative, aphrodisiac, cuts, wounds, etc. (Dhyani et al. 2010). Leaves and rhizomes are also used in phlegm and wind disorder by the ethnic group of Baglung District of Nepal (Boesi 2014). The roots of *P. cirrhifolium* is also used by Naxi People, Northwest Yunnan for antirheumatic and Galactagogue properties (Zhang et al. 2015). The Naxi people of southern China uses the rhizomes as health care food. In Pakistan, the species is used as tonic and vulnerary, and leaves are eaten as vegetables (Ambasta 1986).

4.3 Bioactive Compounds and Pharmacology Activities

Phytochemical analysis of methanolic extracts of rhizomes showed that they are rich in phenolic acid, flavonoids, sterols, fatty acids (Table 3). Among the different flavonoids and phenolic acids, anthraquinones, Caffeic acid, ellagic acid, 3-hydroxy cinnamic acid, and p-coumaric acid are found in the species (Giri et al. 2017). Singh and Patra (2019) reported presence of 62 different bioactive compound in different extract of *P. cirrhifolium* such as antioxidant (sitosterol; hexadecanoic acid, ethyl ester), anti-inflammatory (9,12 Octadecadienoic acid, ethyl ester; 9,12,15 Octadecatrienoic acid,2,3-bis [(trimethylsilyl)-oxy]propyl ester; dodecane; sitosterol; cannabinol, pentafluoropropionate) and adaptogenic compounds (3,5-Dicyclohexyl4-hydroxy-benzoic acid-methyl ester and 6,7-Dichloro-5-[(1-ethylpyrrolidin-2-yl)-methyl-amino]-1,3-dimethyl-pyrido[2,3-d]-pyrimidine 2,4 (1H,3H) dione;13docosenami-de; 1mono-linoleoylglycerol trimethylsilyl ether; 9,12,15Octadecatrienoic acid,2,3bis-[(trimethylsilyl)oxy]propyl ester; sitosterol) (Zhao et al. 2017; Liu et al. 2018).

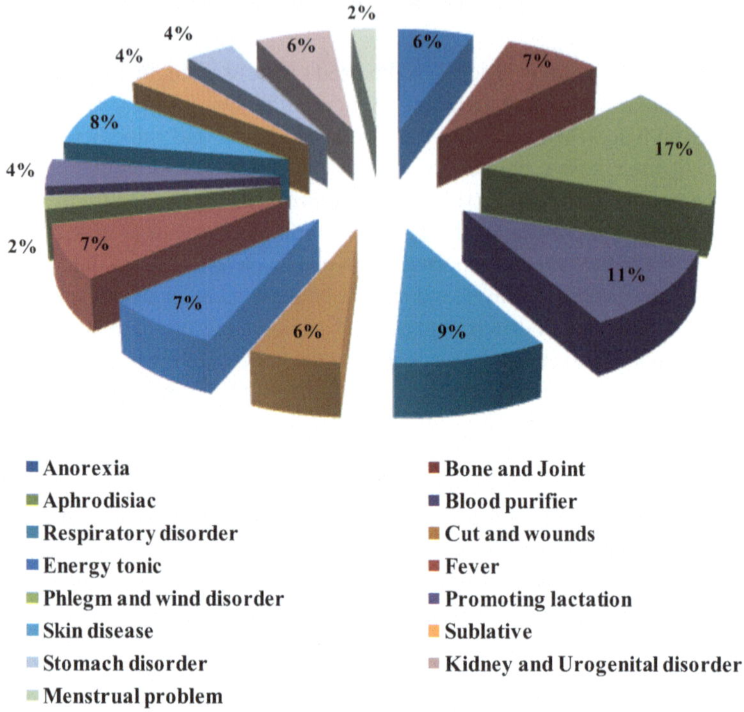

Fig. 3 Reported medicinal uses of *P. cirrhifolium*

Suyal et al. (2022) reported that the rhizomes of species have strong antioxidant potential, as evaluated by various in vitro free radical scavenging and reducing assays such as 2,2-azinobis (3-ethylbenzoline-6-sulfonic acid) radical assay (ABTS: 3.39–5.12 mM AAE/100 g dry weight) and 1,1-diphenyl-2-picrylhydrazyl radical assay (DPPH: 1.09–1.70 mM AAE/100 g dry weight), ferric reducing anti-oxidant properties (FRAPS: 0.58–2.33 mM AAE/100 g dry weight). Singh and Patra (2019) measured the presence free radical scavenging ability of different extracts of *P. cirrhifolium* and reported that the ethanol extract possessed the highest DPPH and ABTS+ scavenging activity (IC50 of 56.16 ± 2.38 and 83.28 ± 1.29 μg/ml, respectively). Giri et al. (2017) reported, the presence of phenolics groups such as caffeic acid (0.09 mg/100 g), 3-hydroxy cinnamic (53.86 mg/100 g), ellagic acid (18.11 mg/100 g) in methanolic extract of the species using HPLC.

Wang et al. (2006) reported that the n-butanol extract of rhizomes of species showed strong anti-microbial activity against *Valsa mali*, *Exserohilum turcicum* and *Verticillium dahliae*. Dongmei et al. (2007) reported that the

Table 2 Traditional methods of application of *P. cirrhifolium*

Condition	Mode of application	References
Phlegm and wind disorder	Cooked leaves and rhizome	Boesi (2014)
Suboleative	Powder of both *P. cirrhifolium* and *P. verticillatum*, *Malaxis musifera* and *Desmodium gangeticum* taken with milk	Dhyani et al. (2010)
Energy tonic	Leaves cooked with onion and potato	Dhyani et al. (2010)
Cuts and wounds	Tuber paste applied to affected area	Dhyani et al. (2010)
Aphrodisiac	Tuber infusion with milk	Dhyani et al. (2010)
Blood purifier	Tuber infusion with milk	Dhyani et al. (2010)
Cuts and wounds	Tuber paste	Nautiyal et al. (2003)
Cough and cold	Root extract	Nautiyal et al. (2003)
Aphrodisiac	Roots are roasted with clarified butter and taken approximately 2 g, three times a day up to 2 months	Rana et al. (2013)
Urogenital disorder, menstrual disorder, Fluid accumulation in joints	Rhizome mixed *Berginia ciliata* extract	Jain (2016)

Table 3 Bioactive compound isolated and identified in *P. cirrhifolium*

Class	Compound	Uses	References
Phenolic acid	Anthraquinones, Caffeic acid, ellagic acid, 3-hydroxy cinnamic acid, and p-coumaric acid	Antioxidant and antimutagenic activity	Giri et al. (2017)
Phytosterols	Diosgenin, γ-Sitosterol, β-sitosterol, Daucosterol	Anti-inflammatory, antioxidant, anticancer agent, antistress, hepatoprotective	Zhao et al. (2017), Wang et al. (2006), Dongmei et al. (2007), Li et al. (2010)
Fatty acid	Linoleic acid; hexadecanoic acid; Octadecadienoic acid; maleamic acid	Antioxidant, anti-inflammatory	Singh and Patra (2019), Virk et al. (2017)
Other compounds	Acetic acids; amino acids; glycosides; proteins; flavonoids; carbohydrates; polysaccharides	Nutritional value	Singh and Patra (2019), Wang et al. (2006)

isolated steroidal saponin compounds from rhizomes, such as (25R)-spirost-5-ene-3β-ol-3-O-α-L-rham-nopyranosy(1 → 2)-[α-L rhamnopyranosyl (1 → 4)]-β-D-glucopyranoside and (25R)-spirost-5-ene-3β-ol-3-O-α-L-rhamnopyranosyl (1 → 4)-β-D-glucopyranoside exhibit higher inhibitory effect on plant pathogenic bacteria.

4.4 Propagation and Conservation Strategies

Propagation and cultivation efforts of *P. cirrhifolium* are at preliminary level. Of the available literature, different methods have been adopted by various conservationists to conserve the species at regional and national levels. Some used nursery techniques (Butola and Badola 2008; Lohani et al. 2011) while, other uses field as well as laboratory conditions (Lattoo et al. 2001; Qazi et al. 2005; Vinay et al. 2011) to propagate the species using different chilling treatments, thermoperiods, photoperiods, organic fertilizers, growth hormones and media. Summary of the propagation techniques reported for *P. cirrhifolium* is given in Table 4.

Lattoo et al. (2001) used different photoperiods (continuous light, light and dark (16 h photoperiod), continuous dark) and thermoperiods: 25/15; 30/20; 20 and 30 °C (±1 °C) to check the phenological patterns of seed germination under laboratory and field condition (sand + soil mixture in pots). Results from this study revealed that germination (97%) occurred only under continuous darkness at a constant temperature of 30 °C and alternating temperature regime of 30/20 °C. Study also reveals that germination is absolutely photosensitive and indicates that seeds of *P. cirrhifolium* exhibits epicotyl dormancy and require two or more growing seasons to complete germination. However, Butola and Badola (2008) reported 71% germination when seeds were incubated at 25 °C under alternate light regimes (16 h light and 8 h dark). Further, that germination in *P. cirrhifolium* is considered as moderate and the reason of less germination is dormancy and hard seed coat. Later, Vinay et al. (2011) studied seed germination in *P. cirrhifolium* to check the influence of different chilling treatments i.e. −20 °C and 4 °C (15, 30, 60, 120 days), 25 °C (control) and photoperiod (light 16 h/dark 8 h) along with 100 ppm of GA_3 on seeds. Study suggested that GA_3 100 ppm at 25 °C maximize germination to 80%.

To study the vegetative propagation, Lohani et al. (2011) used randomized block design to check the effect of different organic fertilizers (forest litter, vermicompost, manure) and nursery beds (plain, rows and furrows) with a spacing of 30 × 30 cm. In all the combinations 100% sprouting were observed. The results show that yield was lower in control beds when compared with forest litter, farmyard manure and vermicompost in all type of beds prepared in rows, furrows and plain. The yield was highest at beds with furrows supplemented with forest litter. A study by Butola and Badola (2008) reported 73% survival of plants propagated through rhizomes in nursery open beds (size 1 × 3 m^2) supplemented with sandy loam soil (pH: 6.7; Moisture content: 40%, water holding capacity: 57%, and bulk density: 0.92 mg/cm^3); farmyard manure and vermi-compost (3 kg m^{-2}).

Table 4 Various research and results of in-vitro propagation conducted in *P. cirrhifolium*

Explant	Method	Response	References
Seed	Photoperiods (continuous light, light and dark (16 h photoperiod), continuous dark) and thermoperiods: 25/15; 30/20; 20 and 30 °C (±1 °C)	97% germination at 30 °C and alternating temperature regime of 30/20 °C	Lattoo et al. (2001)
Seed	Seeds incubated at 25 °C under alternate light regimes (16 h light and 8 h dark)	71% germination	Butola and Badola (2008)
Seeds	Chilling treatments, Gibberellic acid	80% germination at 25 °C in seeds treated with GA100 ppm	Vinay et al. (2011)
Rhizome	Randomized block design, organic fertilizers (forest litter, vermicompost, manure), spacing 30 × 30 cm	100% sprouting	Lohani et al. (2011)
Rhizomes	Nursery open beds (1 × 3 m^2), sandy loam soil (pH: 6.7; Moisture content: 40%, water holding capacity: 57%, and bulk density: 0.92 mg/cm^3); farmyard manure and vermi-compost (3 kg m^{-2})	73% survival	Butola and Badola (2008)
Seeds	GAs (2–50 mg/L); BAP (1–6 mg/L) and NAA (0.1–1.0.1 mg/L) + MS basal culture medium; GA (1–20 mg/L) + 3 mg/L of BAP + 1.0 mg/L of NAA	98.53% germination at 25 mg/L of GA_3; 97.03 of axillary buds producing leaves at 20 mg/L of GA_3	Qazi et al. (2005)

In vitro propagation of seeds of *p. cirrhifolium* was studied by Qazi et al. (2005) using different levels of GAs (2–50 mg/L) and incubated at 30 and 20/30 °C under continuous darkness at a RH of 60% to check its effect on germination percentage and duration of germination; release of epicotyl, coleoptile radicle and days to synchronization, various combinations of BAP (1–6 mg/L) and NAA (0.1–1.0.1 mg/L) supplemented to MS basal culture medium and (v) varying levels of GA. (1–20 mg/L) in combination with 3 mg/L of BAP and 1.0 mg/L of NAA to study the effect of various levels of GAs in combination with 2 mg/L of BAP and 1.0 mg/L of NAA on de novo axillary bud differentiation and the release of foliage leaves. Results revealed (i) 98.53% germination at 25 mg/L of GA_3 with a minimum duration of 7–23 days at 50 mg/L of GA_3 respectively at 30 °C, (ii) in 20/30 °C incubation maximum germination (62.12%) obtained at 50 mg/L of GA_3 (iii) 100 mg/L of GA_3 results in 100%

synchronization after 11 days of culture, (iv) synergistic combination of BAP and NAA is effective at all levels, BAP to NAA ratio of 3:1 yields the optimum response resulting in release of first foliage leaves in 85.15% of secondary explants, (v) 15 mg/L of GA_3 produces 100% axillary bud per explant with 11.67 mean numbers of axillary buds. Maximum percentage (97.03) of axillary buds producing leaves at 20 mg/L of GA_3.

5 Discussion and Conclusion

Medicinal plants play an important role in human life by way of not only functioning as traditional medicines, but also as trade commodities and meeting the pharmaceutical demands (Pant and Pant 2011). The majority of MAPs in the Himalayan region have strong habitat preferences that are governed by limited population sizes and high monetary value (Suyal et al. 2020; Rawat and Jugran 2023). This frequently leads to unrestrained exploitation of MAPs in the region, with serious consequences for their availability, population reduction, and species endangerment (Sekar et al. 2015). During last two decades, there has been a tremendous change in overall number of threatened plant species globally, and over 15,274 plants have now been placed in different threat categories [i.e., Critically Endangered (3229); Endangered (5727); Vulnerable (6818)] (IUCN 2019). With the increasing demand for herbal drugs the use of medicinal plants is growing rapidly all across the world. The global import of medicinal plants is around 660,261 metric tons (MT) having value of US $2.9 billion and export is 5,37,149 MT valued at US $3 billion (Nguyen Viet et al. 2019). This is also true in case of *P. cirrhifolium*, which has been utilized in the traditional and modern medicinal system and in many polyherbal formulations (Balkrishna et al. 2012).

Polygonatum cirrhifolium roots and rhizomes has become an important drug in international market and an estimated annual trade of its rhizome is 100–200 metric tons per year (National Medicinal Plant Board, Government of India, 2017). However, species can be considered neglected by the scientific community. The availability of this plant in the wild is decreasing continuously due to heavy harvesting for different medicinal and commercial purposes. Restricted distribution in small pockets, loss of habitat, and excessive grazing (Balkrishna et al. 2012; Pandey et al. 2019; Lohani 2012; Suyal et al. 2020) are some of the factors threatens the survival of the species in the wild and therefore, there is an urgent need to develop strategy for its conservation.

Tissue culture, seed germination, and vegetative propagation techniques for rapid multiplication are available in the species (Lattoo et al. 2001; Qazi et al. 2005; Lohani et al. 2011; Vinay et al. 2011). However, due to insufficient germination response, cultivation of the species through conventional propagation (from seed) is often not chosen. However, for large scale cultivation of species, a robust and quick propagation technique with improved quality traits need to be developed for commercialization along with the involvement of agronomic practices to help farmers to cultivate this species for its sustainable utilization and reducing pressure on wild populations.

Initiatives for conservation of the targeted species are currently underway at some regional and national organizations such as high-altitude Plant Physiology Research Centre (HAPPRC) and a field demonstration and learning site at Sri Narayan Ashram (SNA)—an initiative of the G.B. Pant National Institute of Himalayan Environment (GBP-NIHE). Therefore, collaborative initiatives at regional, national, and international levels are required to pool resources, expertise, and knowledge to address the conservation challenges faced by *Polygonatum cirrhifolium* and other medicinal plants in the Himalayan region.

Therefore, it can be concluded that species have tremendous therapeutic applications and has proven vitality strengthening and rejuvenating properties (Ghosh and Mallick 2014). In-depth research needs to be carried out in this aspect to promote the species as a nutritive or functional food.

Therefore, conservation efforts must be focused on preserving the natural habitats of these plants, promoting sustainable harvesting practices, and establishing gene banks to safeguard genetic diversity. Additionally, traditional knowledge held by indigenous communities about the uses and cultivation of medicinal plants should be respected and integrated into conservation strategies, as these communities often play a vital role in the preservation of these species.

References

Ambasta SP (1986) The useful plants of India. CSIR, New Delhi

Balkrishna A, Srivastava A, Mishra RK, Patel SP, Vashistha RK, Singh A et al (2012) Astavarga plants—threatened medicinal herbs of the North-West Himalaya. Int J Med Aromat Plants 2(4):661–676. ISSN 2249–4340

Bhatt A, Rawal RS, Dhar U (2005) Germination improvement in *Swertia angustifolia*: a high value medicinal plant of Himalaya. Curr Sci 1008–1012

Bhatt D, Ravi K, Tewari LM, Joshi GC (2014) *Polygonatum cirrhifolium* Royle and *Polygonatum verticillatum* (L.) Allioni: status assessment and medicinal uses in Uttarakhand, India. J Med Plants Res 8(5):253–259. https://doi.org/10.5897/JMPR2013.5234

Boesi A (2014) Traditional knowledge of wild food plants in a few Tibetan communities, pp 1–19

Butola JS, Badola HK (2008) Threatened Himalayan medicinal plants and their conservation in Himachal Pradesh. J Trop Med Plants 9(1):125–142

Chauhan HK, Bisht AK, Bhatt ID, Bhatt A, Gallacher D, Santo A (2018) Population changes of *Trillium govanianum* (Melanthiaceae) amid altered indigenous harvesting practices in the Indian Himalayas. J Ethnopharmacol 213:302–310

Chowdhery HJ, Wadhwa BM (1984) Flora of Himachal Pradesh. Botanical Survey of India, Calcutta, p 163

Dahal S, Sharma TP, Borthakur SK (2017) Database on medicinal plants of Tamze Medicinal Plants Conservation Area (MPCA) of Sikkim Himalaya, India. NeBIO Int J Environ Biodivers 8:45–56

Dhaliwal DS, Sharma M (1999) Flora of Kullu district. Bishen Singh Mahendra Pal Singh, 744p

Dhyani A, Nautiyal BP, Nautiyal MC (2010) Importance of Astavarga plants in traditional systems of medicine in Garhwal, Indian Himalaya. Int J Biodivers Sci Ecosyst Serv Manag 6(1–2):13–19

Dongmei W, Jingfang Z, Xiaoming L (2007) Steroid saponins of *Polygonatum cirrhifolium* root and their antisepsis activity. Scientia Silvae Sinicae 43(8):91

Gaur RD (1999) Flora of the district Garhwal, North West Himalaya. Transmedia, 811pp

Ghosh DK, Mallick JK (2014) Flora of Darjeeling Himalayas and Foothills: angiosperms. Research Circle, Forest Directorate, Government of West Bengal & Bishen Singh Mahendra Pal Singh

Giri L, Belwal T, Bahukhandi A, Suyal R, Bhatt ID, Rawal RS, Nandi SK (2017) Oxidative DNA damage protective activity and antioxidant potential of Ashtvarga species growing in the Indian Himalayan Region. Ind Crops Prod 102:173–179

Gogoi R (2010) The family Liliaceae in Manipur, India. Bull Arunachal For Res 26(1 and 2):73–81

IUCN (2019) IUCN Red list version 2019-3. www.iucnredlist.org. Accessed 10 Dec 2019

Jain AK (2016) Indian ethnobotany: emerging trends. Scientific Publishers

Khan H, Rauf A (2015) Phytochemistry of genus Polygonatum: a review. Am J Biomed Life Sci 3(2–1):5–20

Khan H, Saeed M, Khan MA, Dar A, Khan I (2010) The antinociceptive activity of *Polygonatum verticillatum* rhizomes in pain models. J Ethnopharmacol 127(2):521–527

Khan H, Saeed M, Muhammad N (2012a) Pharmacological and phytochemical updates of genus Polygonatum. Phytopharmacology 3(2):286–308

Khan H, Saeed M, Khan MA, Khan I, Ahmad M, Muhammad N, Khan A (2012b) Antimalarial and free radical scavenging activities of rhizomes of *Polygonatum verticillatum* supported by isolated metabolites. Med Chem Res 21(7):1278–1282

Lattoo SK, Dhar AK, Jasrotia A (2001) Epicotyl seed dormancy and phenology of germination in *Polygonatum cirrhifolium* Royle. Curr Sci 81(11):1414–1417

Li D, Wang D, Zhang J, Liu J, Yu Z, Li J (2010) Effects of water-soluble compounds of *Polygonatum cirrhifolium* rhizomes on plant pathogenic fungi. Allelopathy J 25(2)

Liu Q, Li W, Nagata K, Fu H, Okada S, Tanabe I et al (2018) Isolation, structural elucidation, and liquid chromatography–mass spectrometry analysis of steroidal glycosides from Polygonatum odoratum. J Agric Food Chem 66(2):521–531

Lohani N (2012) Population studies, habitat assessment and conservation strategies of four medicinal herbs of Kumaun (Thesis). Kumaun University, Nainital, Uttarakhand

Lohani N, Kumar R, Tewari LM, Joshi GC (2011) Effect of different organic treatments on *ex situ* conservation of *Polygonatum cirrhifolium* Royle. Int J Biodivers Sci Ecosyst Serv Manag 7(2):134–140

Manikandan R, Srivastava SK (2015) Diversity, medicinal and threatened plants in Govind Pashu Vihar Wildlife Sanctuary, Western Himalaya. Indian For 141(9):966–973

Nair NC (1977) Flora of Bashahr Himalayas. International Bioscience Publishers, 360 pp

Naithani BD (1984) Flora of Chamoli, vol I & II. Botanical Survey of India, Howrah

Nautiyal S, Maikhuri RK, Rao KS, Saxena KG (2003) Ethnobotany of the Tolchha Bhotiya tribe of the buffer zone villages in Nandadevi Biosphere Reserve, India

Negi VS, Giri L, Sekar KC (2018a) Floristic diversity, community composition and structure in Nanda Devi National Park after prohibition of human activities, Western Himalaya, India. Curr Sci 115(6):1056

Negi VS, Maikhuri RK, Maletha A, Phondani PC (2018b) Ethnobotanical knowledge and population density of threatened medicinal plants of Nanda Devi Biosphere Reserve, Western Himalaya, India. Iran J Sci Technol Trans A Sci 43(1):63–73

Nguyen Viet D, Le Ba V, Nguyen Duy T, Pham Thi VA, Tran Thi H, Le Canh VC, Long GB, Kim YH, Tuan Anh HL (2019) Bioactive compounds from the aerial parts of *Hypericum sampsonii*. Nat Prod Res 1–3

Pandey A, Chandra Sekar K, Joshi B, Rawal RS (2019) Threat assessment of high-value medicinal plants of cold desert areas in Johar valley, Kailash Sacred Landscape, India. Plant Biosyst Int J Deal All Aspects Plant Biol 153(1):39–47

Pant S, Pant VS (2011) Status and conservation management strategies for threatened plants of Jammu and Kashmir. J Phytol 3(7):50–56

Qazi GN, Lattoo SK, Dhar AK, Purohit P, Raina RK, Dhar RS (2005) U.S. Patent No. 6,905,876. U.S. Patent and Trademark Office, Washington, DC

Rai ID, Adhikari BS, Rawat GS (2010) A unique patch of timberline ecotone with three species of Lady's slipper orchids in Garhwal Himalaya, India. J Threat Taxa 2(3):766–769

Rana Man S, Samant SS (2011) Diversity, indigenous uses and conservation status of medicinal plants in Manali Wildlife Sanctuary, North Western Himalayas. Indian J Tradit Knowl 10(3):439–459

Rana CS, Ballabha R, Tiwari JK, Dangwal LR (2013) An ethnobotanical study of the plant resources in the Nanda Devi Biosphere Reserve (a world heritage site), Uttarakhand, India. J Ethnobiol Tradit Med 120:591–601

Rawat S, Jugran AK (2023) Bioactive compounds and biological activities of *Roscoea purpurea* Sm. In: Murthy HN et al (eds) Bioactive compounds in the storage organs of plants, reference series in phytochemistry. https://doi.org/10.1007/978-3-031-29006-0_31-1

Rawat DS, Tiwari JK, Tiwari P, Singh H (2016) Floristic diversity of montane zone of Western Ramganga Valley, Uttarakhand, India. J Econ Taxon Bot 40(3–4):104–125

Samant SS (1996) Conservation of rare endangered plants: the context of Nanda Devi Biosphere Reserve. In: Conservation and management of biological resources in Himalaya. Oxford and IBH Publishing Company

Samant SS, Butola JS, Sharma A (2007) Assessment of diversity, distribution, conservation status and preparation of management plan for medicinal plants in the catchment area of parbati hydroelectric project stage—III in Northwestern Himalaya. 4(1):34–56

Sarin YK (1967) A survey of vegetable raw material resources of Lahaul. Indian Forester 93(7):489–499

Sekar KC, Lalit G, Aseesh P, Srivastava SK (2015) A note on distribution of Juniperus semiglobosa in Uttarakhand, India. Indian J For 38(1):79–80

Sharma P, Samant SS (2014) Assessment of fuel resource diversity and utilization pattern in Nargu Wildlife Sanctuary of Himachal Pradesh, NW Himalaya. Int J Biodivers Conserv 6(1):17–27

Sharma PK, Chauhan NS, Lal B (2010) Conservation of phyto-diversity of Parvati Valley in Northwestern Himalayas of Himachal Pradesh-India. In: Medicinal plants of the Himalayas: advances and insights. Global Science Books, UK, p 47

Singh SK, Patra A (2019) Evaluation of adaptogenic potential of *Polygonatum cirrhifolium* (Wall.) Royle: in vitro, in vivo and in silico studies. S Afr J Bot 121:159–177

Son KH, Do JC, Kang SS (1990) Steroidal saponins from the rhizomes of *Polygonatum sibiricum.* J Nat Prod 53(2):333–339

Suyal R, Bhatt D, Rawal RS, Tewari LM (2020) Status of two threatened Astavarga herbs, *Polygonatum cirrhifolium* and *Malaxis muscifera*, in west Himalaya: conservation implications. Proc Natl Acad Sci India Sect B Biol Sci 90(3):695–704

Suyal R, Jugran AK, Bhatt ID, Rawal RS (2022) Assessment of genetic diversity, population structure and phytochemical variations in Polygonatum cirrhifolium (Wall.) Royle: an endangered medicinal herb. Genet Resour Crop Evol 69(7):2383–2397

TPL (2017). The plant list. http://www.theplantlist.org/. Last accessed 22 Aug 2017

Tropicos.org. Missouri Botanical Garden. http://www.tropicos.org/Name/18404756. Last accessed on 04 Jul 2018

Ved DK, Kinhal GA, Ravikumar K, Prabhakaran V, Ghate U, Sankar RV, Indresha JH (2003) CAMP report: conservation assessment and management prioritisation for the medicinal plants of Jammu & Kashmir, Himachal Pradesh & Uttaranchal. In: Workshop, Shimla, Himachal Pradesh. FRLHT, Bangalore, India, p 206

Vinay P, Hemlata B, Nautiyal MC (2011) Seed germination enhancement in high altitude medicinal plants of Garhwal Himalaya by some pre-sowing treatments. Res J Seed Sci 4(4):199–205

Virk JK, Gupta V, Kumar S, Singh R, Bansal P (2017) Astavarga plants—suffering a triple standardization syndrome. J Tradit Complement Med 7(4):392–399

Wang D, Zhu W, Chen G (2006) The preliminary study on antibacterial activity of extracts from *Polygonatum cirrhifolium* Root. J Northwest For Univ 21(1):126

Wang D, Li D, Zhu W, Zhang J, Peng P (2009) Steroidal saponins from the rhizomes of *Polygonatum odoratum.* Nat Prod Res 23(10):940–947

Watt G (1892) A dictionary of the economic products of India, vol 6. Superintendent of Government Printing

Wujisguleng W, Liu Y, Long C (2012) Ethnobotanical review of food uses of Polygonatum (Convallariaceae) in China. Acta Societatis Botanicorum Poloniae 81(4)
Zhang Z, He JS, Li J, Tang Z (2015) Distribution and conservation of threatened plants in China. Biol Conserv 192:454–460
Zhao P, Zhao C, Li X, Gao Q, Huang L, Xiao P, Gao W (2017) The genus Polygonatum: a review of ethnopharmacology, phytochemistry and pharmacology. J Ethnopharmacol 214:274–291

From Wild to Cultivated: Unraveling the Potential of *Swertia chirayita* (Roxb.) H.Karst. for Crop Improvement and Conservation

Bandan Thapa, Kanica Upadhyay, Binoy Chhetri, Natasha Gurung, and Nirmal Kumar Meena

Abstract The current communication sheds light on investigating the possibility of commercialization and conservation of *Swertia chirayita*. It aims to give comprehensive review of the current understanding of the ecological significance, traditional use, and potential commercial worth of *S. chirayita*. By analyzing the available literature and highlighting the challenges and possibilities inherent in production of *S. chirayita*, it hopes to promote sustainable practices that can boost the plant's agricultural and conservation potential. The only viable option for sustainable utilisation is to engage in commercial cultivation, which will also alleviate the strain on its natural population. However, in order to make the cultivation of this plant cost-effective, it is necessary to develop genetically enhanced variants that have a large biomass and active content. It discusses the challenges of achieving sustainable cultivation of *S. chirayita*, emphasises the significance of plant conservation, and presents the most recent advancements in crop enhancement research pertaining to this particular species.

B. Thapa
AICRP on Medicinal and Aromatic Plants and Betelvine, Regional Research Station, Hill Zone, Uttar Banga Krishi Viswavidyalaya, Kalimpong 734301, West Bengal, India

K. Upadhyay (✉)
Department of Forest Products and Utilization, College of Horticulture and Forestry, Jhalawar 326001, Rajasthan, India
e-mail: kanicaupadhyay@gmail.com

B. Chhetri
Regional Research Station, Hill Zone, Uttar Banga Krishi Viswavidyalaya, Kalimpong 734301, West Bengal, India

N. Gurung
ICAR - Indian Agricultural Research Station, Regional Station, Kalimpong 734301, West Bengal, India

N. K. Meena
Department of Fruit Science, College of Horticulture and Forestry, Jhalawar 326001, Rajasthan, India

A. P. Mishra et al. (eds.), *Threatened Medicinal Plants in the Indian Himalayan Region*, World Sustainability Series, https://doi.org/10.1007/978-3-031-73687-2_10

Keywords Conservation · Crop improvement · Cultivation · Medicinal plants · Sustainable agriculture

1 Introduction

1.1 Overview of Swertia chirayita and Its Significance

Swertia chirayita, a temperate Himalayan medicinal plant is widely recognised as the most significant species within the *Swertia* genus in India, owing to its notable therapeutic properties. The medicinal application of this plant has been recorded in various authoritative sources, including the Indian pharmaceutical codex, the British and American pharmacopoeias, as well as esteemed traditional medical systems such as Ayurveda, Unani, and Siddha (Joshi and Dhawan 2005). This species of medicinal plant is also categorised as endangered in the Indian Red Data Book and vulnerable in the Darjeeling and Sikkim Himalaya (Thapa et al. 2021).

Due to the high domestic and international demand for its raw material, the National Medicinal Plant Board (NMPB) has prioritized this species among the 32 medicinal species for intensive cultivation and has also given farmers a 75% subsidy to encourage its cultivation (Kala and Sajwan 2007). According to NMPB, Ministry of AYUSH, the anticipated Annual Trade (MT) of *S. chirayita* is 500–1000 (MT), with the current market price of its rhizome being Rs. 300 per kg.

1.2 Importance of Crop Improvement and Conservation Efforts

Crop improvement and conservation works as hand in glove for the sustainability of endangered and threatened medicinal plant species. The assessment of variability within the existing germplasm of the genus is essential and relies on the examination of morphological and genetic aspects (Samaddar et al. 2014). To effectively enhance resource utilization and mitigate adverse effects on wild populations, it is imperative to assess many chemotypes of *S. chirayita* across diverse phytogeographical locations. By doing thorough biochemical screening, it is possible to identify populations that are superior in meeting the broad commercial demand, revealing the important contribution of species other than *S. chirayita.*

The majority of plants are currently gathered from the wild by local traders. This widespread and unregulated collection has resulted in a decline in plant populations and poses a subsequent risk of extinction (Joshi and Joshi 2008). Cultivation must be stepped up to fulfill the rising demand in domestic and international wholesale markets for raw plant materials. For decades, people have been calling for medicinal plants to be cultivated so that a steady supply of raw materials could be made available for industry and local use, relieving some of the strain on the wild population (Dobremez 1976).

1.3 Scope of the Work

The major objective of this communication is to shed light on exploring the possibility of commercialization and conservation of *S. chirayita.* This article aims to give a comprehensive review of the current understanding of the ecological significance, traditional use, and potential commercial worth of *S. chirayita.* By analyzing the available literature and highlighting the challenges and possibilities inherent in *S. chirayita's* production, it hopes to promote sustainable practices that can boost the plant's agricultural and conservation potential. It talks about how difficult it is to grow *S. chirayita* sustainably, how important it is to preserve the plant, and the latest findings in crop enhancement research for this species and how innovative techniques needs to be adopted. The review is useful for a variety of audiences, including researchers, farmers, and policymakers.

2 Taxonomy, Distribution, and Ecological Status of *S. chirayita*

2.1 Taxonomic Classification and Species Description

Swertia L. (Gentianaceae-Gentianeae-Swertiinae) is a taxonomically distinct yet morphologically varied genus. *S. chirayita* (Roxb.) H.Karst. also known as *Swertia chirata*, Buch.-Ham., *Ophelia chirata* Grisebach., *Agathotes chirayita* Don., *Gentiana chirayita* Roxburgh., and *G. floribunda* Don. in the literature. The genus *Swertia* is named after Dutch gardener and illustrator Emanuel Sweert (sometimes spelt *Swert*, 1552–1612), who published an important florilegium in the early 1600s. The plant, on the other hand, gained its specific name from the '*kirat*' tribals and is also known as '*chirata*' or '*chirayita*'.

Different populations of *S. bimaculata, S. chirayita, S. dilatata, S. nervosa,* and *S. paniculata* were collected and documented from Darjeeling District (West Bengal), Sikkim, and Arunachal Pradesh (Pandey et al. 2012). There is no coherence in the literature citing *S. chirayita's* habit. Some authors refer to it as an annual (Kirtikar and Basu 1984), while others identified it as a biennial or pluri-annual (Raina et al. 2013). It is an erect, thickly branched plant that grows to a height of 3–5.5 feet, and has a dark green to purple stem, and yellow roots (Bentley and Trimen 1880; Joshi and Dhawan 2005). It takes 3–4 months for the seeds to germinate, and then another year and a half or more for the plant to mature through its vegetative stage. Large

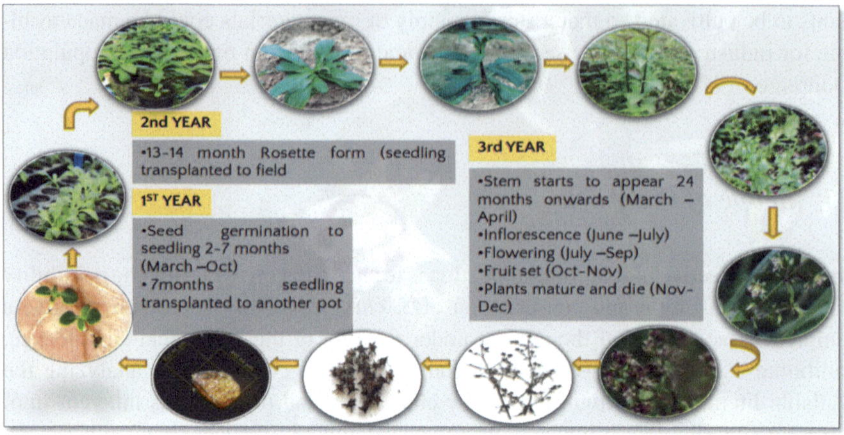

Fig. 1 Pluri-annual life cycle of *Swertia chirayita*

radical rosette leaves (20–25 cm in length) persist throughout the vegetative stage for over 2 years (Fig. 1). The lanceolate, opposite, sessile, 5–7 nerved leaves are 4 cm in length and are acuminate, cordate at the base. These leaves start to degrade and dry out until the stem emerges 3 years after sowing. After the stalk emerges, a green panicle bears little cauline leaves and a profusion of flowers. The root exhibits an easy morphology, characterised by its yellowish coloration, oblique or geniculate shape, tapering and truncated structure, with an average thickness of around half an inch (Bentley and Trimen 1880; Scartezzini and Speroni 2000). The flowers exhibit characteristics such as tiny size, high abundance, tetramerous arrangement, green-yellow coloration, and the presence of purple and green or white hairs on large leafy panicles. Flowers bloom in the third year, from late September to early October, followed by fruits from late October to November. Cross pollination is the mode of pollination for the $2n = 26$ diploid chromosomes (Khoshoo and Tandon 1963; Raina et al. 2013; Samaddar et al. 2014; Chaudhuri et al. 2007; Verma et al. 2020).

2.2 *Geographical Distribution, Habitat and Current Ecological Status*

Swertia chirayita is distributed in the mountains of tropical Asia, Europe, America and Africa. Regionalization of world flora by Takhtajan (1978) suggested *Swertia* to be present in 14 regions of the world, with the major distribution center being in the Himalaya and South-West China. The genus mostly occurs in alpine or temperate habitats in Asia, Africa and North America. The circumscription of the genus has often been debated, resulting disagreement amongst taxonomists due to the morphological similarities (nectariferous and rotate corolla lobes) among the species of

Swertia and the related genera. Over 32 *Swertia* species are found in India at high altitude (1,200–3,000 m) in the temperate Himalayan region stretching from Kashmir to Bhutan, as well as in the Khasi hills and Western Ghats, with 16 being documented from the North West Himalaya. Eleven of the documented northwest Himalayan *Swertia* (Misra et al. 2010) have medicinal characteristics, as do roughly 13–14 species from India's north eastern states.

Swertia chirayita grows best on slopes that face north (Sharma et al. 2011). It grows on a south-facing slope between 1500–3000 m elevation. It falls below 1500 m on the north facing slope whereas in general, a range of 2000 m altitude is preferred. It grows well in acidic soil with a pH of 4.7–5.5. The leading associate species differed between populations, but some of the key associated species of *S. chirayita* were *Anaphalis* sp. and *Anemone* sp. Other associated species with which varied with altitude and with western and eastern Himalayas region were *Desmodium* sp., *Elsholtzia* sp., *Fragaria* sp., *Oxalis corniculata, Stachys sericea, Polygonum amplexicaule, Ageratina adenophora, Gleichenia gigantea,* and *Artemisia vulgaris* (Pyakurel 2008). *S. chirayita* was found to grow most frequently among tree roots, followed by open grassy slopes in the eastern Himalaya, whereas bamboo canopies, rocky slopes, and wet grassy slopes were the least preferred microhabitats, possibly due to the low interspecific competition for sunlight, moisture, and nutrients reported by Kalliovirta et al. (2006). In the western Himalaya, almost all populations of *S. chirayita* studied were discovered growing under the canopy of *Acer* and *Quercus* mixed forests, primarily on a southeast aspect (Bhatt et al. 2006). *S. chirayita* is thought to be endemic to the Himalaya (Samant et al. 1998a, b). According to Balick et al. (1996), the extinction of even one population of the plant can result in the loss of all of its unique genetics, phytochemicals, and associated traits. Increased intensity of forest resource use and grazing leads in trampling, habitat loss during road clearing are among anthropogenic pressures causing damage to *S. chirayta* plants or plant parts, thus compromising the species' survival impacting its supply. Changes in land use, particularly agricultural intensification, are a key cause of falling plant populations, since *S. chirayta* plants and other weed species are uprooted while preparing land for agriculture. In human-modified landscapes, habitat degradation is seen as a primary driver of species extinction (Pradhan and Badola 2015). It's missing a discussion of current and emerging threats to *S. chirayita's* habitat and ecological status. *S. chirayita* is thought to be endemic to the Himalaya (Samant et al. 1998a, b). According to Balick et al. (1996), the extinction of even one population of the plant can result in the loss of all of its unique genetics, phytochemicals, and associated traits. Increased intensity of forest resource use and grazing leads in trampling, habitat loss during road clearing are among anthropogenic pressures causing damage to *chirayita* plants or plant parts, thus compromising the species' survival impacting its supply. Changes in land use, particularly agricultural intensification, are a key cause of falling plant populations, since *chirayta* plants and other weed species are uprooted while preparing land for agriculture. In human-modified landscapes, habitat degradation is seen as a primary driver of species extinction (Yokogawa et al. 2013; Pradhan and Badola 2015).

As the entire plant is used for medical purposes (Kirtikar and Basu 1984; Nadkarni 1998), individuals are uprooted indiscriminately, resulting in low densities in each population. Furthermore, removing the entire plant prior to seed maturation diminishes the likelihood of seed development for future regeneration (Sheldon et al. 1997). The severely uncommon status of *S. chirayita* is reflected in its low population density and localized distribution in distinct pockets (habitats). This has conservation significance since species with specialized habitat requirements are more likely to become extinct than species with a wide range of habitat requirements (Samant et al. 1996).

3 Traditional Uses and Medicinal Properties of *S. chirayita*

3.1 Indigenous Knowledge and Historical Use

Swertia chirayita's therapeutic efficacy is formally registered and recognized in the Indian pharmaceutical codex, the British pharmacopoeia, and other Indian traditional systems of medicine, including Ayurveda, Unani, Siddha, and Tibetan medicine (Joshi and Dhawan 2005). Traditional medicine uses the entire plant of *S. chirayita* to cure fever, malaria, cough, cold, stomach ache, skin ailments, intestinal worms, and diarrhea (Qayum et al. 2016; Pradhan and Badola 2008). The species has long been utilized as medicine, both in formal and informal institutions. The species is advised for lowering fever (jvaraghna) and purifying breast milk (stanyasodhana) in the Charaka Samhita Sutra (third to second century BCE). During pregnancy, the herb reduces inflammation and relaxes the uterus (Kirtikar and Basu 1984). The herb's root is traditionally thought to be the most potent part, and it is used to treat fever, cough, and cold (Joshi and Dhawan 2007). Water taken from the entire plant is drunk to treat liver problems, persistent fever, bronchial asthma, dyspepsia, and debility. It is also used to cure diarrhea and as a febrifuge, laxative, stomachic, anthelmintic, tonic, and febrifuge. Fever is treated with a fresh-shoot decoction paste. As a general health tonic, dried-shoot powder is ingested after being dissolved in warm milk (Singh et al. 2002; Joshi and Joshi 2008).

Ayurvedic pharmacology (Joshi 2000) describes chirata as having a bitter taste (rasa). *Chirata's* thermal activity (virya) is classified as cooling (shita). Chirata is light (laghu), which means it is easily digestible, and ruksha (dry). These properties remove heat from the blood and liver. Its application is also stated in Unani medicine (Mukherji 1953). Chirata with cardamom, turmeric, and kutki is used to treat gastrointestinal infections, and it is also used to treat fever (Kirtikar and Basu 1984). It treats numerous skin disorders when used with neem, manjishta, and gotu kola. In situations of scorpion bite, it is administered in conjunction with other medications (Nadkarni 1976). Diabegon, D-400, Chirayita pill, GlucoBuster, DBCare, Himoliv, and Melicon V are all branded herbal medications made from the plant. It is one of the ingredients in Ayurvedic formulations like Kabdeen (for treating viral hepatitis), Sudarshan, Mahasudarsana, Kiratatiktadai, Bhunimbadi, Kiratadi taila (all for treating fever), Chandra Prabati (for cancer), Palas abijadi churna (anthelmintic), Dermafex oil (for treating skin issues), and as hair vitalizers (Pradhan and Badola 2015).

3.2 Medicinal Properties and Bioactive Compounds

The pharmacological properties of *S. chirayita* extracts, including aqueous, alcoholic, and methanolic extracts, have been seen and documented in laboratory research. The comprehensive antimicrobial properties of the complete *S. chirayita* plant have been documented in previous studies (Alam et al. 2009; Laxmi et al. 2011; Rehman et al. 2011). Recent studies have provided evidence that extracts derived from *S. chirayita* possess a diverse array of biological properties, including antibacterial, antifungal, antiviral, anticancer, anti-inflammatory, antidiabetic, antioxidant, and anti-hepatitis B virus (anti-HBV) activities (Verma et al. 2008; Arya et al. 2011a, b; Chen et al. 2011; Zhou et al. 2015).

Amarogentin, Amaroswerin, Swertiamarin, are members of the secoiridoid glycoside family, and Swerchirin, a prominent xanthone with therapeutic properties, are the four primary bioactive ingredients responsible for *S. chirayita's* medicinal and pharmacological significance (Fig. 2). Even after being diluted to a ratio of 1:58,000,000, amarogentin still has a very bitter flavour. It has been documented to have antileishmanial effects (Medda et al. 1999) and chemo-preventive qualities (Saha et al. 2005). Swertiamarin has been discovered to have analgesic characteristics, while amaroswerin has been reported to have a gastro-shielding action (Niiho et al. 2006). Antimalarial, hypoglycemic, hepatoprotective, and pro-hematopoietic properties of swerchirin (Table 1) have been documented.

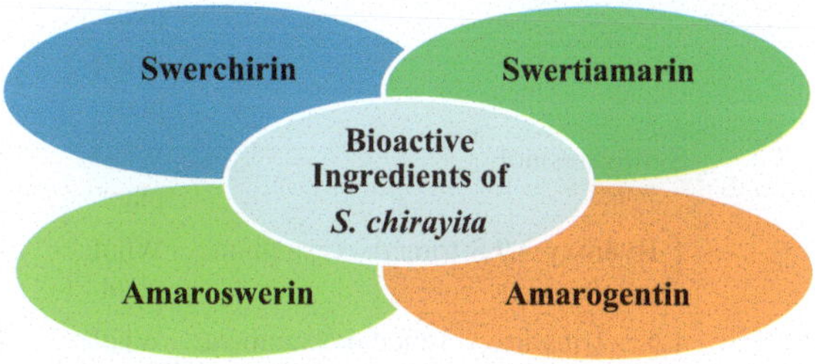

Fig. 2 Major bioactive components of *Swertia chirayita*

Table 1 Biological activity and important chemical compounds of *Swertia chirayita*

S. No.	Chemical constituent	Part Used	Biological activity	References
1	Mangiferin	Plant aqueous decoction	Anti-viral, antiparkinsonian	Kavitha and Dattatri (2013)
2	Swertiamarin	Dry plant decoction	CNS depressant, anti-diabetic, anti-arthritic	Saravanan et al. (2014)
3	Amaroswerin	Whole plant decoction	Gastroprotective	Niiho et al. (2006)
4	Amarogentin	Whole plant	Antileishmanial, anticancer, anti-diabetic, gastroprotective, anthelmintic	Sankar et al. (2017)
5	Swerchirin	Whole plant	Chemo preventive	Hirakawa et al. (2005)
6	Ursolicacid	Whole plant	Antitumor, antimicrobial	Soica et al. (2014)
7	Sweroside	Seeds	Hepatoprotective, antibacterial, hyperpigmentation	Jeong et al. (2015)
8	Swertanone		Anti-inflammatory	Tabassum et al. (2012)
9	Gentianine	Whole plant decoction	Antimalarial, anti-hepatitis B virus, antipsychotic	Natarajan et al. (1974)
10	Bellidifolin	Whole plant	Hypoglycemic	Basnet et al. (1995)
11	Oleanolicacid	Whole plant	Antimicrobial	Jesus et al. (2015)
12	Syringaresinol	Whole plant	Hepatoprotective	Sankar et al. (2017)
13	1-Hydroxy-3,7,8-trimethoxyxanthone	Whole plant	Antiulcerogenic	Ateufack et al. (2014)
14	1,5,8-Trihydroxy-3-methoxyxanthone	Whole plant	Blood sugar lowering	Ghosal et al. (1973)
15	β-Amyrin	Roots paste and whole plant	Anti-inflammatory	Holanda et al. (2008)
16	Chiratol	Roots paste and whole plant	Anti-inflammatory	Banerjee et al. (2000)

4 Domestication and Cultivation Potential of *S. chirayita*

Plant domestication is the evolutionary process whereby a population of plants becomes accustomed to human provision and control (Pourkheirandish and Komatsuda 2007). For many authors (Harlan 1992; Zohary and Hopf 1993), domestication is generally considered to be the end-point of a continuum that starts with exploitation of wild plants, continues through cultivation of plants selected from the wild but not yet genetically different from wild plants and ends with the adaptation to the agro ecology through conscious or unconscious human morphological selection, and hence genetic differences distinguishing the domesticated species from its wild progenitor. The domestication of a plant starts, when its usefulness is proved, its demand is confirmed and regular, its availability around dwellings is seriously decreasing and when getting the desired quantity on time for use becomes problematic.

4.1 Advantages of Domestication and Cultivation

Domestication and cultivation of plants ensures consistent botanical identification, whereas gathering plants from their natural habitats often leads to a mix of unwanted or even harmful species. The amount of wild harvest is influenced by a variety of uncontrollable causes, and supply fluctuations are a common occurrence. A consistent supply of raw materials is ensured by cultivation. Through agreements with growers, wholesalers and pharmaceutical companies can determine volumes and prices over time. Selecting and developing genotypes with desirable traits from wild or managed populations can promote the economic growth of medicinal plant species as crops. Furthermore, cultivation enables controlled post-harvest handling, ensuring quality controls and adjusting product standards to meet regulations and consumer preferences. Additionally, cultivated materials can easily obtain organic or biodynamic certification (Palevitch 1991; Uniyal et al. 2000; Pierce et al. 2002; Leaman 2004; Peter et al. 2005).

4.2 Challenges of Domestication and Cultivation of Medicinal Plants

Since 1976, Dobremez have been advocating the cultivation of medicinal plants due to its potential to provide a constant supply of raw materials to the industry and local communities while reducing pressure on the wild population. While it is reasonable to expect that cultivated plants will have attributes that differ from those found in their natural environments, it is also evident that certain values in plants can be purposely boosted under controlled cultivation circumstances (Palevitch 1991; Uniyal et al. 2000).

However, studies show that farming does not always allow for the domestication of wild medicinal plants. Many species are difficult to cultivate due to biological or ecological limitations (slow growth rate, particular soil requirements, low germination rates and pest vulnerability). The fundamental reason for bringing a species into cultivation is economic feasibility, but it is also a significant limitation as long as sufficient volumes of material can still be obtained at a lower price from wild gathering.

The cultivated plants will be in direct competition with wild collections that is harvested by commercial gatherers and sold without incurring any cultivation-related costs (Cunningham 1994). It takes several years of research (e.g., 12 years for *Alchemilla alpina*; Schneider et al. 1999) as well as a substantial financial expenditure to domesticate wildly collected species.

4.3 Domestication and Cultivation Efforts of S. chirayita

Despite having a well-established market in India and abroad (Badola and Pal 2002; Joshi and Dhawan 2005), *S. chirayita* is still collected from the wild and rarely cultivated, and little effort has been made to develop suitable agro-techniques for its domestication (Joshi and Dhawan 2005). Plants of *S. chirayita* are removed in natural environments before the seeds mature, reducing the chances of regeneration in the wild and increasing the risk of extinction (Bhattarai and Shrestha 1996). Given the high pharmaceutical demand for *S. chirayita* and the resulting decrease in the natural environment, cultivation of the species is required (Badola and Pal 2002).

Swertia chirayita plants are harvested commercially during the erect stage. Natural regeneration potential has been diminished due to early harvesting before seed development. The problems with *S. chirayita* are low seed viability, slow growth, a three-year gestation period, poor germination of seed, and difficulty in handling of seedlings due to their small size, leading to high mortality rates, and high cultivation efforts on the production process, which may lead to decreased productivity. According to our observations, this herb is highly site and temperature sensitive, and can grow appropriately only above 1200 m msl altitude, making it difficult to acclimatize in lower elevations (Basnet 2001; Badola and Pal 2002; Raina et al. 1994; Joshi and Dhawan 2005; Chaudhuri et al. 2007). The main barriers to the commercialization of cultivation are a lack of quality planting material, inadequate development and extension support in cultivation and processing, and unorganized marketplaces.

According to Rai et al. (2000), the number of *S. chirayita* plants taken from the Sikkim Himalaya is exceptionally large, totaling to 3,000,000 plants per year. Ex-situ farming was cited by Pradhan and Badola as an appropriate tool for in-situ conservation of endangered species. However, it is widely held that plants grown outside of their natural environment lose their therapeutic properties.

5 Genetic Improvement Strategies for *S. chirayita*

5.1 Phenology of S. chirayita

Phenological studies impart knowledge on the morphological and functional traits of plants, providing clues about adaptability mechanisms of plants under varying and periodically changing climatic conditions in which they grow, and thus aid in the development of management and conservation plans for valuable plant species. Climate parameters, such as temperature, water availability, photoperiodism, dry spell duration, variation in day length and temperature, and so on, have been demonstrated in studies to be primarily responsible for vegetative and reproductive phenology at both the community and species levels (Bhat and Murali 2001; Badola 1994, 2010).

Particularly in the case of herbs, various species might respond to a growing environment in different ways (Butola and Badola 2008). Transplanted genotypes perform better in their origin location than at foreign sites (Van 1998; Hufford and Mazer 2003), provided adequate substrates are available, which plays a substantial impact in the species' improved performance. According to investigations on the phenology, growth, and development of *S. chirayita*, which is also a pluri-annual crop, the third growing season is the most critical phase in the life cycle of *S. chirayita*, because major phenological events occur in this phase, from shoot commencement through plant senescence. *S. chirayita* plants develop in a regular seasonal pattern regardless of growing conditions, even when housed in a controlled environment. Seedlings/plants of *S. chirayita* that have adapted to the microclimate of the site after transplantation perform similarly to those naturally established in wild environments in terms of phenology, growth, and development.

5.2 Reproductive Biology of S. chirayita

For any genetic improvement strategy to be successful, an understanding of the breeding system (self or cross) is essential. Few studies on *S. chirayita's* cytology, genetics, and reproductive biology have been conducted thus far. Based solely on floral traits, such as a vibrant corolla and the presence of nectaries, it was claimed to be a cross-pollinated species without any experimentations (Khoshoo and Tandon 1963).

In a field experiment, Chakraborty et al. (2009) established and shown that natural pollination of seed development of *S. chirayita* plant is the result of self-pollination, with 16–20% likelihood of cross pollination. Raina et al. (2013), on the other hand, observed that *S. chirayita* had protogynous flowers and the existence of various cytotypes. The species appears to prefer cross pollination, with approximately 37–48% self-pollination potential. Because of its endangered status, this is extremely beneficial to this species. Out crossing would ensure the development of new gene combinations, assisting its stock in overcoming the vagaries of rapid climatic change (Shah et al. 2011). The species' self-pollinating ability is especially valuable since it may be used to maintain the purity of any enhanced strain that is established, as this species is not receptive to vegetative propagation. Rao et al. hypothesized that producing inbred seed is a "feel safe" approach for producing seeds when pollinators are scarce.

5.3 Diversity and Variability Among Populations of *S. chirayita*

Geographically influenced diversity in plant morphology is a consequence of phenotypic changes in response to local environmental conditions, variation in genes and evolution among populations, and an individual species' biogeographic history (Ellison et al. 2004). Morphological variation and geographical separation between populations are both required for the establishment of subspecies and species (Losos and Glor 2003). An evaluation of germplasm in nature often provides an insight about ideal climatic conditions for domestication and cultivation of an important plant species, and elite germplasm can be chosen for crop improvement based on phenotypic features (Vashistha et al. 2006). Assessing genetic diversity and natural distribution patterns is thought to be essential for developing conservation strategies for endangered species like *S. chirayita* since the survival index of a species is significantly influenced by the existing polymorphisms and gene flow between populations. Chakraborty et al. (2016) found morpho variant diversity in populations of Neora Valley and Lava area in eastern Himalaya. The authors found 5 morpho-variants and 24 subvariants based on NBPGR's designated characters' minimal descriptors. Some population morpho-variant parameters revealed critical diversity. They suggested categorization and subcategorization in the grouping rule as the sole approach to conserve this plant (Wang et al. 2011).

The availability of a pool of genetic diversity is a necessary prerequisite for a species' survival in the face of environmental adverse conditions (Frankel 1993). Genetic diversity is also associated to phytochemical distinctiveness (Arya et al. 2011a, b). In the context of global climate change, the very high genetic diversity seen could be the synergistic result of cross pollination between individuals in the population and the act of evolutionary processes (natural selection, genetic drift, and mutation) (Slatkin 1987). This could be due to the species' high survival capacity in its current environment, as well as a high level of adaptability and acclimatization to the new selection pressure caused by environmental changes. In addition to disease resistance, greater genetic diversity leads to greater productivity in plant communities, greater nutrient retention as well as ecosystem stability (Tilman 2000).

The extinction of the species' native environment as a result of human activity has also been linked to genetic erosion of the species. As a result, there is an urgent need to catalogue and preserve *S. chirayita's* present diversity. Genetic diversity studies aid in the development of pragmatic conservation strategies and the creation of core collections for endangered plant species by identifying regions with the greatest diversity and thus conserving them with the smallest population size (Shanker and Ganeshaiah 1997; Rao and Hodgkin 2002). Morphological, biochemical (Karan et al. 1997; Bhatia et al. 2003), and isozyme markers (Verma and Kumar 2001) have been employed to distinguish various species in the genus *Swertia.*

Due to the commercial relevance of *S. chirayita*, immediate cultivation efforts are required, and superior lines must be developed. Because wild stocks are essentially non-existent, there is no way to conduct selection breeding after large population surveys; the only choice is to search the existing limited genepool for measurable and provable diversity. Patil (2011) uncovered chemical variations in *S. chirayita* that can be linked to specific morphological characteristics. In the lack of evidence that superior lineages exist in the species, such variation must be conserved and multiplied. Because this species only reproduces sexually, these variants cannot be multiplied by traditional vegetative means; nevertheless, they can be multiplied by tissue culture, for which protocols are currently available (Ahuja et al. 2003). Raina et al. (2016) found that gamma rays can increase therapeutically essential amarogentin and amaroswerin in critically endangered temperate *S. chirayita*, which is on the brink of extinction. Since *S. chirayita* is being grown commercially due to high demand for its raw material, strains with improved bio compounds are needed.

The influence of height on population differentiation and genetic diversity is a crucial ecological element. According to Royer et al. (2005), the presence of high altitude directly impacts the growth and development of plants. Based on the findings of a separate study conducted by Pradhan and Badola (2015), it was observed that factors such as elevation, latitude, longitude, soil pH, species richness, species diversity, and degree of slope exhibited minimal or no influence on *S. chirayita.* There are very limited studies on assessing the genetic diversity and population genetic structure using molecular markers in *S. chirayita.* Joshi and Dhawan (2007), Chhipi et al. (2013) and Neupane et al. (2017) observed extremely high levels of polymorphism, demonstrating that substantial levels of genetic variation are still retained within the declining populations of *S. chirayita*, indicating good survival capacity and adaptation in shifting environmental scenarios. As a result, it is critical to protect this valuable germplasm and its habitat from further devastation.

Selecting and developing genotypes with favorable features from wild or established populations can help medicinal plant species flourish economically as crops. A diverse set of accessions must be identified to serve as the foundation for future breeding programmes, genetic improvement, and commercial cultivation, as well as potential sources of high-quality germplasm to increase the quality of herbal products and relieve pressure on threatened wild populations. In the last decade only, handful of studies have been conducted for its genetic improvement. In six explorations and collections from Darjeeling and Kalimpong districts of West Bengal and Sikkim, Thapa et al. (2023) identified twenty-five diverse accessions based on morphometric traits as prospective sources of high-quality germplasm for genetic improvement programmes and commercial cultivation. This will improve herbal goods and reduce pressure on endangered species. In Uttarakhand, Verma et al. (2020) tried to acclimatize and identify genotypes from the eastern and western Himalayas with improved results over three altitudes. A specific accession from Kalimpong at 1247 m had improved performance in plant height, number of leaves per plant, fresh weight, and dry weight and can perform better in low altitudes.

6 Conservation Approaches for *S. chirayita*

There has been widespread international apprehension regarding the decline in populations, reduction in genetic variety, and local extinction of significant medicinal plant species (Badola and Aitken 2003, 2010; Canter et al. 2005). There are other factors contributing to the issues at hand, namely unsustainable harvesting, habitat degradation, overgrazing, diseases, and herbivores, among others. The global recognition of the significance of *S. chirayita* is widely acknowledged. However, there has been a lack of initiatives to restore its population, resulting in the species being classified as critically endangered by the International Union for Conservation of Nature (IUCN), endangered according to the Indian Red Data Book, critically rare and virtually endemic to the Himalayas as reported by Samant et al. (1998a, b).

Without the implementation of significant conservation efforts, it is likely that the species may face extinction in several areas of the Himalayan region. A conservation program's success hinges on its capacity to accurately depict species during storage and retain germplasm for future recovery attempts. In order to ensure the successful preservation of endangered medicinal plants, it is crucial to use conservation techniques that transcend traditional limitations. This entails integrating various in situ and ex situ measures for conservation, known as Trans situ conservation approaches which have been framed out for *S. chirayita.* These approaches should be complemented by research and educational initiatives, and should aim to expand their scope from local to national and ultimately global levels. Additionally, these efforts should specifically target taxa that need conservation.

6.1 Conservation Through Habitat Preservation

Before initiating the conservation efforts for any given species, it is vital to first conduct a thorough mapping of its geographical distribution, followed by an inventory of its current populations. These first steps are essential in order to accurately determine and define the conservation status of the species (Volis 2015). The decline in species and biodiversity can be attributed to the driving factors of habitat destruction, degradation, and fragmentation. Various human activities can have substantial adverse effects on ecosystems, including the clearance of forests for agricultural purposes, urban development, industrialization, and the installation of transportation infrastructure. Consequently, the initial phase in the preservation of a species involves the implementation of habitat conservation strategies, with particular emphasis on the preservation of micro-habitat niches as valuable indicators. The presence of *S. chirayita* in close proximity to human settlements or in areas affected by human activity, as well as in open spaces resulting from other causes, suggests the species' ability to adjust to slight disruptions and its habitat preferences. The objective of this intervention is to evaluate the invasion of shrubs and other undesirable or invasive plant species that present a risk to the survival of *S. chirayita.* The effective conservation of *S. chirayita* habitats within their distinct ecosystems can be facilitated through the mobilization and delegation of responsibility to local communities. The individuals under consideration demonstrate a clear association with the natural environment, thereby facilitating the application of their traditional knowledge and expertise to promote habitat conservation.

6.2 Conservation Through Good Agricultural Collection and Cultivation Practices

The protection of *S. chirayita* and the implementation of training and conservation programmes have become imperative, serving as essential requirements for all stakeholders involved. It is imperative to establish and deliver comprehensive guidelines and training programmes pertaining to Good Agricultural Collection Practices for the local residents residing in close proximity to the forest, as well as those individuals whose livelihoods rely on sustainable harvesting. By doing so, we can enhance the long-term value of the harvest and work towards the preservation of natural resources. This is essential for preventing damage to young plants and limiting the local environment's temperature and humidity swings. The assessment of economic viability plays a significant role in determining the feasibility of engaging in the production of medicinal plants (Badola and Pal 2002; Schippmann et al. 2002; Badola and Butola 2005; Badola 2010). In this particular setting, it is imperative to cultivate awareness among the relevant stakeholders, mostly farmers or entrepreneurs, periodically regarding the market potential and market-related concerns associated with *S. chirayita.* Scalable agriculture is considered a highly effective strategy for addressing the conservation needs while simultaneously meeting the substantial demands of both national and international markets. The development of Good Agricultural Practices for *S. chirayita* with significant commercial demand has become imperative and urgent. Their cultivation would alleviate the strain on natural resources and the practice of gathering from the wild, as well as hinder the utilization of counterfeit alternatives and contaminants. Consequently, this will lead to deterioration in the quality of pharmaceuticals used in Indian systems of Medicine.

6.3 Conservation Through Inter-Situs Approach

This pertains to an off-site collection that is managed inside a natural habitat, and can be seen as an in-situ reintroduction technique (Husband and Campbell 2004). The recommended strategy for the preservation of plant species that have become extinct in their original habitat or are in danger of extinction is by reintroduction, as proposed by Maunder (1992), Falk et al. (1996), IUCN (1998), Godefroid et al. (2011). Additionally, this technique does not work for rare species whose natural populations are not protected and are in peril, even while protected areas within the known range of the species can be utilized to establish new populations. Reintroduction of populations to areas beyond the present distribution but within the historical range of a species has been observed (Burney and Burney 2007). The site where reintroduction takes place is subject to rigorous restoration efforts, which encompass horticultural and agricultural management, invasive species removal, and protection measures. Over time, the level of care provided to the reintroduced plants is gradually reduced.

Furthermore, when considering reintroduction efforts, it is crucial to possess a comprehensive understanding of the magnitude and arrangement of genetic diversity, the capacity for dispersal through seeds and pollen, and the significance of local adaptation. This knowledge plays a pivotal role in informing decisions regarding the reintroduction of gene flow among fragmented populations or the preservation of locally adapted populations.

Severe rainfall causes erosion or deep percolation, fungal infestation, and seed predation are all factors that impede the development of *S. chirayita* seedlings in their native settings. Because of these factors, *S. chirayita* is critically endangered in its natural range. *S. chirayita* populations in their natural habitat are also threatened by unsustainable harvesting practices. Seedlings can be grown at a nursery close to the population's location under certain circumstances. These plants can be propagated from natural seeds collected from populations with comparable genetic makeup. These young plants can be reintroduced to areas where numbers are decreasing. However, population size should be considered before performing a transplant to ensure that genetic diversity is not lost to genetic erosion. Only those who have had proper education and experience should collect seeds for reproduction.

6.4 Circa Situm Conservation

Hawkes et al. (2001) define *circa situm* conservation as a middle ground between in-situ and ex-situ conservation methods. Farmland where natural forest or woodland containing the same species was formerly found, but has been destroyed or drastically transformed due to agricultural expansion, may benefit from the preservation of planted and/or remnant plants and wildings. When clearing land or eradicating weeds for agricultural or gardening purposes, this strategy involves actively encouraging and protecting the presence of *S. chirayita* inside and around agricultural fields or home gardens.

6.5 Conservation Through Field Gene Bank (FGB) and Seed Bank

Plant genetic resources in FGB are conserved as living plants, necessitating ongoing growth and maintenance efforts. Other ex situ techniques are frequently employed in situations when it is challenging or unfeasible to preserve germplasm as seeds. This may occur when seeds are not generated, when seeds exhibit recalcitrance, or when seed generation requires a significant amount of time. Seed banks are the most effective approach to ex-situ conservation of plant species. Currently, it is exclusively regarded as applicable solely to agricultural crops. The preservation of crucial medicinal plants, such as *S. chirayita*, can be achieved through their storage in seed banks. This practice ensures the possibility of reintroducing these species to their native habitats in the event of their imminent risk of extinction or endangered status.

Nevertheless, the efficacy of conserving a species in a field and seed bank and subsequently reintroducing it to its original ecosystem is significantly constrained unless the underlying reasons contributing to its vulnerability or potential extinction are predominantly eradicated.

6.6 Conservation Through Capacity Building, Communication and Cooperation

Public support is an essential component in the conservation of medicinal plants that are in danger of extinction. To ensure the successful initiation and implementation of conservation efforts, it is crucial to develop a fundamental basis of trust and collaboration between governmental entities and community members involved in the process. The attainment of this objective can be realized by formulating a comprehensive communication strategy and executing information and education initiatives. The involvement of the community is of utmost importance in the preservation of endangered medicinal plants through in-situ conservation efforts. The challenge of properly monitoring is attributed to the inherent difficulty encountered by government officials and allied entities.

6.7 Conservation Through Biotechnological Approaches

The propagation of *S. chirayita* on a large scale by seed is challenging due to the limited quantity of seeds caused by the premature harvesting of plants before seed maturation. Traditional techniques of protecting germplasm are insufficient to ensure the successful restoration of this endangered species to its natural environment. To encourage large-scale production for ex-situ conservation and to meet the medicinal needs, an in vitro regeneration protocol for *S. chirayita* is urgently needed. Mass propagation, which can be employed in conjunction with conventional breeding for sustainable use of this ancient medicinal plant, requires the adoption of alternative repeatable micropropagation technologies Plant tissue culture provides a means by which commercially harvested medicinal plants can be preserved for future use. Micropropagation, cell and organ culture allow for the efficient multiplication and ex situ conservation of plant biodiversity from a minimal amount of available plant material.

Among the many practical applications of current plant biotechnology is synthetic seed technology, which has great promise for facilitating micropropagation, preserving plant germplasm through cryopreservation, and general seed management. Research efforts for in vitro propagation and in vitro conservation have been done by many scientists showing positive and reproducible results. In a study by, Wawrosch et al. (1999) they devised a protocol aimed at the micropropagation of *S. chirayita.* Their investigation revealed that the most favourable approach for achieving adventitious shoot regeneration was by the use of root explants. Chaudhuri et al. (2007) successfully generated genetically uniform plants using nodal explants. Joshi and Dhawan (2007) provided a detailed account of axillary shoot multiplication. Bisht et al. developed a protocol for plant regeneration through indirect organogenesis. Balaraju et al. (2009) established a rapid system for this purpose. Wang et al. (2009) investigated the effects of phytohormones on shoot regeneration. Pant et al. (2010) developed an efficient protocol for the same purpose. Lastly, Kumar and Chandra (2013) reported on the production of synthetic seeds and subsequent plant regeneration. These techniques hold practical significance as they can be utilised for many purposes such as large-scale multiplication, germplasm conservation, and extraction of important secondary metabolites from *S. chirayita.*

7 Future Directions and Research Priorities

7.1 Investigating the Mechanisms of Action of the Plant's Active Compounds

Such studies would aim to examine the underlying mechanisms of action associated with the bioactive chemicals derived from the plant. This research endeavor will aid in the development of new medication therapies enhancing the comprehension of the mechanisms behind the efficacy of *S. chirayita* and facilitate the development of novel pharmacological interventions utilizing its bioactive components.

7.2 Conducting Clinical Trials to Assess Its Efficacy

This is essential to test the effectiveness of *S. chirayita* in humans to see if it can be used to treat various ailments. It is imperative to ascertain whether the plant possesses the qualities necessary to serve as a secure and efficacious remedy for various ailments.

7.3 Studying Its Safety for Long-Term Use

Ensuring the prolonged safety of the plant is paramount in its utilization. A comprehensive study is imperative to confirm that extended use poses no potential harm to consumers. Rigorous assessment of its long-term effects becomes crucial to guarantee the plant's sustained application without adverse consequences. This scrutiny is essential for establishing the plant's safety profile and instilling confidence in its extended use, aligning with the principles of responsible and secure botanical utilization.

8 Conclusion

8.1 Recap and Insights

Swertia chirayita is a temperate Himalayan medicinal plant in India, known for its therapeutic properties and high domestic and international demand for its raw material. Traditional medicine and indigenous peoples have used the plant to cure a wide range of conditions, including the flu, malaria, coughs, colds, stomachaches, skin problems, worms in the intestines, and diarrhea. The plant has been used in various traditional systems of medicine, including Ayurveda, Unani, Siddha, and Tibetan medicine. Medicinal properties and bioactive compounds of *S. chirayita* include antibacterial and antifungal action, anticancer, anti-inflammatory, and antioxidant activities. The National Medicinal Plant Board has prioritized this species among the 32 medicinal species for intensive cultivation and given farmers a 75% subsidy to encourage its cultivation. Crop improvement and conservation efforts are crucial for the sustainability of endangered and threatened medicinal plant species. Assessing variability within the existing germplasm of the genus is essential and relies on the examination of morphological and genetic aspects. Through rigorous biochemical screening, we can uncover the significant contribution of species other than *S. chirayita* in meeting the extensive commercial demand. The majority of plants are currently gathered from the wild by local traders, leading to a decline in plant populations and posing a risk of extinction. To meet the increasing demand in national and international trade markets for raw plants, cultivation needs to be intensified. The entire plant is used for medicinal purposes, leading to low population density and localized distribution in distinct pockets. This has conservation significance since species with specialized habitat requirements are more likely to become extinct than those with a wide range of habitat requirements.

Domestication and cultivation of *S. chirayita* are essential for conservation and meeting the high pharmaceutical demand. Domestication ensures consistent botanical identification, reducing the risk of unwanted or harmful species. It is collected from the wild and rarely cultivated, and little effort has been made to develop suitable agro-techniques for domestication. However, early harvesting before seed development reduces natural regeneration potential and increases the risk of extinction. The main barriers to commercialization of cultivation include lack of quality planting material, inadequate development and extension support in cultivation and processing, and unorganized marketplaces.

Due to the commercial relevance of *S. chirayita*, immediate cultivation efforts are required, and superior lines must be developed. High altitude directly impacts population differentiation and genetic diversity, with factors such as elevation, latitude, longitude, soil pH, species richness, species diversity, and degree of slope exhibiting minimal or no influence on the species. Selecting and developing genotypes with favorable features from wild or established populations can help medicinal plant species flourish economically as crops. A diverse set of accessions must be identified to serve as the foundation for future breeding programs, genetic improvement, and commercial cultivation, as well as potential sources of high-quality germplasm to increase the quality of herbal products and relieve pressure on threatened wild populations.

Conservation through inter-situs approach involves off-site collection managed within a natural habitat, such as reintroduction. *Circa situm* conservation is a middle ground between in-situ and ex-situ conservation methods. It involves actively encouraging and protecting the presence of plants and wildings in farmland where natural forests or woodlands contain the same species were once found but have been destroyed or drastically transformed due to agricultural expansion. Field gene banks (FGB) and seed banks are effective approaches for ex-situ conservation of plant species. Capacity building, communication, and cooperation are essential components in the conservation of medicinal plants in danger of extinction. A comprehensive communication strategy and information and education initiatives are essential for community involvement. Indian Council of Agricultural Research has been networking with All India Co-ordinated Research Project on Medicinal and Aromatic focusing on aspects of *S. chirayita*'s conservation, cultivation and imparting training to farmers with centers from Western to Eastern Himalayas at

Kashmir, Himachal Pradesh, Uttarakhand and Northern part of West Bengal. Biotechnological approaches are needed for large-scale propagation of *S. chirayita.* Mass propagation requires the adoption of alternative repeatable micropropagation technologies. Synthetic seed technology has great promise for facilitating micropropagation, preserving plant germplasm through cryopreservation, and general seed management. Research on in vitro propagation and in vitro conservation has shown positive results, with some studies showing promising results.

References

Ahuja A, Kool S, Kaul BL, Verma NK, Kaul MK, Raina RK, Quazi GN (2003) Media composition for faster propagation of *Swertia chirayita.* WO 03/045132 AL. US. Patent 7238527

Alam KD, Ali MS, Parvin S, Mahjabeen S, Akbar MA, Ahamed R (2009) In vitro antimicrobial activities of different fractions of *Swertia chirata* ethanolic extract. Pak J Biol Sci 12:1334–1337

Arya R, Sharma SK, Singh S (2011a) Antidiabetic effect of whole plant extract and fractions of *Swertia chirayita* Buch.-Ham. Planta Medica 77(05):138

Arya V, Yadav S, Yadav JP (2011b) Intra-specific genetic diversity of different accessions *of Cassia occidentalis* by RAPD markers. Genet Eng Biotechnol J 22:1–8

Ateufack G, Nguelefack TB, Wabo HK, Tane P, Kamanyi A (2014) Antiulcerogenic activity of 1-Hydroxy-3,7,8-trimethoxyxanthone isolated from the methanol extract of *Anthocleista vogelii* PLANCH. in rats. Ulcers 172096. https://doi.org/10.1155/2014/172096

Badola HK, Pal M (2002) Endangered medicinal plants in Himachal Pradesh. Curr Sci 83(7):797–879

Badola HK (1994) Bud phenological studies as an aid to forestry research: an overview. In: Forestry research and education. Diamond Jublee Publication, pp 142–153

Badola HK (2010) Phenology and climate responses in Himalayan rhododendrons. In: Rhododendrons: conservation and sustainable use. Proceedings of international conference. Printed at Concept, pp 48–59

Badola HK, Aitken S (2003) The Himalayas of India: a treasury of medicinal plants under seize. Biodiversity 4(3):3–13

Badola HK, Aitken S (2010) Potential biological resources for poverty alleviation in Indian Himalaya. Biodiversity (3–4):8–18

Badola HK, Butola JS (2005) Effect of ploughing depth on the growth and yield of *Heraceum candicans*: a threatened medicinal herb and a less explored potential crop of the Himalayan Region. J Mount Sci 2:173–180

Balaraju K, Agastian P, Ignacimuthu S (2009) Micropropagation of *Swertia chirata* Buch. Hams. ex Wall: a critically endangered medicinal herb. Acta Physio Planta 31(3):487–494

Balick MJ, Elinabotsky E, Laird SA (1996) Medicinal resources of the tropical forest. Columbia University Press, New York

Banerjee S, Sur TP, Das PC, Sikdar S (2000) Assessment of the anti-inflammatory effects of *Swertia chirata* in acute and chronic experimental models in male albino rats. Indian J Pharmacol 32:21–24

Basnet P, Kadota S, Shimizu M, Takata Y, Kobayashi M, Namba T (1995) Bellidifolin stimulates glucose-uptake in rat-1 fibroblasts and ameliorates hyperglycemia in streptozotocin (stz)-induced diabetic rats. Planta Med 61:402–405

Basnet DB (2001) Evolving nursery practices and method of cultivation of high value medicinal plant *Swertia chirata* Ham. Environ Ecol 19:935–938

Bentley R, Trimen H (1880) Medicinal plants. In: Bentley R, Trimen H (eds) J and A Churchill, London, p 183

Bhat DN, Murali KS (2001) Phenology of under storey species of tropical moist forest of Western Ghats region of Uttara Kannada district in South India. Curr Sci 81(7):799–805

Bhatia A, Karan M, Vasisht K (2003) Morphological and chemotypic comparison of certain Indian species of *Swertia*. J Med Aroma Plant Sci 25:336–343

Bhatt A, Rawal RS, Dhar U (2006) Ecological features of a critically rare medicinal plant, *Swertia chirayita*, in Himalaya. Plan Species Biol 21:49–52. https://doi.org/10.1111/j.1442-1984.2006.00150

Bhattarai KR, Shrestha S (1996) Ecological study on Chiraito (*Swertia chirayita*, Roxb. ex Fleming (Karsten) in Northern Gorkha. J Nat Hist Museum 15:13–16

Burney DA, Burney LP (2007) Paleoecology and "inter-situ" restoration on Kauai, Hawaii. Front Ecol Environ 5:483–490

Butola JS, Badola HK (2008) Threatened medicinal plants and their conservation in Himachal Pradesh. J Trop Med Plants 9(1):125–142

Canter PH, Thomas H, Erns E (2005) Bringing medicinal plants into cultivation: opportunities and challenges for biotechnology. Trends Biotechnol 23(4):180–185

Chakraborty S, Mukherjee D, Dasgupta T (2009) Cytological study on chromosome behaviour and new report on nature of mode of pollination of *Swertia chirayita*, a high value endangered medicinal plant of North Eastern Himalayan Region. Caryologia 62(1):43–52

Chakraborty S, Mukherjee D, Baskey S (2016) Morphological diversity and nomenclature of *Swertia chirayita* (Gentianaceae)—recovery of endangered medicinal plant population in North Eastern Himalaya. Am J Plant Sci 7:741–755

Chaudhuri RK, Pal A, Jha TB (2007) Production of genetically uniform plants from nodal explants of *Swertia chirata* Buch. Ham. ex Wall—an endangered medicinal herb. In Vitro Cell Dev Biol Plant 43:467–472

Chen Y, Huang B, He J, Han L, Zhan Y, Wang Y (2011) In vitro and in vivo antioxidant effects of the ethanolic extract of *Swertia chirayita*. J Ethnopharmacol 136:309–315. https://doi.org/10.1016/j.jep.2011.04.058

Chhipi JK, Shrestha T, Bhattarai J, Sijapati N, Rana J, Maharjan DS, Rawal BB, Raskoti S, Shrestha (2013) Assessment of genetic diversity in nepalese populations of *Swertia chirayita* (Roxb. ex Fleming) H. Karst using RAPD-PCR technique. Am J Plant Sci 4:1617–1628

Cunningham AB (1994) Management of medicinal plant resources. In: Proceedings of the 13th plenary meeting of AETFAT173-189

Dobremez JF (1976) Le Nepal: Ecologieet Biogeographie [Ecology and biogeography of Nepal]. Centre Nationale de la Recherche Scientifique, Paris, France

Ellison AM, Buckley HL, Miller TE, Gotelli NJ (2004) Morphological variation in *Sarracenia purpurea* (Sarraceniaceae): geographic, environmental, and taxonomic correlates. Am J Bot 91(11):1930–1935

Falk DA, Millar CI, Olwell ME (1996) Restoring diversity: strategies for reintroduction of endangered plants. Island Press, Washington, DC

Frankel OH (1993) The place of management in conservation. In: Genetics and conservation: a reference manual for managing wild animals and plant population, pp 1–14

Ghosal S, Sharma PV, Chaudhuri RK, Bhattacharya SK (1973) Chemical constituents of the Gentianaceae V: tetraoxygenated xanthones of *Swertia chirata* Buch.-ham. J Pharm Sci 62:926–930

Godefroid S, Piazza C, Rossi G, Buord S, Stevens AD, Aguraiuja R, Cowell C, Weekley CW, Vogg G, Iriondo JM, Johnson I, Dixon B, Gordon DR, Magnanon S, Valentin B, Bjureke K, Koopman R, Vicens M, Virevaire M, Vanderborght T (2011) How successful are plant species reintroductions? Biol Coserv 144:672682

Harlan JR (1992) Crops and man, 2nd edn. American Society of Agronomy/Crop Science Society of America, Madison (WI)

Hawkes JG, Maxted N, Ford Lloyd BV (2001) The ex situ conservation of plant genetic resources. Kluwer Academic Publishers, London, UK

Hirakawa K, Yoshida M, Nagatsu A, Mizukami H, Rana V, Rawat MSM (2005) Chemo preventive action of xanthone derivatives on photosensitized DNA damage. Photochem Photobiol 81:314–319

Holanda SA, Pinto LM, Cunha GM, Chaves MH, Santos FA, Rao VS (2008) Anti-inflammatory effect of alpha, beta-Amyrin, a pentacyclic triterpene from *Protium heptaphyllum* in rat model of acute periodontitis. Inflammopharmacology 16:48–52

Hufford KM, Mazer SJ (2003) Plant ecotypes: genetic differentiation in the age of ecological restoration. Trends Ecol Evol 18:147–155

Husband BC, Campbell LG (2004) Population responses to novel environments: implications for ex situ plant conservation. Ex situ plant conservation: supporting species survival in the wild. Island Press, Washington, pp 231–266

IUCN (1998) Guidelines for re-introductions. IUCN/SSC Reintroduction Specialist Group, IUCN, Gland, Switzerland

Jeong YT, Jeong SC, Hwang JS, Kim JH (2015) Modulation effects of sweroside isolated from the *Lonicera japonica* on melanin synthesis. Chem Biol Interact 238:33–39

Jesus JA, Lago JHG, Laurenti MD, Yamamoto ES, Passero LFD (2015) Antimicrobial activity of oleanolic and ursolic acids: an update. Evid Based Complement Altern Med 2015:620472

Joshi K (2000) Medicinal plant-lore in some hilly villages of the Central Development Region, Nepal. Bio Notes 2(4):69–71

Joshi K, Joshi A (2008) *Swertia* L. (Gentianaceae) in Nepal Himalaya: checklist, phytogeography, ethnobotany and conservation status. Ethnobot Leaflets 12:361–372

Joshi P, Dhawan V (2005) *Swertia chirayita*—an overview. Curr Sci 89(4):635–640

Joshi P, Dhawan V (2007) Axillary multiplication of *Swertia chirayita*: a critically endangered medicinal herb of temperate Himalayas. In Vitro Cell Dev Biol Plant 43(6):631–638

Kala CP, Sajwan BS (2007) Revitalizing Indian systems of herbal medicine by the national medicinal plants board through institutional networking and capacity building. Curr Sci 93:797–806

Kalliovirta M, Ryttari T, Hiekinen RK (2006) Population structure of a threatened plant, Pulsatilla patens, in boreal forests: modelling relationships to overgrowth and site closure. Biodivers Conserv 15:3095–3108

Kavitha KN, Dattatri AN (2013) Experimental evaluation of antidiabetic activity of *Swertia chirata*—aqueous extract. J Pub Health Med Res 1:71–75

Karan M, Vasisht K, Handa SS (1997) Morphological and chromatographic comparison of certain Indian species of *Swertia*. J Med Aromat Plant Sci 19:955–963

Khoshoo TN, Tandon SR (1963) Cytological and pollination studies on some Himalayan species of *Swertia*. Caryologia 16:445–477

Kirtikar KR, Basu BD (1984) Indian medicinal plants, vol III. LM Basu Publishers, Allahabad

Kumar V, Chandra S (2013) Efficient regeneration and antioxidant activity of the endangered species *Swertia chirayita*. Int J Pharma Biol Sci 4(4):823–833

Laxmi A, Siddhartha S, Archana M (2011) Antimicrobial screening of methanol and aqueous extracts of *Swertia chirata*. Int J Pharm Sci 3:142–146

Leaman DJ (2004) Sustainable wild collection of medicinal and aromatic plants: practice standards and performance criteria. Med Plant Conserv 9(10):7–8

Losos JB, Glor RE (2003) Phylogenetic comparative methods and the geography of speciation. Trends Ecol Evol 18(5):220–227

Maunder M (1992) Plant reintroduction an overview. Biodivers Conserv 1:5161

Medda S, Mukhopadhyay S, Basu MK (1999) Evaluation of the in vivo activity and toxicity of amarogentin, an antileishmanial agent, in both liposomal and niosomal forms. J Antimicrob Chemother 44:791–794

Misra A, Shasany AK, Shukla AK, Darokar MP, Singh SC, Sundaresan V, Singh J, Bagchi GD, Jain SP, Saikia D, Khanuja SPS (2010) AFLP markers for identification of *Swertia* species (Gentianaceae). Genet Mol Res 9:1535–1544

Mukherji B (1953) Indian pharmaceutical codex, indigenous drugs, vol I. CSIR, New Delhi, pp 64–65

Nadkarni KM (1998) Indian plants and drugs with their medicinal properties and uses. Asiatic Publishing House, Delhi

Nadkarni KM (1976) Indian materia medica, Bombay popular Prakashan, vol I. Elsevier, Bombay, pp 1184–1186

Natarajan PN, Wan AS, Zsaman V (1974) Antimalarial, antiamobeic and toxicity tests on gentianine. Planta Med 25:258–260. https://doi.org/10.1055/s-0028-1097940

Neupane S, Sijapati J, Bhattarai T, Shrestha S (2017) Genetic diversity in Nepalese population of *Swertia chirayita* (Roxb. ex Fleming) H. Karst based on inter-simple sequence repeats (ISSR) markers. Afr J Biotechnol 16:895–907

Niiho Y, Yamazaki T, Nakajima Y, Yamamota T, Ando H, Hirai Y (2006) Gastroprotective effects of bitter principles isolated from Gentian root and *Swertia* herb on experimentally-induced gastric lesions in rats. J Nat Med 60:82–88

Palevitch D (1991) Agronomy applied to medicinal plant conservation. In: Akerele, Heywood OV, Synge H (eds) Conservation of medicinal plants. University Press, Cambridge, UK, pp 168–178

Pandey DK, Basu S, Jha TB (2012) Screening of different East Himalayan species and populations of *Swertia* L. based on exomorphology and mangiferin content. Asian Pac J Trop Biomed 2:1450–1456

Pant M, Bisht P, Gusain MP (2010) In vitro propagation through axillary bud culture of *Swertia chirata* Buch. Ham. ex Wall: an endangered medicinal herb. Int J Integr Biol 10(1):48–53

Patil P (2011) Studies on morphological variation and breeding system in *Swertia chirayita* (Roxb. ex Flem.). Karst Thesis Master of Science in forestry (forest products) college of forestry Dr. Yashwant Singh Parmar University of Horticulture and Forestry, Nauni-173 230, Solan, (H.P.) India

Peter HC, Howard T, Edzard E (2005) Bringing medicinal plants into cultivation: opportunities and challenges for biotechnology. J Trends Biotechnol 23(4):45–49

Pierce A, Laird S, Malleson R (2002) Annotated collection of guidelines, standards and regulations for trade in non-timber forest products (NTFPs) and botanicals. Version 1.0. Rainforest Alliance, New York. [http://www.rainforest-alliance.org/news/2002/botanicals-standards.pdf]

Pourkheirandish M, Komatsuda T (2007) The importance of Barley genetics and domestication in a global perspective. Ann Bot 100:999–1008

Pradhan BK, Badola HK (2008) Ethnomedicinal plant use by Lepcha tribe of Dzongu valley, bordering Khangchendzonga Biosphere Reserve, in North Sikkim, India. J Ethnobiol Ethnomed 4:22

Pradhan BK, Badola HK (2015) *Swertia chirayta*, a threatened high-value medicinal herb: microhabitats and conservation challenges in Sikkim Himalaya, India. Mt Res Dev 35(4):374–381

Pyakurel D (2008) Resource assessment of Chiraito (*Swertia chirayita*) in Langtang National Park Buffer Zone Area. LNPBZ-SP/WWF, Rasuwa

Qayum A, Arya R, Lynn AM (2016) Ethnobotanical perspective of antimalarial plants: traditional knowledge-based study. BMC Res Notes 9:67. https://doi.org/10.1186/s13104-015-1827-z

Rai LK, Prasad P, Sharma E (2000) Conservation threats to some medicinal plants of the Sikkim Himalaya. Biol Cons 93:27–33. https://doi.org/10.1016/S0006-3207(99)00116-0

Rao VR, Hodgkin T (2002) Genetic diversity and conservation and utilization of plant genetic resources. Plant Cell Tissue Organ Cult 68:1–19. https://doi.org/10.1023/A:1013359015812

Raina R, Johri AK, Srivastava LJ (1994) Seed germination studies in *Swertiachirata* L. Seed Res 22:62–63

Raina R, Patil P, Sharma YP, Rana RC (2013) Reproductive biology of *Swertia chirayita*—a temperate critically endangered medicinal plant. Caryologia 66(1):12–20

Raina R, Kumar J, Sharma YP, Bandana K (2016) Impact of gamma rays on amarogentin and amaroswerin content in M1 generation of *Swertia chirayita* (Roxb. ex Fleming) H. Karst. Med Plants Int J Phytomed 8:46–51

Rehman S, Latif A, Ahmad S, Khan AU (2011) In vitro antibacterial screening of *Swertia chirayita* Linn. against some gram-negative pathogenic strains. Int J Pharm Res Dev 4:188–194

Royer DL, Wilf P, Janesko D, Kowalski EA, Dilcher DL (2005) Correlations of climate and plant ecology to leaf size and shape: potential proxies for the fossil record. Am J Bot 92(7):1141–1151

Saha P, Das S, Bitter (2005) Fraction of *Swertia chirata* prevent carcinogenic risk due to DMBA exposure. Indian J Med Res 22–27 Poster Presentation

Samaddar T, Jha S, Jha T B (2014) Indian *Swertia* from Eastern Himalaya: strategies of conservation and biotechnological improvement. In: The Gentianaceae: characterization and ecology, vol 1. https://doi.org/10.1007/978-3-642-54010-3-11

Samant SS, Dhar U, Rawal RS (1996) Conservation of rare endangered plants: the context of Nanda Devi Biosphere Reserve. In: Ramakrishnan PS, Purohit AN, Saxena KG, Rao KS, Maikhuri RK (eds) Conservation and management of biological resources in Himalaya. Oxford & IBH, New Delhi, India, pp 521–545

Samant SS, Dhar U, Palni LMS (1998a) Medicinal plants of Indian Himalaya: diversity, distribution potential value. Himavikas Occasional Publication No. 13. Gyanodaya Prakashan, Nainital, India

Samant SS, Dhar U, Palni LMS (1998b) Medicinal plants of Himalaya: diversity, distribution and potential values. Gyanodaya Prakashan, Nainital

Sankar ASS, Sugathan NV, Mohan M (2017) A review on synergistic and hormetic action of *Swertia chirata* from homoeopathic perspective. Int J Ayurvedic Med Res 2456–6209

Saravanan S, Hairul Islam VI, Prakash Babu N, Pandikumar P, Thirugnana Sambantham K, Chellappandian M (2014) Swertiamarin attenuates inflammation mediators via modulating NF-kB/I kB and JAK2/STAT3 transcription factors in adjuvant induced arthritis. Eur J Pharm Sci 56:70–86

Scartezzini P, Speroni E (2000) Review on some plants of Indian traditional medicine with antioxidant activity. J Ethnopharmacol 71:23–42. https://doi.org/10.1016/S0378-8741(00)00213-0

Schippmann U, Leaman DJ, Cunningham AB (2002) Impact of cultivation and gathering of medicinal plants on biodiversity: global trends and issues. In: Biodiversity and the ecosystem approach in agriculture, forestry and fisheries. Satellite event on the occasion of the Ninth Regular Session of the Commission on Genetic Resources for Food and Agriculture. Inter-Departmental Working Group on Biological Diversity for Food and Agriculture, Rome

Schneider E, Stekly G, Brunner P (1999) Domestikation von Bergfrauenmantel (*Alchemilla alpine* agg.). Zeitschrift für Arznei und Gewürzpflanzen 4:134–140

Shah IA, Sharma YP, Raina R, Rana R (2011) Pollination studies in *Swertia chirayita*—a critically endangered medicinal plant of Western Himalayas. Open Access J Med Aromat Plant 2(1):14–17

Shanker RU, Ganeshaiah KN (1997) Mapping genetic diversity of *Phyllanthus emblica*: forest gene banks as a new approach for in situ conservation of genetic resources. Curr Sci 73:163–168

Sharma V, Srivastava N, Kamal B, Dobriyal AK, Jadon VS (2011) *Swertia chirayita*: a review to revitalize its importance in pharmaceutical arena. J Pharm Res 4(6):1784–1787

Sheldon JW, Balick MJ, Laird SA (1997) Medicinal plants: can utilization and conservation coexist? Adv Econ Bot 12:1–104

Singh HB, Prasad P, Rai LK (2002) Folk medicinal plants in the Sikkim Himalayas of India. Asian Folklore Stud 61:295–310

Slatkin M (1987) Gene flow and geographic structure of natural populations. Science 236(4803):787–792. https://doi.org/10.1126/science.3576198

Soica C, Oprean C, Borcan F, Danciu C, Trandafirescu C, Coricovac D (2014) The synergistic biologic activity of oleanolic and ursolic acids in complex with Hydroxypropyl-γ-Cyclodextrin. Molecules 19:4924–4940

Tabassum S, Mahmood S, Hanif J, Hina M, Uzair B (2012) An overview of medicinal importance of *Swertia chirayita*. Int J Appl Sci Technol 2:298–304

Takhtajan AL (1978) The floristic regions of the world. Soviet Sciences Press, Moscow

Thapa B, Sharma P, Baskey S (2021) *Swertia chirayita*: an endangered medicinal herb of the Himalayas. Agri-India Today Monthly e-Newsletter 01(2)

Thapa B, Chhetri B, Mahato SK, Sharma P, Gurung S, Baskey S (2023) Exploration, collection and conservation of genetic resources of *Swertia chirayita* in Darjeeling and Sikkim Himalayas. Med Plants Int J Phytomed 5(3):511–520

Tilman D (2000) Causes, consequences and ethics of biodiversity. Nature 405:208–211. https://doi.org/10.1038/35012217

Uniyal RC, Uniyal MR, Jain P (2000) Cultivation of medicinal plants in India. A reference books. TRAFFIC India and WWF India, New Delhi, India

Van AJ (1998) Intraspecific variability in the context of ecological restoration projects. Perspect Plant Ecol Evol Syst 1:221–237

Vashistha RK, Nautiyal BP, Nautiyal MC (2006) Conservation status and morphological variation between populations of *Angelica glauca* Edgew and *Angelica archangelica* Linn. in Garhwal Himalaya. Curr Sci 91(11):1537–1542

Verma NK, Kumar A (2001) Isozyme polymorphism and genetic diversity among Swertia species—endangered medicinal plants of Himalayas. Indian J Plant Gen Resour 14:74–77

Verma H, Patil PR, Kolhapure RM, Gopalkrishna V (2008) Antiviral activity of the Indian medicinal plant extract, *Swertia chirata* against herpes simplex viruses: a study by in-vitro and molecular approach. Indian J Med Microbiol 26:322–326

Verma PPS, Nashra A, Singh V, Shanker K, Kumar B (2020) Survival, performance and adaptation of *Swertia chirayita* genotypes at Western Himalayan zone of Uttarakhand. J Med Aromat Plants 42(3–4):220–228

Volis S (2015) Species targeted plant conservation: time for conceptual integration. Isr J Plant Sci 63:232–249

Wang BY, Shi L, Ruan ZY, Deng J (2011) Genetic diversity and differentiation in *Dalbergia sissoo* (Fabaceae) as revealed by RAPD. Genet Mol Res 10(1):114–120

Wang L, An L, Hu Y, Wei L, Li Y (2009) Influence of phytohormones and medium on the shoot regeneration from leaf of *Swertia chirata* Buch-Ham. ex Wall. In vitro. Afr J Biotechnol 8:2513–2517

Wawrosch C, Maskay N, Koop B (1999) Micropropagation of the threatened Nepalese medicinal plant *Swertia chirata* Buch-Ham. ex Wall. Plant Cell Rep 18(12):997–1001

Yokogawa M, Kaneko S, Takahashi Y, Isagi Y (2013) Genetic consequences of rapid population decline and restoration of the critically endangered herb *Polemonium kiushianum*. Biol Conserv 157:401–408

Zhou NJ, Geng CA, Huang XY, Ma YB, Zhang XM, Wang JL (2015) Anti-hepatitis B virus active constituents from *Swertia chirayita*. Fitoterapia 100:27–34. https://doi.org/10.1016/j.fitote.2014.11.011

Zohary D, Hopf M (1993) Domestication of plants in the Old World. Clarendon Press, Oxford

Taxonomy, Demography and Status of Threatened *Phlomoides superba* (Royle ex Benth.) Kamelin & Makhm.

Arti Garg and S. K. Singh

Abstract This study provides a comprehensive taxonomic, demographic, and status assessment of *Phlomoides superba* (Royle ex Benth.) Kamelin & Makhm., a perennial herb belonging to the tribe *Phlomideae Mathieses.* The taxonomic history, including the resurrection of the genus *Phlomoides* based on morphological and molecular studies is elucidated. Also, the growth and flowering phenology, pollination, and breeding mechanisms of *P. superba* are meticulously described. The species is characterized by its unique morphological features, including thick tuberous roots, bilipped yellow flowers, and trigonous nutlets. The study highlights the challenges faced by the species, including anthropogenic threats like over-exploitation for medicinal purposes, habitat loss due to road widening, and the impact of biotic factors such as insect infestation and grazing by wild animals. The threat status of *P. superba* is discussed, emphasizing the critical need for conservation efforts. The demography of *P. superba* is explored, with insights into its reproductive biology, seed set, and germination. The species exhibits poor sexual reproduction and seed germination in its natural habitat, prompting the exploration of alternative vegetative propagation methods. The study presents the successful use of tissue culture techniques for plant propagation and multiplication. Current population status reveals the existence of multiple populations, mitigating initial concerns about its extinction. The article concludes with the recommendation to reevaluate the conservation status of *P. superba* and underscores the importance of ongoing efforts to preserve this species.

Keywords Taxonomic history · Reproductive biology · Demography · Conservation · Threat status

A. Garg (✉) · S. K. Singh
Botanical Survey of India, Northern Regional Centre, 192, Kaulagarh Road, Dehradun, Uttarakhand 248195, India
e-mail: artibsi.garg@gmail.com

A. P. Mishra et al. (eds.), *Threatened Medicinal Plants in the Indian Himalayan Region*, World Sustainability Series, https://doi.org/10.1007/978-3-031-73687-2_11

1 Introduction

Phlomoides superba (Royle ex Benth.) Kamelin & Makhm., belongs to the tribe *Phlomideae* Mathieses. The tribe *Phlomideae* is represented by 250–278 species in the world under six genera viz. *Eremostachys* Bunge, *Lamiophlomis* Kudo, *Notochaete* Benth., *Phlomis* L., *Phlomoides* (L.) Moench and *Pseudoeremostachys* Popov, (Scheen et al. 2010; Kamelin and Makhmedov 1990; Govaerts et al. 2010). The members of this tribe are mainly distributed in sub-alpine and alpine regions from Europe to Mongolia, China and India, with some xeric species. The plants are non-aromatic herbs, subshrubs or shrubs, usually with the characteristic thyrsoid, or rarely with racemoid inflorescence bearing single to many-flowered cymes. The flowers have typically bilipped, zygomorphic corolla, with posterior lip characteristically elongated, hooded and densely to sparsely bearded outside and/or at margins, stamens variously exerted, unequally or equally lobed styles and bearded or glabrous nutlets.

The new, monotypic genus *Phlomoides* was described by Moench in 1794 on the basis of *Phlomis tuberosa* L., after segregating it from the genus *Phlomis* L. The genus *Eremostachys* was established by Bunge (1830), as intermediate genus between *Molucella* L. and *Phlomis* L., with four species, two transferred from other genera and two new species (Regel 1884) viz., *Eremostachys laciniata* (L.) Bunge, *E. moluccelloides* Bunge, *E. Phlomoides* Bunge and *E. tuberosa* (Pall.) Bunge. However, since these species had close affinity with the genera *Phlomis* and *Molucella*, Bunge (1873) again divided the genus *Eremostachys* into two sections, sect. *Phlomoides* and sect. *Molucelloides*. Later, Briquet (1896) added the third section *Metaxoides* Briq. under *Eremostachys*. While the genus *Phlomoides* of Moench (1794) remained overshadowed for about two centuries since its establishment, it was resurrected in 1986 by Adylov et al. (1986) which was subsequently followed by Adylov and Makhmedov (1987) to accommodate the species with open hooded (not laterally compressed) upper corolla lip, the galea, and tuberous roots. Thus, the genus *Eremostachys* bearing the aforesaid characters were placed under section *Phlomoides,* while the species assigned to sect. *Metaxoides* of Briquet were kept under genus *Paraeremostachys* Adylov, Kamelin & Makhm. (Adylov et al. 1986; Adylov and Makhmedov 1987). Being ‘illegitimate’ homotypic synonym, the genus *Paraeremostachys* was treated under synonymy of *Eremostachys* (Hegde 1990). Later, the molecular taxonomic studies conducted by Mathiesen, Scheen and Lindqvist (2011) proved the validity of genus *Phlomoides* of Moench (1794), already endorsed by Adylov et al. (1986) with the subsumed genera *Pseudoeremostachys* Popov, *Notochaete* Benth. and *Lamiophlomis* Kudo. Considering paraphylletic origin, the remaining two genera, *Phlomis* sensu stricto and *Eremostachys* were treated independently. Azizian and Culter (1982) concluded that the genera *Eremostachys* and *Phloimoides* were congeneric by their close affinities in Karyology, morphology and occurrence of similar glycosides, chrysoeriol and luteolin, hence based on principle of priority, the name *Phlomoides* (L.) Moench was accepted over *Eremostachys* Bunge. The molecular chloroplast DNA studies conducted by Salmaki et al. (2012) strengthened the aforesaid view, and as such the two genera viz., *Phlomis* L. and *Phlomoides* (L.) Moench. were accepted under the tribe.

The genus *Phlomoides* (L.) Moench, with about 173 species worldwide (POWO) is globally distributed from western limits of Czechoslovakia and Austria up to the eastern boundaries of China and Korea and from northern Russia to southern India. India has poor representation of the genus, only 8 species, mainly occurring in West Himalayan states, Himachal Pradesh, Jammu and Kashmir, Uttarakhand and Uttar Pradesh (Dash and Mao 2020). The species *Phlomoides superba* (Royle ex Benth.) Kamelin & Makhm. (syn. *Eremostachys superba* Royle ex Benth.), known as the 'Golden Himalayan Spike', occurs in foot-hills of Garhwal Himalaya in India, and Peshawar in Pakistan (Hooker 1885; Srivastava et al. 2017). It inhabits loamy soil on calcareous rocks, in partially exposed places along forest edges (Karavelioouilary et al. 2008), mostly in association with herbs of Asteraceae, Euphorbiaceae and Fabaceae. The species was described as new species by Royle in 1832 based collection from Kheeree Pass (now known as Mohand pass) in lower Shiwalik hills of Dehradun (Bentham 1833). Almost a century later, it was again collected from Himachal Pradesh by Mukherjee (1940). Severe threats both natural and anthropogenic were imposed to the species, hence it was treated vulnerable by Jain and Sastry (1980) and 'endangered' by Jain and Sastry (1984) and finally siphoned to 'critically endangered' with a single population of ca. 100 plants in the type locality at Mohand in Uttar Pradesh that further declined to merely 25 extant individuals (Babu 1977; Rao and Garg 1994; Garg and Rao 1997a). It is interesting to note that this population size was less than 50% strength of minimum viable population (Shaffer 1981) and much below (almost 5%) the strength of minimum number of individuals suggested for long term evolutionary flexibility (Menges 1991). The alarming collapse invoked investigations on demography and breeding mechanism of the species (Garg and Rao 1997a, b, c), coupled with extensive surveys to underexplored areas (Koul et al. 1997) in search of any other unnoticed or eclipsed populations, if any, which might generate a new ray of hope to salvage of the species from extinction. Fortunately, Botanical Survey of India, Northern Regional Centre, Dehradun took up initiative to conserve and safeguard the species and multiplied a huge germplasm (Panwar et al. 2015). Since then, the species is thriving well within the associated garden.

2 Methodology

The research methodology involved extensive field surveys in various regions of India, particularly in Jammu and Kashmir, Himachal Pradesh, and Uttarakhand. Taxonomic investigations included morphological characterization and molecular studies to validate the taxonomic status of *Phlomoides superba*. Demographic studies were conducted to assess population sizes, distribution, and threats faced by the species. In-depth analyses of growth and flowering phenology, pollination, and breeding mechanisms were undertaken through field observations and experiments.

Vegetative propagation methods, including tissue culture techniques, were explored to enhance plant multiplication. Seed set, germination, and seed dormancy were studied under controlled conditions. The current population status was determined through systematic field surveys, aiming to locate new populations and assess the overall health and distribution of *P. superba*. The findings provide a holistic understanding of the taxonomic, demographic, and conservation aspects of this endangered species.

3 Results

3.1 Taxonomic Description

Phlomoides superba (Royle ex Benth.) Kamelin & Makhm., Bot. Zhurn. (Moscow & Leningrad) 75: 248. 1990. *Eremostachys superba* Royle ex Benth., Hooker's J. Bot. Kew Gard. Misc. 3: 381. 1833. *Eremostachys laciniata* var. *superba* (Royle ex Benth.) Regel., Trudy Imp. S.-Peterburgsk. Bot. Sada 9: 554. 1886 (Figs. 1 and 2).

Fig. 1 *P. superba* habit

Fig. 2 *P. superba* flowering spike

Vernacular names: 'Gajar-moola' (Dogri—Jammu & Kashmir); 'Gorein' 'Goreyan' (Pahari, Kangri—Himachal Pradesh); Ban muli (Hindi).

Perennial herb, up to 50 cm high. Root tuberous, thick, hard, growing vertically down. Stem reduced, up to 3 cm high, sometimes 2–3 stems arising from single root. Leaves radical, forming a rosette at stem apex, petiolate; lamina oblong-ovate or ovate-lanceolate, 7–35 cm long, 5–14 cm broad, at base deeply lobed or pinnatipartite, basal pinnae in 1–3 pairs, lower pinnae subopposite remote, shortly stalked; stalk 0.8–1.8 cm long, pubescent, leaf margins cilialte, lobed; lobes 9–12, lobulate; surface scabrid with sparse indumentum of erect hairs on both surfaces; petioles 8–23 cm long villous, canaliculated dorsally. Inflorescence with 3–5 pairs of cauline leaves, terminal, 70–180 cm long; spike single, or up to 3, 25–50 cm long, bearing successive whorls of 10 flowers each in bracteate verticillasters of five on either side of the axis; bracts ovate-lanceolate to lanceolate, 6–10 mm long, almost 1/2 to 2/3 of the calyx length, margins spinulose, surface villous. Flowers yellow, bisexual, zygomorphic. Calyx subcampanulate, broadly 5-lobed, 15–20 mm long, mouth truncate densely woolly, subcoriaceous, with indumentum of simple non-glandular hairs, sometimes also with some shorter glandular hairs; teeth spinulose apically with 1–2 mm long spine. Corolla woolly-pubescent, bright yellow, 2.5–3.0 cm long with short 0.5–1 cm long tube, bilipped; upper lip 2-lobed, erect, open-hooded forming open galea, villous, white-bearded within; lower lip longer than the upper one's, 3-lobed; lobes subequal, midlobe largest. Stamens 4, didynamous, studded on corolla tube, all ascending; filaments lower pair longest, glabrous, upper pair fimbriate at base; anther thecae dark brown, divaricate, filled with pollens; pollen tricolpate, reticulate, homobrochate, 36–37.5 × 30–32.5 μm, subprolate, P/E ratio: 1.14–1.33. Ovary glabrous, bicarpellary with tuft of long erect hairs at distal end; each carpel deeply lobed, appearing as 4; style simple, 4 cm long, bifid at apex; stigma delicate, bifid. Nutlets 4, at the base of persistent calyx, trigonous, golden brown to dark brown, 6.5–7 × 3–3.5 mm, with a tuft of pappus-like hairs at the distal end. Seeds triangular, triquetrous, 8–9 × 4–4.3 μm, apex with tuft of stiff, multicellular hairs.

Flowering & Fruiting: March–May.

Distribution: India (Jammu & Kashmir, Himachal Pradesh, Uttarakhand), Eastern Afghanistan, Pakistan.

Notes: The spikes of the *ex-situ* plants were bigger 25–50 cm long, with 100–130 flowers each, whereas those under *in-situ* conditions were smaller, 15–25 cm, with 30–60 flowers (Garg and Rao 1996).

Uses: The root tubers have medicinal property. The native peoples and Gujjars use root tuber for curing mastitis and restoration of milking in cattle, including cows and buffaloes. Its bioactive components are utilized as antidepressants and antioxidants (Samant et al. 2007).

Growth and Flowering Phenology: The basal leaves emerge from the tuberous root stock, just after monsoon season by the end of September and mature by the mid December forming a rosette. The inflorescence emerges by the end of January to elongates up to 120–180 cm by mid-February, it's stalk bearing 3–5 pair of cauline leaves, which terminates in to an erect, densely wooly spike which is up to 30 cm long. The spike bears 100–130 flowers in ex-situ condition in contrast to natural populations with 30–60 flowers. Flowers appear in dense whorls of 8–10 flowers each, arranged in verticillate of 5, on either side of the axis, in the axil of a densely wooly bract. Bud initiation takes place as soon as the inflorescence emerges. Flowering occurs under dry weather conditions usually by mid-March when temperature reaches around 30–35 °C.

Flowers mature is acropetal and centrifugal succession within an inflorescence, and is simultaneous in the opposite bracts of each whorl. Peak flowering occurs during 3rd week of March when more than70% flowers bloom at a time and the spikes look handsome. Each flowers bloom for about 4–5 days after which the corolla withers and falls off leaving behind the persistent calyx enclosing four nutlets. Anthesis takes place within the buds and anthers dehisce longitudinally and release cream-colored pollen dust on lower corolla lip with some pollen adhering loosely on anthers. The flowers produce about 1,44,560 pollen (SD = 1200) against 4 ovules with mean pollen production being 33,600 (SD = 300) per anther and 95% viability (Garg and Rao 1997a, b, c). All the four ovules are capable of developing into seeds. The seeds produced either all healthy 12–13 mm long with an average weight 31 mg

(28–32.5 mg), or all sterile, 4–5 mm long, shriveled, empty and almost weightless (0.002–0.01 mg), or both types. Post pollination, the fertilization leading to seed maturation, and dispersal takes 20–30 days.

3.2 Pollination, Plant-Pollinator Co-Adaptation and Breeding Mechanism

Although Koul et al. (1997) reported bumble bee (Hymenoptera) as pollen carriers and probable pollinators of *Phlomoides superba* flowers, studies conducted by Garg & Rao from 1994 to 1997 conducted series of experiments and field observations and confirmed that among many insect visitors, only two Hymenoptera bees *Ceratina heiroglyphica* Sm. and *Nomia rustica* West. were active and effective pollinators those visit the flowers for collecting pollen rewards, both in experimental garden as well as in the natural habitat in type locality (Garg and Rao 1996). A symbiotic relationship was observed between flowers and pollinating bees, in which the bees received pollen rewards, and in return the flowers were benefitted by pollen transfer to the stigma which is brought about by the specific foraging pattern of bees (Garg and Rao 1996). The unique pollen collection by the bees were well designed during which they alight on the lower corolla lip, crawl inwards and climb up the stamens to reach anthers and stigma. This process is repeated several times during which the bees work in inverted position (sternotribic pollination). The pollen entangled in their body hairs are packed in pollen baskets and transferred to the sticky stigma during their flights bringing about pollination. In between, the bees took short flights on same and different inflorescences causing autogamous, geitonogamous as well as xenogamous pollination (Garg and Rao 1997a).

In the entire event of pollination process, while the handsome flowering spikes draw attention of pollinators, the nectarless flowers produce abundant pollen rewards, ca. 1,44,560 pollen per flower. The bilabiate corolla provides suitable landing platform on which the bee alights and then proceed inward and upward to access the elevated anthers, in an inverted position for pollen collection. The monolectic, oligolectic and 'pollinivorous' pollinators therefore visit the flowers ethodynamically and sternotribically exhibiting perfect floral fidelity behavior, their visits corresponding to anthesis (Garg and Rao 1997b, c) of the oligophilic flowers, indicating successful pollination. Pollinator limitation therefore, is only an outcome of scanty pollinator population (Garg and Rao 1996) resulting in only one or two insect pollinators, and this being after effect of dwindled plant population with limited number of blossoms. As a result, the plants face inbreeding depression, in the small isolated population leading to reproductive bottleneck. Natural pollination in the population of the type locality at Mohand, Dehradun, produced 6.9% seeds, but was increased to 12.5% in ex situ geitonogamous hand pollinated plants, and 44.4%by xenogamy. Autogamy, by forced selfing reared only 2.8–3.8% seeds, suggesting that the species was a normal out-breeder (Garg and Rao 1997a) but is forced to inbreed for want of another population in vicinity of the pollinators.

Vegetative propagation: The species was found to have a poor sexual reproduction and seed germination in natural habitat, but perennated by means of it's thick, deeply penetrating rhizome. Hence the population devoid of genetic variation, required variability by outcrossing with another population. The only known population existed at Domel in Peshawar, Pakistan, a distance beyond the normal foraging range of the pollinators. Hence for mass propagation, especially for the horticultural purposes, vegetative methods of propagation were adopted and also found suitable (Panwar et al. 2015; Srivastava 2020). IBA treatment of the foliage crown, buds arising from the sides of the main plant as well as from basal rosette of well-developed mature plants yielded good results. The plants propagated fast under treatment with 200 ppm IBA and showed 95% root initiation. Treatment with 100 ppm IBA produced 55% and 500 ppm produced 70% roots. Propagation through seeds was not so effective and took much more, nearly three years' time to bloom.

Seed set and germination: Under natural conditions, seed failed to germinate, however germination occurred under control conditions in MS medium with growth regulators, hence Rao and Garg (1994), suggested tissue culture techniques for plant propagation and multiplication. Later, Panwar et al. (2015) found that under storage conditions, the seeds tend to lose moisture and viability, the loss being directly proportional to the storage time. It was noticed that germination was ameliorated up to 6 months storage time, after which germination get slumping with further storage. Seeds stored under refrigeration at 0–4 °C showed better viability compared to the seeds stored under room temperature. Furthermore, fresh seeds portrayed maximum viability, but low germinability which could not even be enhanced by GA3 application. However, the seeds stored at room temperature as well as at 0–4 °C for 6-months exhibited high germinability, which gets further enhanced by GA3 treatment. This suggests intrinsic dormancy prevalent in fresh seeds which breaks when stored for about 6 months, confirming the observations of Garg and Rao, (1997a). Tricotoledony was also observed in about 2% germinated seeds (Panwar 2014).

Seed germination experiments revealed that presoaking the seeds stored in refrigeration at 0–4 °C for 6 months, in 400 ppm of GA, initiated germination within 17 days and shoot multiplication increased upto 77.8% in medium with BAP growth regulators (Panwar et al. 2015) which proved superior over kinetin (Sunnichan and Shivanna 1998). The micropropagated plants showed good performance under greenhouse conditions with high survival rates even after 6 weeks of transplantation, organogenesis from shoot tip explants was established as the most comprehensive and direct regeneration and micro-propagation method for effective shoot proliferation, rhizogenesis and acclimatization of *P. superba* in vitro conditions. The seedling reared by this method resulted in successful establishment, acclimatization and survival in natural conditions, and prove the best for species multiplication and germplasm conservation for improved vigor and adaptability.

3.3 Threat Status

Phlomoides superba faced several threats operative on the population in type locality—at Mohand due to over exploitation-the shephards, Gujjars and Bakerwals dig out the medicinal root stock in bulks during spring season and store for future use to feed cows, buffaloes and goats for inducing lactation in non-lactating cattle which damages the plant population on the whole. The roots are also used for curing human ailments related to liver, stomach, gout etc. As the plants grow along roadsides of highway habitat loss during road broadening is a potential threat for habitat loss. Elephant stampede and grazing is common in the area which results in breakage and crushing of the plants. Biotic factors, insect infestation and forage of tubers by beetles and wild animals (porcupines, boars and rats) coupled with low regeneration potential, result in complete loss of the plant. The plants are adapted for out -crossing, but in absence of other nearby populations within the normal pollinator range of the specific pollinators, they are forced to inbreed which leads to inbreeding depression in the normal out-breeders and loss of genetic variability and progeny vigor. The low population leads to decline in pollinator population causing pollinator scarcity, low seed set and strong genetic drift leading to pollen competition on stigma surface for good quality pollen (Garg and Rao 1997b). Seed dormancy under natural conditions is detrimental as the seeds are eaten away by beetles, pests, rodents and birds, and also tend to get fungal infection wiping out the chances of population multiplication. The seeds exhibit inherent dormancy, reduced viability and germinability under natural conditions and hence succumb to environmental pressures.

3.4 Current Population Status

Following the alarm raised by demographic studies, scientific surveys were conducted extensively to unearth other eclipsed, under-evaluated or ignored populations of *Phlomoides superba*. The survey teams were sent in March, 1996, by Koul and party to Domel and spotted a population of about 400 plants, which gave respite to the botanical alarm of the species extinction (Rao and Garg 1994). Subsequently, three more populations in District Jammu, one each at village Suketor, Domel, and Khandel, together with about 700 individuals was unearthed (Koul et al. 1997).

A decade later, Verma et al. (2007) reported two populations at Sunderbani and Rajouri in Jammu & Kashmir and Uniyal et al. (2012) reported two populations in Una and Kangra in Himachal Pradesh and in Udhampur District of Jammu & Kashmir by Srivastava et al. (2017) of nearly 500 individuals (200 mature plants and 300 saplings) in about 45 m^2 area. With these populations, and in addition to the *ex-situ* plants surviving in the experimental garden of the Botanical Survey of India, Northern Regional Centre, Dehradun, the present populations of the species *P. superba* scores to a total of about 12 healthy populations comprising an estimated at least 1600 individuals. We therefore conclude that *Phloimoides superba* is one such species which has endured the test of time and can now be restored from the RET category.

Acknowledgements The authors are thankful to Director, Botanical Survey of India, Kolkata and H.o.O., Botanical Survey of India, Northern Regional Centre, Dehradun for facilities and encouragements.

References

Adylov TA, Kamelin RV, Makhmedov AM (1986) Zametki o semeistve Lamiaceae 1 (Notes on Lamiaceae 1). Novosti Sist Vyssh Rast 23:110–114

Adylov TA, Makhmedov AM (1987) *Phlomoides* Moench. In: Adylov TA (ed) Conspectus florae Asiae Mediae, vol 9. Izdatel'stvo Akademii Nauk SSSR, Tashkent, pp 82–107

Azizian D, Cutler DF (1982) Anatomical, cytological and phytochemical studies on *Phlomis* L. and *Eremostachys* Bunge (Labiatae). Bot J Linn Soc 85:249–281

Babu CR (1977) Herbaceous flora of Dehra Dun, Council of Scientific and Industrial Research, New Delhi

Bentham G (1833) Account of Indian Labiatae in the collection of J. F. Royle. Bot Misc 3:370–384

Briquet J (1896) Labiatae. *Eremostachys* Bunge, *Phlomis* L. In: Engler A, Prantl K (eds) Die natürlichen Pflanzen familien, IV (3a, 134). Engelmann, Leipzig, pp 246–249

Bunge A (1830) *Eremostachys* Bunge. In: Ledebour CF, Meyer CA, Bunge A (eds) Flora altaica, vol 2. G. Reimer, Berlin, pp 414–416

Bunge A (1873) Labiatae persicae. Mém Acad Imp Sci Saint Petersbourg ser 7, 21:1–84

Dash SS, Mao AA (2020) Flowering plants of India: an annotated checklist, vol 2: Dicotyledons. Botanical Survey of India, Kolkata

Garg A, Rao RR (1996) Pollination ecology of endangered *Eremostachys superba* (Labiatae). Taiwania 41(4):309–321

Garg A, Rao RR (1997a) Observations on the reproductive biology of endangered *Eremostachys superba* (Labiatae). Proc Ind Nat Sci Acad B 63(142):79–88

Garg A, Rao RR (1997b) Pollen limitation in endangered *Eremostachys superba*. Taiwania 42(4):307–315

Garg A, Rao RR (1997c) Anthecological relationship between *Eremostachys superba* and its pollinator. Taiwania 42(2):99–103

Govaerts R, Paton A, Harvey Y, Navarro T (2010) World checklist of Lamiaceae & Verbenaceae. The Board of Trustees of the Royal Botanic Gardens, Kew. http://www.kew.org/wcsp/Lamiaceae

Hedge IC (1990) Eremostachys. In: Ali SI, Nasir YJ (eds) Flora of Pakistan, vol 192. University of Karachi, Karachi, pp 137–147

Hooker JD (1885) The flora of British India (Labiatae). L. Reeve & Co., London

Jain SK, Sashtry ARK (1984) The Indian plant red data book—1. Botanical Survey of India, Howrah-64

Jain SK, Sastry ARK (1980) Threatened plants of India. A state-of the-art report. Botanical Survey of India, Howrah (India)

Kamelin RV, Makhmedov AM (1990) Sistema roda Phlomoides (Lamiaceae) [The system of the genus Phlomoides (Lamiaceae)]. Bot Zhurn (Moscow & Leningrad) 75:241–250

Karaveliooullary FA, Ekici M, Celik C, Baser B, Uzunhisarcykly ME, Ocak A, Bani B (2008) The revision of the genus *Eremostachys* (Labiatae) in Turkey. Acta Bot Hungar 50(3–4):000–000. https://doi.org/10.1556/abot.50.2008.3-4.Xx

Koul AK, Mangotra R, Verma S (1997) Is *Eremostachys superba* Royle ex. Benth. really at the verge of extinction? Curr Sci 73:313–314

Mathiesen C, Scheen A-C, Lindqvist C (2011) Phylogeny and biogeography of the Lamioid genus *Phlomis* (Lamiaceae). Kew Bull 66:1–17

Menges ES (1991) In genetics and conservation of rare plants. In: Falk DA, Holsinger KE (eds). Oxford University Press, New York, Oxford, pp 43–61

Moench C (1794) Methodus planta shorti botanicietagri Marburgensis, a staminum situ describendi. Officina Nova Libraria Academiae, Marburg

Mukherjee SK (1940) Records of the botanical survey of India, vol XIV. Botanical Survey of India, Calcutta

Panwar GS, Srivastava SK, Uniyal PL (2015) In vitro propagation of *Eremostachys superba* Royle ex Benth.—an endangered, medicinal and ornamental herb of North-West Himalaya. Med Pl 7(4):264–271

Panwar GS (2014) Tricotyledony in critically endangered herb *Eremostachys superba* Royle ex Benth. (Lamiaceae). Indian For 140(12):1265–1266

Rao RR, Garg A (1994) Can *Eremostachys superba* be saved from extinction. Curr Sci 67(2):80–81

Regel E (1884) Monographia generis *Eremostachys*. Acta Horti Petropolitani 9:529–571

Salmaki Y, Zarre S, Ryding O, Lindqvist C, Scheunert A, Bräuchler C, Heubl G (2012) Phylogeny of the tribe Phlomideae (Lamioideae: Lamiaceae) with special focus on Eremostachys and Phlomoides: new insights from nuclear and chloroplast sequences. Taxon 61(1):161–179

Samant SS, Pant S, Singh M, Lal M, Singh A, Sharma A, Bhandari S (2007) Medicinal plants in Himachal Pradesh, northwestern Himalaya, India. Int J Biodivers Sci Manag 3:234–251

Scheen A-C, Bendiksby M, Ryding O, Mathiesen C, Albert VA, Lindqvist C (2010) Molecular phylogenetics, character evolution and suprageneric classification of Lamioideae (Lamiaceae). Ann Missouri Bot Gard 97:191–219

Shaffer ML (1981) Minimum population sizes for species conservation. Bioscience 31(2):131–134

Srivastava A, Sharma YP, Sharma OP, Vidyarthi, Srivastava SK (2017) Conservation status assessment and new population record of the threatened Golden Himalayan Spike *Phlomoides superba* (Royle ex Benth.) Kamelin & Makhm. from Jammu & Kashmir. Indian J Threatened Taxa 9(4):10089–10095

Srivastava A (2020) Development of propagation protocol for the endangered species Phlomoidessuperba (Royle ex Benth.) Kamelin & Makhm. Trop Plant Res 7(3):541–546

Sunnichan VG, Shivanna KR (1998) Micropropagation of *Eremostachys superba*-an endangered, endemic species from India. Curr Sci 74:699–702

Uniyal SK, Prakash O, Lal B (2012) Relocating Eremostachys superba Royle ex Benth. In Himanchal Pradesh. Curr Sci 103(5):467–468

Verma S, Karihaloo JL, Tiwari SK, Magotra R, Koul AK (2007) Genetic diversity in Eremostachys superba Royle ex Benth. (Lamiaceae), an endangered Himalayan species, as assessed by RAPD. Genet Resour Crop Evol 54:221–229

Spatial Extent of Threatened Medicinal and Aromatic Plants in Alpine Zone of Kumaun, Western Himalaya

Naveen Chandra, Gajendra Singh, Dhani Arya, Saumya Agnihotri, Vinod Chandra Joshi, Arun Pratap Mishra, Ravindra Kumar, and Lalit Mohan Tewari

Abstract The utilization of wild Medicinal and Aromatic Plants (MAPs) has been integral to healthcare practices since the inception of human civilization. The escalating demand for herbal healthcare systems has exerted immense pressure on wild resources, leading to the depletion of several MAP populations. In this context, comprehending the environmental suitability and potential distribution of these species becomes imperative for effective conservation and management. This study seeks to evaluate the potential niche of 25 MAPs in the alpine region of the Kumaun Himalaya through species distribution modelling. The study area encompasses 4617 km^2 area between 3000 and 5300 m elevation range. Topographic variables derived from Digital Elevation Models, vegetation types, and climatic variables are considered for assessing habitat suitability. Distribution records of the species are obtained through extensive field surveys conducted in 30 alpine meadows of the Kumaun Himalaya, employing 63 transects and approximately 1120 quadrats following the rapid mapping exercise. Six dominant alpine habitats, including tall forbs, mixed herbaceous, *Danthonia* grassy slopes, moist alpine scrubs, *Kobresia* sedge meadows, and marshy meadows, with *Danthonia*-dominated grassy slopes as the most preferred by MAPs. Analysis of habitat suitability reveals that ~309 km^2 area is highly suitable for species reintroduction. Among the MAPs, *Picrorhiza kurroa* exhibits the maximum potential distribution area (230 km^2), while *Aconitum balfourii* displays the lowest (58 km^2) in the region. The MaxEnt model results indicate that altitude,

N. Chandra · D. Arya · R. Kumar
Department of Botany, Soban Singh Jina University Campus, Almora, India

G. Singh
Uttarakhand Space Application Centre, Upper Aamwala, Nalapani, Dehradun, India

S. Agnihotri · L. M. Tewari
Department of Botany, D.S.B Campus, Kumaun University, Nainital, India

V. C. Joshi · A. P. Mishra (✉)
Department of Forestry and Remote Sensing, Earthtree Enviro Private Limited, Shillong, Meghalaya 793012, India
e-mail: arunpratap7371@gmail.com

A. P. Mishra et al. (eds.), *Threatened Medicinal Plants in the Indian Himalayan Region*, World Sustainability Series, https://doi.org/10.1007/978-3-031-73687-2_12

aspect, vegetation type, precipitation during the wettest months, and annual precipitation are the most influential factors determining the potential distribution of MAPs across various habitats.

Keywords Alpine region · Medicinal and Aromatic Plants (MAPs) · Species distribution modelling · Habitat suitability

1 Introduction

The Himalayan region is one of the most biodiverse areas globally, harbouring a diverse range of plant and animal species including several rare and threatened species (Dar and Khuroo 2020; Mishra et al. 2023a). The region is home to several high use medicinal plant species, many of which possess significant medicinal properties. The Indian Himalayan Region (IHR) is repository of medicinal and aromatic plants (MAPs) having over 1748 species of medicinal importance; some of them are categorized under various threatened categories of IUCN (Kandari et al. 2011). In the IHR, a total of 713 species, including 112 MAPs have been listed as threatened (Mehta et al. 2020). Whereas, in the alpine region of Uttarakhand, out of the 1400 species of angiosperm plants, 400 species are medicinal in nature (Rawat 2005) and about 40 MAPs are under various threatened categories (Ved et al. 2005). Threatened medicinal plants have garnered international attention due to their rapid population loss in the wild and high demand as a commodity for health care (Kala 2003; Kumar et al. 2023). MAPs face risk of extension across the IHR states, due to low population, over-exploitation, habitat degradation, invasive species, livestock grazing, and other anthropogenic activities (Kala 2000; Chandra et al. 2021c). The alpine region has a huge variety of medicinal plants and endemism because of the broad range of habitats that help to retain high levels of species richness and variation (Suissa et al. 2021). This makes the alpine region extremely significant from an ecological point of view. However, human-induced activities such as overgrazing, unorganized and over harvesting, habitat destruction and climate change have led to rapid decline in the populations of these medicinal plant species, leading to their threatened status (Mishra et al. 2023a). These threats have led to the decline and even extinction of several plant species, which has further exacerbated the challenges faced by local communities (Kala 2000). Efficient information on their distribution, which is generally obtained over extended periods of time and has low spatial accuracy (Lomba et al. 2010; Chandra et al. 2023a). Similarly, locating the populations of rare, endangered, and threatened (RET) species is crucial for biodiversity conservation and the development of successful in-situ and ex-situ conservation (Menon et al. 2010; Mishra et al. 2023b; Mishra 2023).

However, owing to rugged terrain and low accessibility, the available information on the spatial extent, distribution pattern and population status of threatened MAPs of the alpine region is fragmented (Kala 2000; Chandra et al. 2023b). The niche assessment of the threatened medicinal plants and the prioritisation of management areas has received inadequate attention (Chandra et al. 2021a, b; Rawat et al. 2022). Hence, the analysis of the niche dynamics of vulnerable MAPs is essential for future monitoring and their management. Considering the aforesaid in view, the current study elucidated the potential habitat niche of MAPs in Kumaun Himalaya in the absence of any such previous attempt.

2 Study Area

The alpine region of the Kumaun region was selected for the intensive study (Fig. 1). A total 4617 km^2 between 3000 and 5300 m (above mean sea level) elevation was proposed for present study. The local inhabitants above 3000 m are basically agro-pastoralists and extensively use the alpine areas for livestock grazing during summer. The intensive study area has diverse timberline communities within a short distance. According to Champion and Seth's classification (1968), vegetation in the study area are falls under sub-alpine, moist alpine and dry alpine scrub. Physiographically the study area is characterized by undulating topography with gentle slopes on northern, north-eastern and north-western aspects and steep slopes (>60°) on southern and south-western faces. At the higher elevations, micro-topography is more diverse and reflected in the plant community composition. Topography range from flower laden lush green meadows to extreme xeric stony deserts and ending into stupendous snow-clad mountain peak. The precipitous mountain slopes of alpine region offer little room for cultivation but wherever the little land is available, high altitude village settle down. In Kumaun Himalaya upper limit of plant growth has been recorded upto 5400 m elevation whereas, it is up to 5600 m in Norh-West Himalaya (Rawat 2007).

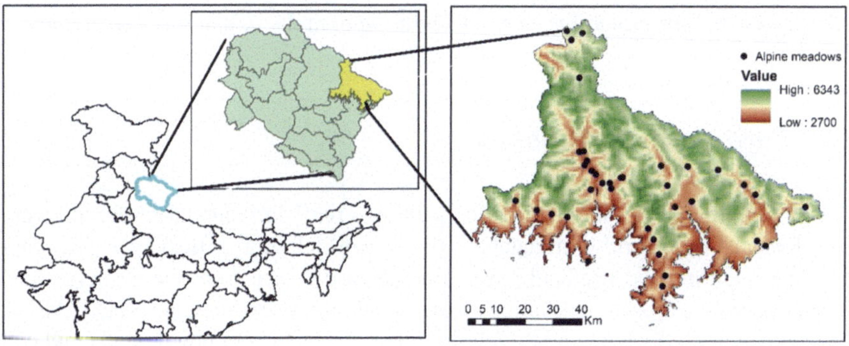

Fig. 1 Map showing distribution of surveyed meadows in Kumaun region of Uttarakhand

3 Methodology

3.1 Collection of Occurrence Data of MAPs

Distribution data or occurrence points of plants was collected from following sources (i) Primary information was collected based on the extensive field work and collection of various ground truth points, (ii) Questionnaire survey, (iii) Secondary information was collected from different floral literature, and (iv) Consulting herbarium records.

3.2 Satellite Data Used

Landsat-8 satellite data was used to generate various geo-spatial layers of the region (Mishra et al. 2021). Satellite data of September–October 2017 was downloaded in GEOTIFF format with UTM WGS84 Zone 44 projection from USGS (https://earthexplorer.usgs.gov). The swath of Landsat 8 data was 185 km (115 miles) with 30 m multi-spectral spatial resolutions and 16 days temporal resolution. The path and row of the Landsat-8 satellite data were 146 and 39, respectively.

3.3 Topographical Data from DEM

The ASTER Digital Elevation Model with spatial resolution of 30 m was downloaded from http://earthdata.nasa.gov (accessed on November 2017) to generate various topographical layers (Slope, Aspect, Elevation). These layers were used to study the variations in habitat suitability of MAPs with topography (Chandra et al. 2023c).

3.4 Environmental Variables

The 19 bioclimatic variables (bio1-19) with 30 s (ca.1 km) spatial resolution were downloaded from World-clim database (www.worldclim.org). Besides these, other variables such as land use/land cover, Euclidean distance from the drainage line, vegetation type map, geomorphology and degradation maps were generated. All the environmental layers have to be with the same extent (geographic bounds and cell size) as MaxEnt model requires all the information (environmental layers) in raster format having same cell size, extent and projection system (e.g., geographic or UTM) in order to execute the prediction (Rawat et al. 2021; Mishra et al. 2023c). All the geo-spatial environment variable layers were created and modified in the Arc GIS 10.2 version platform. Environmental variables used for habitat suitability assessment are presented in Plate 4.1.

3.5 Spatial Extend Assessment

Maximum Entropy (MaxEnt) or ecological niche modelling (Phillips et al. 2006) technique was used for predicting the potential distribution of MAPs and key factors were identified associated with its distribution. All the species occurrence records and environmental variables were incorporate in to population distribution model. MaxEnt algorithm was used for generating an estimate of probability of presence of the species that varies from 0 and 1, where 0 being the lowest and 1 is the highest probability. The maximum number of background points was 10,000 and Linear, quadratic and hinge features were used as auto feature. We selected 70% data for training and the rest 30% for testing. The Jackknife procedure was used to assess the importance of the variables. The area under the receiving operator curve (AUC) was used to evaluate model's goodness-of-fit and model with highest AUC value was considered as the best performer. The performance of the model was validated by applying the receiver operating characteristic (ROC) curves, which is generally used for assessing the accuracy of any diagnostic test (Swets 1988). We used area under ROC curves (AUC) that is a base for assessing accuracy of the model. The AUC value ranges between 0 and 1, of which AUC value of 1 indicates the ideal model i.e. good predictive power of model. In contrast, AUC value ≤ 0.5 indicate the inaccurate model i.e. the results of the model outcome are poor (Phillips et al. 2006).

4 Results

4.1 MAPs with General Habitat Preferences and Wide Environmental Limits

The species of this category were more generalised in their habitat preferences as they are recorded from different habitat types. *Aconitum heterophyllum* showed preference in terms of its distribution in sub-alpine forest, grassy slopes, and herbaceous meadows. In some areas, it was found distributed in glacial riverine, rocky moist areas, alpine dry scrub, shady moist alpine slopes usually in humus-rich soil between 2800 and 5000 m. It is mainly associated with *Danthonia cachemeriana* dominated grassy slope in association with *Potentilla atrosanguinea*, *Geum elatum* in herbaceous meadows and shruberries. *Aconitum balfourii* found to be distributed in moist shady subalpine slope under canopy of *Rhododendron campanulatum, Taxus wallichiana* and *Quercus semecarpifolia,* moist shady open meadows and moist shady places. High potential zone was predicted near sub-alpine forest (Darma and Ralam valley) margin and open moist meadows. Least potential zone in grassy slope and *Kobresia* meadows. *Aconogonum rumicifolium* observed in shrubberies and alpine slopes. At places, dominant along treeline gaps and in herbaceous meadows. Highly preferable areas were under the treeline gap and tall herbaceous formations, while least was observed in grassy slope and shrubberies habitat. *Euphrasia himalaica* was reported another generalist species which occur on moist slopes, herbaceous meadows and grassy slope. *Jurinea dolomiaea* was a common species in gently undulating alpine herbaceous slopes and grassy slope dominated by *Danthonia cachmeriana*. The observed and predicted sites were mostly in grassy slope (70%) and alpine scrub (30%) and rarely in *Kobresia* sedge meadows. *Malaxis muscifera* mainly prefer herbaceous meadows, alpine grassy slopes and sparsely distributed in subalpine forest of Himalayan birch (*Betula utilis*) and *Rhododendron campanulatum* krummholz. *Ophiocordyceps sinensis* was found to prefer in all kinds of alpine habitats including herbaceous, *Danthonia* grasslands and *Kobresia* sedge meadows. It also occurs in open dwarf vegetation formed by *Rhododendron anthopogon* in moist slopes. Major vegetation type encountered for their potential sites were mixed herbaceous formations, *Kobresia* sedge meadows and gaps of alpine moist scrub. *Sinopodophyllum hexandrum* was observed to prefer sub-alpine forest margin (*Rhododendron campanulatum, Betula utilis*), alpine scrub (*Berberis jaeschkeana, Juniperus indica*), rocky areas and grassy slope.

4.2 MAPs with Special Habitat Preferences and Narrow Environmental Limits

Allium stracheyi prefers moist herbaceous meadows and rocky slopes in association with *Allium* spp., *Rheum webbianum* in rocky slopes and *Danthonia cachmeriana* in grassy slopes. Natural habitats of *Angelica glauca* are moist shady sub-alpine slopes mostly found in the shady canopies of *Rhododendron companulatum, Betula utilis, Quercus semecarpifolia* and *Acer* forests between 3000 and 3700 m. The observed and predicted *A. glauca* sites were mostly in forest fringes (60%) followed by scrubland (30%) and least in grassy slopes. Natural habitats of *Allium wallichii* are open alpine slopes and shrubberies, although it is found near sub-alpine forest and herbaceous meadows. *Fritillaria cirrhosa* occur on alpine grassy slopes and in moist herbaceous meadows. The observed and predicted sites were mostly in *Danthonia* grassy slope (50%) in west facing slope. *Picrorhiza kurroa* was mostly found along forest fringe and shady moist and well- drained soil in association with *Salix lindleyana* and *Gaultheria tricophylla*. The observed and predicted *P. kurroa* sites were mostly in forest fringes (42%) followed by grassy (30%) and rocky slopes (23%), with slopes between 28° and 36° in the south-west and north-west aspects being highly preferred. *Dactylorhiza hatagirea* mostly prefers mesic and moist alpine meadows associated with tall forbs, mixed herbaceous meadows, and *Danthonia* grassy slopes for its growth, although it has more suitable habitats near the treeline under Himalayan birch (*Betula utilis*) forest and *Rhododendron campanulatum* krummholz. The species is mostly found in mixed herbaceous meadows (45%) followed by *Danthonia* grassy slopes (30%) and tall forbs (25%) areas having >34° slopes and North, North-East aspects.

Arnebia benthami grows well on the open slopes and shrubberies. Its natural habitats are near open alpine slopes and shrubberies. Scattered population was found near the sub-alpine forest and herbaceous meadows. *Bergenia stracheyi* grows gregariously in rocky and bouldery slopes. Its population observed on bouldery/rocky moist slopes. *Ephedra gerardiana* occur in alluvial fans, gravel terraces and rocky slopes. *Hippophae salicifolia* occur on moist riverine area in the inner valleys and drier alpine slope. *Meconopsis aculeata* shows preferred and potential sites in rocky alpine slope, bouldery and rock ledges. *Megacarpaea polyandra* mainly confined to moist scrub habitat. Its potential sites were encountered in open slopes, understory of *Rhododendron anthopogon* scrub and shady moist localities. *Nardostachys jatamansi* occur on shady moist rocks and boulder area often associated with *Rhododendron anthopogon* and *Juniperus communis*. *Rheum australe* thrives in rocky and bouldery slopes. Scattered potential habitat suitability was predicted in dry rocks, moist rocks, scree slope, grassy slope, under shrub, undulating meadows and rarely in grassland habitat. *Rheum moorcroftianum* mainly confined to alpine stony and scree slopes between 3500 and 4500 m. *Trillidium govanianum* prefer under the shade of rocks or dense canopy of trees. Habitat suitability was also predicted in temperate and subalpine mixed forest (*Betula utilis, Abies pindrow, Cedrus deodara* and *Juniperus indica*). Krummholtz (*R. campanulatum*) zones were also hospitable for persistence of species. The observed and predicted sites were mostly in subalpine forest (60%), and Krummholtz zone (40%).

4.3 Environmental Variables Influencing Species Distribution

The Jackknife test showed the training gain of each variable if the model was run in isolation, and compared it to the training gain with all the variables. This is useful to identify which variables contribute the most individually. The environmental variable when used in isolation, with the highest gain is vegetation type followed by soil type and drainage for *Aconitum heterophyllum*, drainage for *Aconitum balfourii*, aspect (*Adenia rumicifolia, Angelica glauca, Allium wallichii,* Arnebia benthamii, *Bergenia stracheyi, M. polyandrea, Fritillaria cirrhosa*) slope (*Allium stracheyi, Dactlorhizea hatagirea*), degradation (*Ephedra gerardiana, Hyoscyamus niger*), bio 12 (*E. himalaica, Jurinea dolomiaea*, bio 13 (*Hippophae salicifolia, Malaxis muscifera, Nardostachys jatamansi, Trillium govanianum, Picrorhiza Kurroa*). Vegetation types for *Meconopsis aculeata*, bio 2 for *Ophiocordyceps sinensis*, bio 2, bio 13, bio 12 and altitude for *Sinopodophyllum hexandrum*, soil type (*R. australe, R. moorcroftianum*) which therefore, appears to have the most useful information by itself. Altitude of the area also structure the distribution of MAPs (*M. aculetea, M. polyandra, H. salicifolia, D. hatagirea, A. benthamii, A. glauca* and *B. stracheyi*), when used individually. The environmental variable that decreases the gain the most when it is omitted is precipitation seasonally (bio 15), drainage, degradation, geomorphology appears to have the most information that is not present in the other variables.

4.4 Potential Habitat Suitability of MAPs

The MaxEnt model generated a map of predicted probabilities of occurrence showing areas with high potential to low potential (Figs. 2, 3, 4and5). High potential habitat of MAPs encompasses grassy slope, shrubberies and herbaceous meadows. The maximum potential zone was recorded for *P. kurroa* (230 km^2) followed by *A. stracheyi* (210 km^2) and *B. stracheyi* (200 km^2), while it was lowest for *Aconitum balfourii* (53 km^2) and *Malaxis muscifera* (57 km^2). Detailed of habitat suitability of MAPs in (Table 1).

5 Discussion

Due to sparse and clustered data on distribution of Himalayan threatened plants, it is arduous to model their suitable habitat distribution using commonly used modelling approaches. There are many combinations of predictors that can explain distributions equally well, especially when environmental factors are correlated with the geographical locations of the species. In this study, the total geographical area estimated between 3000 and 5300 m elevation zone was 4617 km^2, of which a high potential habitat for 10 MAPs was recorded only 515 km^2. Most of the moderate to high potential habitats of the threatened medicinal plants predicted and evaluated through field surveys were recorded in tussock grassy slopes (*Danthonia* dominated), herbaceous meadows (understory of *Betula*, *Rhododendron* and *Quercus* forests), shrubberies (*Berberis*, *Cotoneaster* and *Rhododendron*). These trees, shrubs and tussock forming grasses form microclimate and shelter to the tiny herbaceous species and also supports to eliminate from grazing and trampling.

Species namely, *A. balfourii*, *A. glauca* and *Sinopodophyllum hexandrum* show their maximum potential zone under the forest fringes *(R. campanulatum*, *Betula* and *Abies* forest). *Aconitum heterophyllum* mainly scattered in grassy slope, its distribution was also determined by vegetation type and soil in the region. *Dactylorhiza* thrives in marshy meadows and topography of the region determines its potential site for suitability. However, mean diurnal range (bio 2), aspect and slope are responsible for the persistence of *Dactylorhiza* populations in Western Himalaya (Chandra et al. 2021a, b, c). *A. glauca*, a native and endemic species of the Himalayan region that is distributed along the sub-alpine and alpine regions mainly grows in grassy, moist rocky slopes, near water springs, humus rich soil and forest edges. It has about 332 km^2 highly suitable area which is reported in alpine region of Pithoragarh district (66.76 km^2), Kumaun Himalaya and Kedarnath wildlife sanctuary (53.53

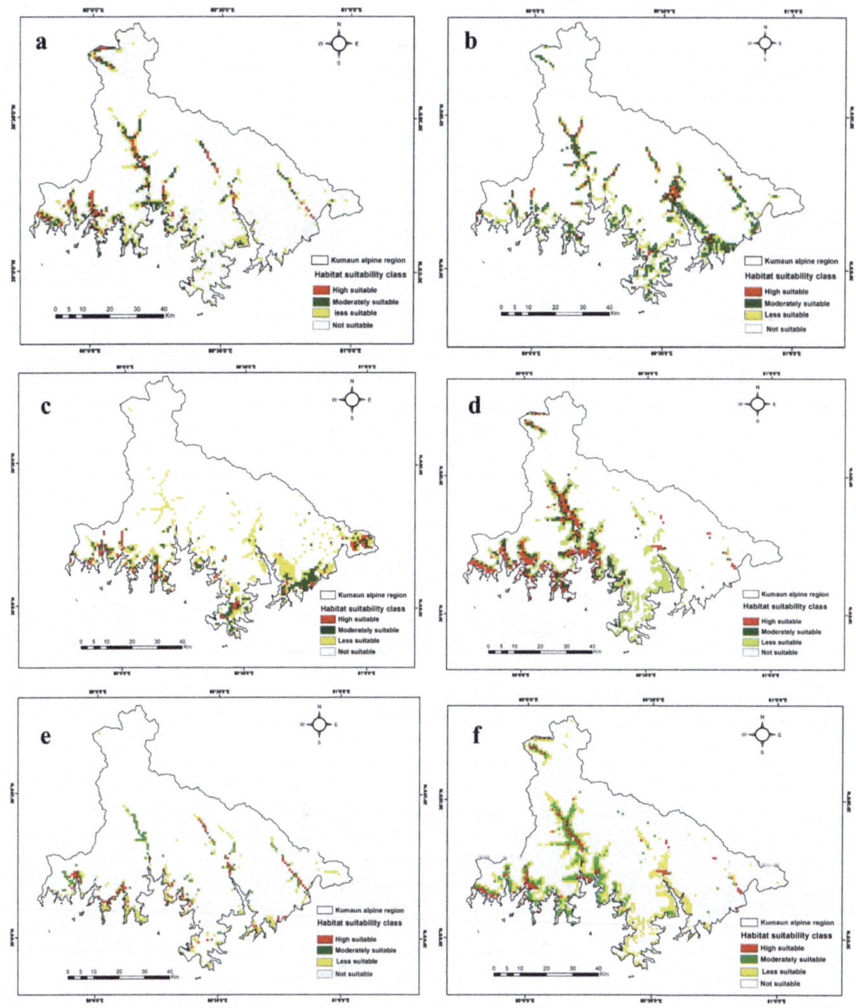

Fig. 2 Predicted habitat suitability of MAPs **a** *Aconitum heterophyllum*, **b** *Aconitum balfourii*, **c** *Anthurium rumicifolium*, **d** *Allium stracheyi*, **e.** *Angelica glauca*, **f** *Allium wallichii*

km^2). Precipitation played a key role in determining the distribution of potential habitats of *A. glauca* in its native range in Western Himalaya (Singh et al. 2020). Among all the selected variables, bio 12 (annual precipitation) has highest contribution (72%) for *Euphrasia himalayica* followed by bio 13 (precipitation of wettest month) for *Hippophae salicifolia* and aspect for *B. stracheyi*. Precipitation of wettest month has also the maximum contribution to determine the suitability of *N. jatamansi*, *P. kurroa*, *T. govanianum* and *Euphrasia himalayica*. Owing to their habitat in rocky and boulder areas, *R. australe* and *R. moorcroftianum* sites were determined

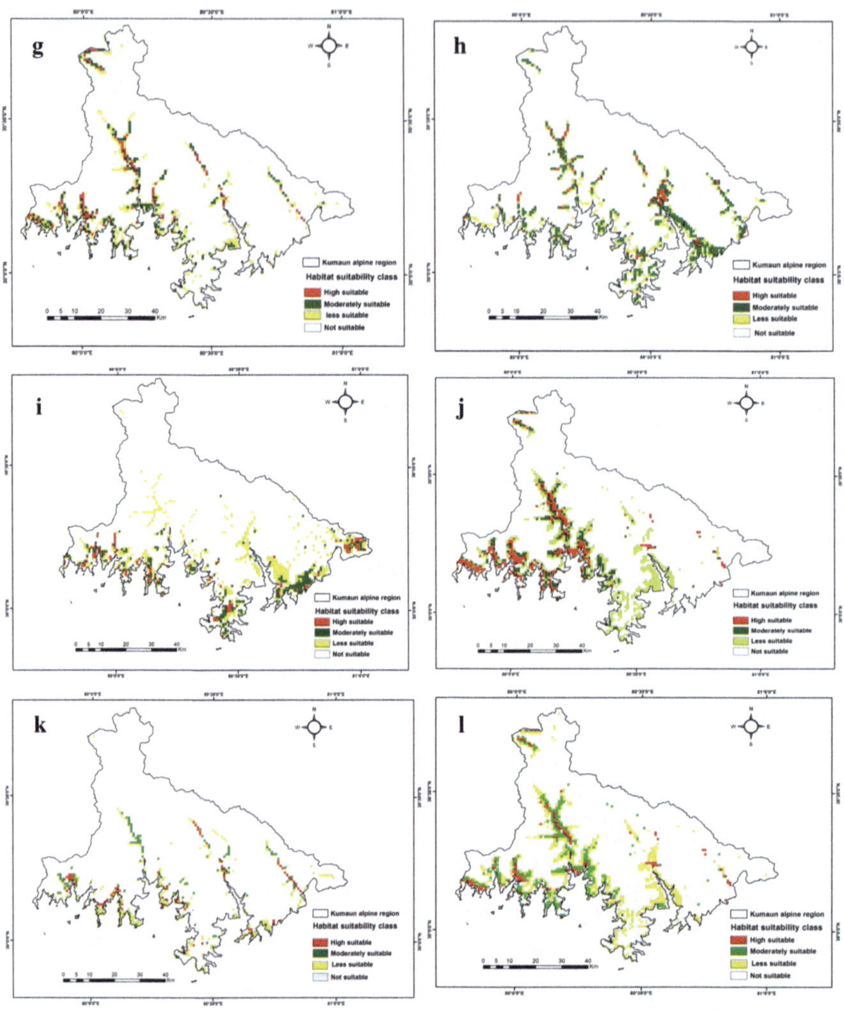

Fig. 3 **g** *Arnebia benthamii*, **h** *Bergenia stracheyi*, **i** *Dactylorhiza hatagirea*, **j** *Ephedra gerardiana*, **k** *Euphrasia himalayica*, **l** *Fritillaria cirrhosa*

by existing soil type and slope of the area. *M. polyandra* require optimum topography for its persistence in the region. Among the topographic factors, aspect has minimum contribution to determine the potential sites of *Malaxis muscifera*, *B. stracheyi*, *A. wallichi*, *F. roylei* and *A. rumicifolium*. However, land use and land cover and precipitation were the most influential environment predictors that determine its distribution in Western Himalaya (Chandra et al. 2021a, b, c; Singh et al. 2023). Disturbance regime and drainage are hospitable for *E. gerardiana* and *Jurinea dolomiaea* distribution in the landscape. Aspect also plays an important role to structure the potential habitats of threatened species followed by slope, whereas altitude has a

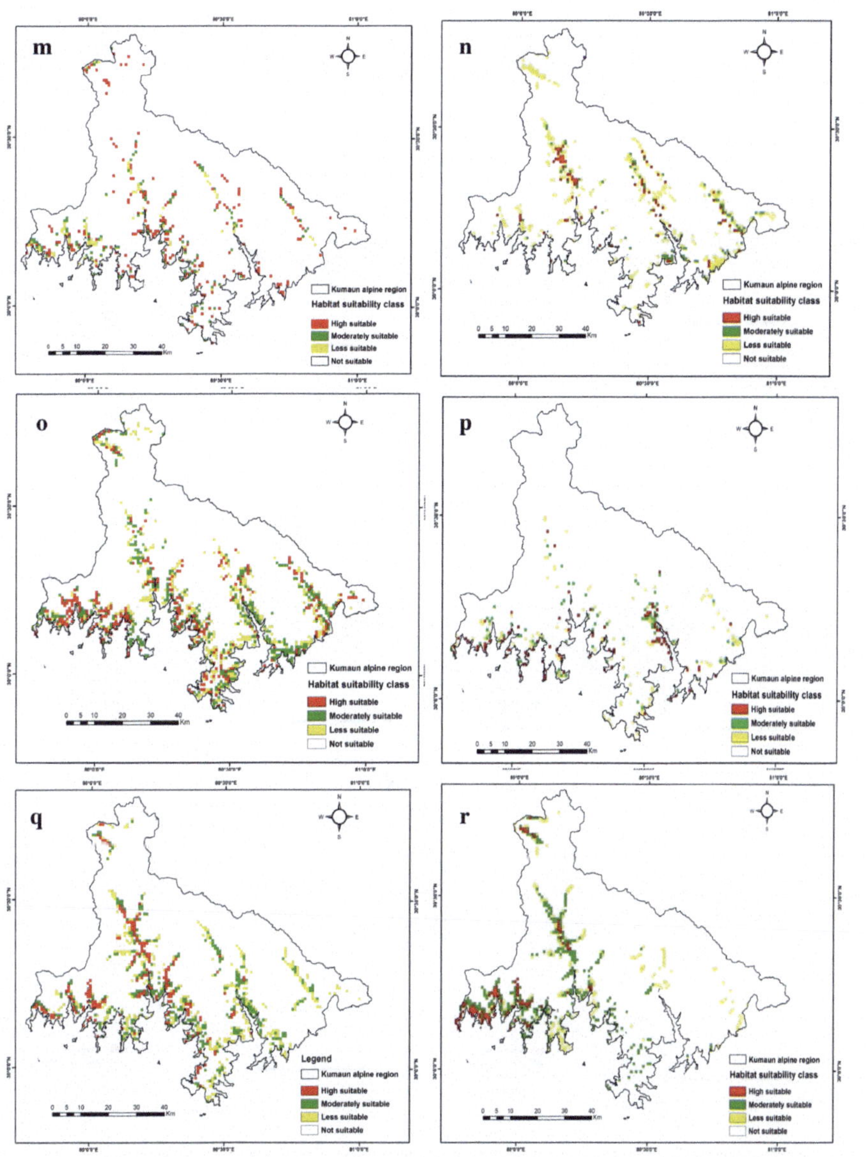

Fig. 4 **m** *Hippophae salicifolia*, **n** *Hyoscyamus niger*, **o** *Jurinea dolomiaea*, **p** *Malaxis muscifera*, **q** *Meconopsis aculeata*, **r** *Megacarpaea polyandra*

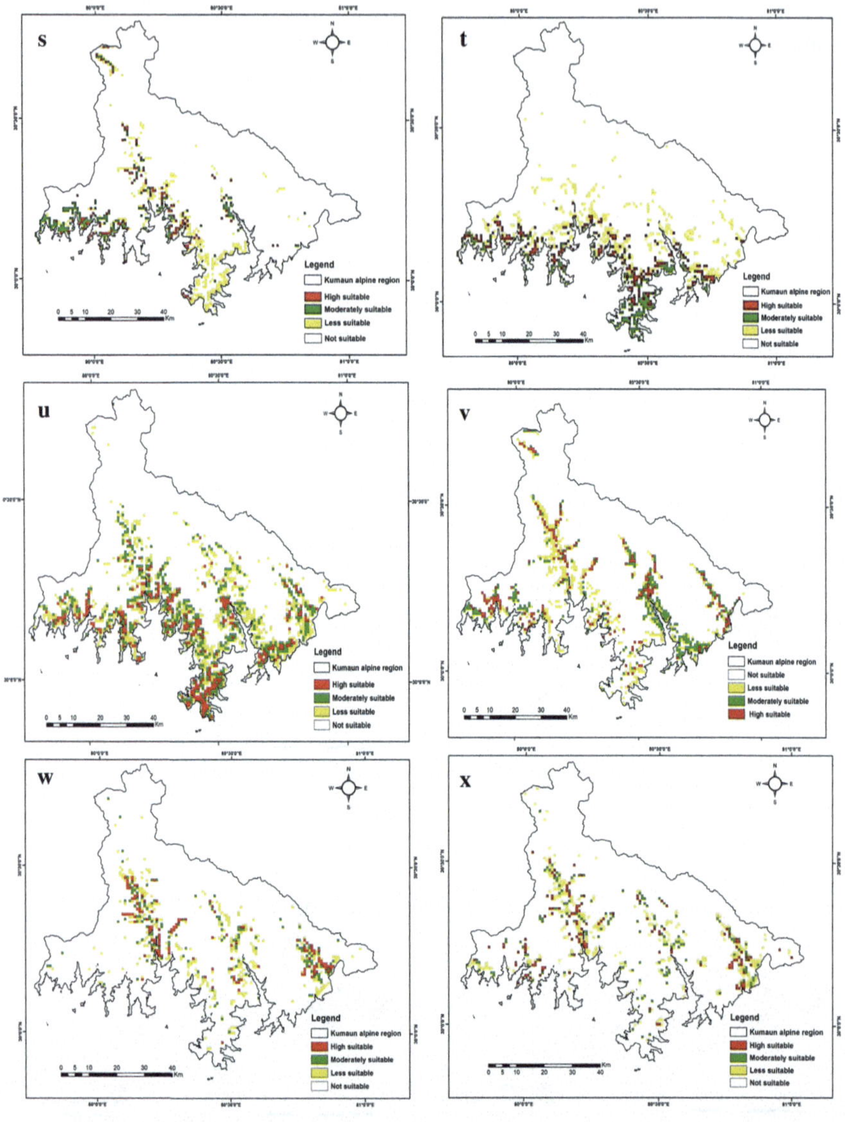

Fig. 5 Predicted habitat suitability of MAPs **s** *Nardostachys jatamansi*, **t** *Ophiocordyceps sinensis*, **u** *Picrorhiza Kurroa*, **v** *Sinopodophyllum hexandrum*, **w** *Rheum australe*, **x** *Rheum moorcroftianum*

vital role in local distribution of a species (Adhikari et al. 2012). The combination of all the three factors creates a microclimate for rare species those are habitat specific in nature. Topography, precipitation, temperature and vegetation has vital importance for site suitability assessment of MAPs in the region. The distribution of medicinal plants is also regulated significant by soil moisture, soil nutrient availability,

Table 1 Potential habitat suitability of MAPs in Kumaun Himalaya

MAPs name	Highly suitable (km^2)	Moderately suitable (km^2)	Less suitable (km^2)	Not suitable (km^2)
Aconitum heterophyllum	106 (2.3)	190 (4.1)	310 (6.1)	2556 (55.5)
Aconitum balfourii	53 (1.12)	106 (2.29)	240 (5.19)	3000 (64.29)
Anthurium rumicifolium	102 (2.2)	165 (3.5)	363 (7.8)	4000 (82.2)
Allium stracheyi	210 (4.6)	170 (3.6)	419 (9.1)	3800 (79.2)
Angelica glauca	78 (1.6)	180 (3.8)	360 (7.7)	2268 (49.2)
Allium wallichii	130 (2.16)	315 (6.9)	560 (12.3)	3540 (76.2)
Arnebia benthamii	110 (2.4)	140 (3.1)	270 (6.7)	3950 (85.5)
Bergenia stracheyi	200 (4.3)	347 (7.8)	423 (9.1)	921 (19.3)
Dactylorhiza hatagirea	59 (1.27)	101 (2.1)	165 (3.57)	1581 (34.2)
Ephedra gerardiana	100 (2.1)	178 (3.8)	210 (4.5)	1444 (31.2)
Euphrasia himalayica	184 (3.9)	264 (5.7)	200 (4.3)	1456 (31.5)
Fritillaria cirrhosa	130 (2.8)	197 (4.3)	200 (4.4)	1444 (31.7)
Hippophae salicifolia	70 (1.51)	110 (2.38)	160 (3.4)	1596 (34.5)
Hyoscyamus niger	61 (1.32)	65 (1.6)	236 (5.1)	1510 (32.7)
Jurinea dolomiaea	203 (4.3)	305 (6.6)	315 (6.8)	1050 (22.7)
Malaxis muscifera	57 (1.2)	68 (1.6)	90 (1.9)	1724 (37.3)
Meconopsis aculeata	146 (3.16)	252 (5.4)	270 (5.8)	1236 (26.7)
Megacarpaea polyandra	80 (1.7)	235 (5)	123 (3.9)	1477 (31.9)
Nardostachys jatamansi	154 (3.3)	170 (3.4)	247 (5.3)	1389 (30)
Ophiocordyceps sinensis	159 (3.4)	190 (3.6)	281 (6)	1297 (29)
Picrorhiza Kurroa	230 (4.9)	386 (8.3)	484 (10.4)	1823 (39.4)
Sinopodophyllum hexandrum	156 (3.3)	236 (5.1)	315 (6.8)	1185 (25.6)
Rheum australe	100 (2.1)	113 (2.3)	209 (4.5)	1493 (32.3)
Rheum moorcroftianum	165 (3.5)	180 (3.8)	140 (3)	1430 (30)
Trillium govanianum	90 (1.9)	110 (2.3)	265 (5.7)	1460 (30.9)

precipitation (Mueller et al. 2016). Analysis of habitat suitability reveals that overall suitable area for species reintroduction is 713 km^2 of which 404 km^2 is less suitable and 309 km^2 is highly suitable. The predicted suitable areas compose a mosaic of habitats including *Danthonia* grassy slopes, moist meadows, marshy meadows, *R. campanulatum* and *Betula utilis* patches are among high probability areas for the species, hence, these areas could be used for *in-situ* conservation of the species.

The present study explains the role of habitat suitability modelling in identifying the habitats for reintroduction of threatened and endemic medicinal plant species. Reintroduction of the target species in the identified potential habitats would help in rehabilitating the species population and in improving management status hence in conserving the overall biodiversity of the region. Some studies have used modelling approaches to successfully map distributions of species with fewer occurrences, for instance MaxEnt modelling of Geckos in Madagascar by using five species occurrence records (Pearson et al. 2007). In present study, models calibrated with good number of occurrence records predicted comparatively larger areas as suitable in comparison with known distribution. It is also found that, many areas predicted by models as suitable were without the target species. Possible reasons for this could be either due to commission error by the model or because of the inability of the species to disperse to these locations.

6 Conclusion

In the realm of biodiversity conservation, niche modelling emerges as a pivotal tool, especially when considering the reintroduction or natural assisted regeneration of threatened species. The identification of potential habitats through this study serves as a foundational step for in-situ conservation and eco-restoration efforts directed towards the focal species. The delineation of potential sites provides essential baseline information crucial for informed decision-making in conservation strategies.

An innovative stride towards the conservation of medicinal plants, particularly those facing threats, involves the establishment of Medicinal Plant Conservation Areas (MPCA). This proactive measure can serve as a sanctuary for these valuable plant species, ensuring their survival, and fostering ecological balance. By designating specific areas for the conservation of medicinal plants, we create controlled environments conducive to their growth and propagation, aiding in the preservation of biodiversity.

Additionally, empowering the biodiversity management committees at the Gram Panchayat level assumes paramount significance in the conservation and management of natural resources, particularly the rich diversity of medicinal plants in remote regions. These committees, equipped with local insights and community engagement, play a vital role in implementing sustainable practices, monitoring biodiversity, and safeguarding the fragile ecosystems. The findings of this study not only contribute to the scientific understanding of potential habitats through niche modelling but also provide practical insights for on-the-ground conservation efforts. The establishment of Medicinal Plant Conservation Areas and the reinforcement of local governance structures exemplify actionable steps towards the preservation of medicinal plant diversity, ensuring the sustainability of these invaluable resources for generations to come.

References

Adhikari D, Barik SK, Upadhaya K (2012) Habitat distribution modelling for reintroduction of *Ilex khasiana* Purk., a critically endangered tree species of northeastern India. Ecol Eng 40:37–43. https://doi.org/10.1016/j.ecoleng.2011.12.004

Chandra N, Singh G, Rai ID, Mishra AP, Kazmi MY, Pandey A, Abdo HG (2023c) Predicting distribution and range dynamics of three threatened *Cypripedium* species under climate change scenario in Western Himalaya. Forests 14(3):633. https://doi.org/10.3390/f14030633

Chandra N, Singh G, Lingwal S, Bisht MPS, Tewari LM (2021a) Population assessment and habitat distribution modelling of the threatened medicinal plant *Picrorhiza kurroa* Royle ex Benth. in the Kumaun Himalaya, India. J Threatened Taxa 13(7):18868–18877. https://doi.org/10.11609/jott.5603.13.7.18868-18877

Chandra N, Singh G, Lingwal S, Bisht MPS, Tewari LM, Joshi VC (2021b) Ecological status of alpine medicinal and aromatic plants of Western Himalaya. J Herbs, Spices Med Plants 28(1):73–88. https://doi.org/10.1080/10496475.2021.1963903

Chandra N, Singh G, Lingwal S, Jalal JS, Bisht MS, Pal V, Tiwari LM (2021) Ecological niche modelling and status of threatened alpine medicinal plant *Dactylorhiza hatagirea* D.Don in Western Himalaya. J Sustain For 1–17

Chandra N, Kumar A, Mishra AP, Singh G, Rawat GS (2023a) *Wigandia urens* (Ruiz & Pav.) Kunth: a new emergent invasive alien species in the Western Himalaya. Int J Environ Stud 1–14. https://doi.org/10.1080/00207233.2023.2194157

Chandra N, Rai ID, Mishra AP, Dwivedi SK, Kotiya A, Tiwari UK, Singh G (2023b) Assessing potential habitats and populations of selected medicinal herbs in Alpine areas of Uttarakhand, Western Himalaya. Indian J For 45(3):144–155. https://doi.org/10.54207/bsmps1000 2023-ITRFNL

Dar GH, Khuroo AA (2020) An introduction to biodiversity of the Himalaya: Jammu and Kashmir State. In: Dar G, Khuroo A (eds) Biodiversity of the Himalaya: Jammu and Kashmir State. Topics in biodiversity and conservation. Springer, Singapore, p 18. https://doi.org/10.1007/978-981-32-9174-4_1

Kala CP (2000) Status and conservation of rare and endangered medicinal plants in the Indian trans-Himalaya. Biol Cons 93(3):371–379. https://doi.org/10.1016/S0006-3207(99)00128-7

Kala CP (2003) Commercial exploitation and conservation status of high value medicinal plants across the borderline of India and Nepal in Pithoragarh. Indian For 129(9):80–84

Kandari LS, Rao KS, Maikhuri RK, Kharkwal G, Chauhan K, Kala CP (2011) Distribution pattern and conservation of threatened medicinal and aromatic plants of Central Himalaya, India. J For Res 22:403–408. https://doi.org/10.1007/s11676-011-0110-y

Kumar A, Adhikari BS, Mishra AP (2023) Status of threatened medicinal and Biosphere Reserve, Uttarakhand. A report (TR No./2022/35) submitted to International Union for Conservation of Nature—India, p 103

Lomba A, Pellissier L, Randin C, Vicente J, Moreira F, Honrado J, Guisan A (2010) Overcoming the rare species modelling paradox: a novel hierarchical framework applied to an Iberian endemic plant. Biol Cons 143(11):2647–2657. https://doi.org/10.1016/j.biocon.2010.07.007

Mehta P, Sekar KC, Bhatt D, Tewari A, Bisht K, Upadhyay S, Negi VS, Soragi B (2020) Conservation and prioritization of threatened plants in Indian Himalayan Region. Biodivers Conserv 29:1723–1745. https://doi.org/10.1007/s10531-020-01959-x

Menon S, Choudhury BI, Khan ML, Peterson AT (2010) Ecological niche modelling and local knowledge predict new populations of *Gymnocladus assamicus* a critically endangered tree species. Endangered Species Res 11(2):175–181. https://doi.org/10.3354/esr00275

Mishra AP, Rai ID, Pangtey D, Padalia H (2021) Vegetation characterization at community level using sentinel-2 satellite data and random forest classifier in western Himalayan Foothills, Uttarakhand. J Indian Soc Remote Sensing 49:759–771. https://doi.org/10.1007/s12524-020-01253-x

Mishra AP, Kumar A, Yadav SN (2023a) Ecology and conservation of threatened medicinal plants in the Trans-Himalayan region of Nanda Devi Biosphere Reserve. Western Himalaya. https://doi.org/10.1016/j.tfp.2023.100451

Mishra AP, Chandra N, Mandy JJ, Dwivedi SK, Alruzuq A, Pande CB (2023b) Analytical hierarchy process (AHP) based on the spatial assessment of an endangered alpine medicinal herb *Aconitum heterophyllum* in the Western Himalayan environment. In: Climate change impacts on natural resources, ecosystems and agricultural systems. Springer International Publishing, Cham, pp 579–594. https://doi.org/10.1007/978-3-031-19059-9_25

Mishra AP, Chandra N, Srivastava A, Singh SK (2023c) An updated checklist of the vascular flora of Asan Conservation Reserve, Uttarakhand, India. Indian J For 46(1):12–21. https://doi.org/10.54207/bsmps1000-2023-7P1KSV

Mishra AP (2023) Understanding the decline of *Butea monosperma* (Lam.) Kuntze in Gorakhpur District, India: an ecological investigation. J For Geosci 1:2

Mueller M, Schickhoff U, Scholten T, Drollinger S, Böhner J, Chaudhary RP (2016) How do soil properties affect alpine treelines? General principles in a global perspective and novel findings from Rolwaling Himal, Nepal. Prog Phys Geogr 40(1):135–160

Pearson RG, Raxworthy CJ, Nakamura M, Townsend Peterson A (2007) Predicting species distributions from small numbers of occurrence records: a test case using cryptic geckos in Madagascar. J Biogeogr 34(1):102–117

Phillips SJ, Anderson RP, Schapire RE (2006) Maximum entropy modelling of species geographic distributions. Ecol Model 190(3/4):231–259. https://doi.org/10.1016/j.ecolmodel.2005.03.026

Rawat GS (2007) Pastoral practices wild mammals and conservation status of alpine meadows in Western Himalaya. J Bombay Nat Hist Soc 104(1):251–257

Rawat N, Purohit S, Painuly V, Negi GS, Bisht MPS (2022) Habitat distribution modeling of endangered medicinal plant *Picrorhiza kurroa* (Royle ex Benth) under climate change scenarios in Uttarakhand Himalaya, India. Ecol Informatics. https://doi.org/10.1016/j.ecoinf.2021.101550

Rawat N, Mishra AP, Chandra N, Painuly V, Negi GS, Bisht MPS, Purohit S (2021) Maxent modeling for predicting the potential distribution of *Valeriana jatamansi* jones. in chakrata forest division of Garhwal Himalaya. Indian J Ecol 48(2):367–372

Rawat GS (2005) Alpine meadows of Uttaranchal: ecology, landuse and status of medicinal & aromatic plants. Bishen Singh Mahendra Pal Singh, Dehradun

Singh G, Chandra N, Lingwal S, Bisht MPS, Tiwari LM (2020) Distribution and threat assessment of an endemic and endangered species *Angelica glauca* in high ranges of western Himalaya. J Herbs Spices Med Plants 26(4):394–404

Singh R, Mishra AP, Kumar M, Pande CB (2023) Classification of vegetation types in the mountainous terrain using random forest machine learning technique. In: Climate change impacts on natural resources, ecosystems and agricultural systems. Springer International Publishing, Cham, pp 615–628. https://doi.org/10.1007/978-3-031-19059-9_27

Suissa JS, Sundue MA, Testo WL (2021) Mountains, climate and niche heterogeneity explain global patterns of fern diversity. J Biogeogr 48(6):1296–1308. https://doi.org/10.1111/jbi.14076

Swets JA (1988) Measuring the accuracy of diagnostic systems. Science 240:1285–1293

Ved DK, Kinhal GA, Ravikumar K, Sankar RV, Haridasan K (2005) Conservation assessment and management prioritisation (CAMP) for wild medicinal plants of North-East India. Med Plant Conserv 11:40–44

Harmony in Habitats: Unveiling the Conservation Insights Through MaxEnt Modeling of *Withania somnifera* (L.) Dunal in Uttarakhand

Rishabh Kumar, Himanshu Sahu, Salam Bhopen Singh, Naveen Chandra, Sipu Kumar, and Kusum Arunachalam

Abstract *Withania somnifera* (L.) Dunal, a pivotal perennial plant with extensive therapeutic applications in traditional and modern medicine was investigated for its potential distribution in Uttarakhand, India. Employing WorldClim bioclimatic variables such as slope, aspect and elevation (derived from a Digital Elevation Model), and spatially dispersed species occurrence points, the study predicted the potential distribution within the 53,483 km^2 study area. The highly suitable habitat covered 8,557.7 km^2. Key environmental variables such as annual mean temperature (bio1), mean temperature of the warmest quarter (bio10), precipitation of the wettest month (bio13), precipitation of the wettest quarter (bio16), mean temperature of the wettest quarter (bio8), and elevation significantly influenced the species distribution. This insight is crucial for understanding habitat suitability, with 16% highly suitable, 32% suitable, 22% moderately suitable, and 30% least suitable. The study underscores the importance of considering multiple environmental variables for accurate species distribution predictions. The Jackknife test assessed the importance of these variables, revealing a highly accurate Maxent model with a statistically significant AUC value of 0.89. The study identified districts such as Dehradun, Udham Singh

R. Kumar · K. Arunachalam (✉)
School of Environment and Natural Resources, Doon University, Dehradun 248001, India
e-mail: ka@doonuniversity.ac.in

S. Kumar
Department of Habitat Ecology, Wildlife Institute of India, Dehradun 248001, India

H. Sahu · N. Chandra · S. Kumar
Bhomya Foundation, Monal Enclave, Dehradun 248001, Uttarakhand, India

S. B. Singh
School of Biological Sciences, Doon University, Dehradun 248001, Uttarakhand, India

N. Chandra
Department of Botany, Soban Singh Jeena University, Almora 263601, India

H. Sahu
Department of Animal Ecology and Conservation Biology, Wildlife Institute of India, Dehradun 248001, Uttarakhand, India

A. P. Mishra et al. (eds.), *Threatened Medicinal Plants in the Indian Himalayan Region*, World Sustainability Series, https://doi.org/10.1007/978-3-031-73687-2_13

Nagar, lower elevations of Chamoli and Pauri have about 60% cultivation potential. The results have implications for species restoration, conservation planning, and localization of *W. somnifera*. The findings have relevance in species restoration, conservation planning, and localization of *W. somnifera*. This comprehensive approach holds promise for predicting the potential distribution of medicinal plant species, providing valuable ecological insights and aiding in conservation planning and medicinal applications of *W. somnifera* in Uttarakhand.

Keywords Conservation · Distribution · Habitat · Modeling · Planning · Suitability

1 Introduction

The intricate dynamics of plant-environment interactions have captivated researchers for decades, leading to a nuanced understanding of their profound impact on plant growth (Elith and Leathwick 2009; Mishra et al. 2023a; Chandra et al. 2023a). Environmental factors, encompassing temperature, precipitation, humidity, and soil composition, have emerged as pivotal influencers not only in shaping the distribution of medicinal plants but also in determining the composition of their active ingredients (Jochum et al. 2007; Mishra 2023). This realization comes at a critical juncture when the globalization of herbal medicine, combined with unregulated exploitation and inadequate conservation measures, has placed many medicinal plant species on the precipice of extinction. In response, the imperative for habitat restoration and strategic cultivation has gained prominence as a vital ecological measure for the rehabilitation and conservation of these invaluable plants (Barnosky et al. 2011; Mishra et al. 2023b).

Our research advocates the application of ecological modeling, specifically species distribution models (SDMs), to delineate suitable habitat distributions and identify environmental factors critical for both habitat suitability and optimal production of active ingredients in medicinal plants. The fusion of statistical modeling and Geographic Information System (GIS) techniques has significantly advanced our comprehension of ecological phenomena, contributing to the fields of ecology, conservation, and utilization (Guisan and Zimmermann 2000; Warren et al. 2008; Brito et al. 2009; Chandra et al. 2022; Mishra et al. 2023c). In this context, MaxEnt stands out as a standard model for accurately predicting species distribution, employing entropy to generalize specific observations of species presence without necessitating the incorporation of absence points within its framework (Phillips et al. 2006).The significance of medicinal plants extends beyond ecological considerations to their integral role in human civilization, particularly in regions where herbal medicines remain a primary healthcare resource (Owolabi et al. 2007; Mishra et al. 2021, 2023d). Recognizing the therapeutic potential of secondary metabolites present in these plants, they are increasingly utilized as raw materials for essential drug manufacturing globally. Approximately 80% of the world's population relies on traditional medicines for primary healthcare, with more than 25% of modern medicines and their analogues derived from effective medicinal plants (Chandra et al. 2020). India, with its rich biodiversity of over 45,000 medicinal plant species, including regions like the Eastern Himalayas and Western Ghats as hotspots, serves as a poignant example (Ahmedullah and Nayar 1986; Batugal 2004).Uttarakhand, a recent addition to the Indian landscape, grapples with challenges of unemployment and underproductive agriculture. In light of its abundant biodiversity, particularly in fragrant and medicinal plants, herbal plant farming emerges as a viable economic revival strategy. The establishment of herbal gardens, nurseries, and labs, supported by the Indian government, aligns with efforts to create a state-wide export zone for medicinal plants, adding a new dimension to Uttarakhand's agricultural landscape (Gaur 1999; Bhatt 1999; Tiwari et al. 2015; Chandra et al. 2023b; Joshi et al. 2024).

Despite constituting only 2% of the nation's land area, Uttarakhand contributes significantly, encompassing 25% of the nation's biodiversity (Purohit 2000; Samant and Pal 2003). The Himalayan region alone is a source of over half of the 1600 medicinal plant species historically used in India (Uniyal et al. 2002; Adhlkari et al. 2003). The focal point of this study is the exploration of habitat preferences and environmental factors influencing the growth of Withania somnifera, commonly known as Ashwagandha, utilizing MaxEnt modeling. The overarching objective is to unveil conservation insights crucial for the sustainable management of this medicinal plant in Uttarakhand.

2 Materials and Methods

2.1 Study Area

This research was conducted across Uttarakhand, a northern Indian state spanning latitudes 28.43° N to 31.27° N and longitudes 77.34° E to 81.02° E. Encompassing diverse elevations from the Terai plains to the Greater Himalaya, Uttarakhand offers a rich ecological tapestry with subtropical, temperate, and alpine climates. Renowned for its botanical significance, cultural heritage, and varied landscapes, Uttarakhand provides a unique setting for investigating the habitat suitability of *Withania somnifera*.

2.2 Studied Species

W. somnifera Dunal is an erect, stellate-tomentose, grayish, under shrub with 30–75 cm in height having long tuberous roots. Leaves are sub-opposite, broadly ovate to oblong, alternate, petiolate, entire and sub-acute with lamina (5–10 × 2.5–7) cm. Flowers are bisexual, greenish, small, solitary, axillary, or in few-flowered cymes.

The calyx is gamosepalous possesses five 3–5 mm lobes, inflated and acrescent in a fruit. The corolla is greenish-yellow of five 5–8 mm lobes and campanulate. Stamens are five numbers. The ovary is glabrous, globose and ovoid containing many ovules. The filiform style as well as 2-lobed stigma, fruit is a globose/berry, orange-red in color when ripen and covered in the enlarged calyx. Seeds are many, discoid, yellow and reniform.

2.3 Occurrence Data

Through a multifaceted approach encompassing field surveys, online databases, and published literature, we gathered a total 270 occurrence records for *Withania somnifera*, colloquially known as Ashwagandha. A total of 43 sites were systematically sampled during field surveys across diverse regions of Uttarakhand (Fig. 1). Fifteen precise occurrence points were collected using a handheld Garmin GPS device, ensuring accurate spatial documentation. Additionally, we incorporated 28 occurrence records obtained from various published literature sources.

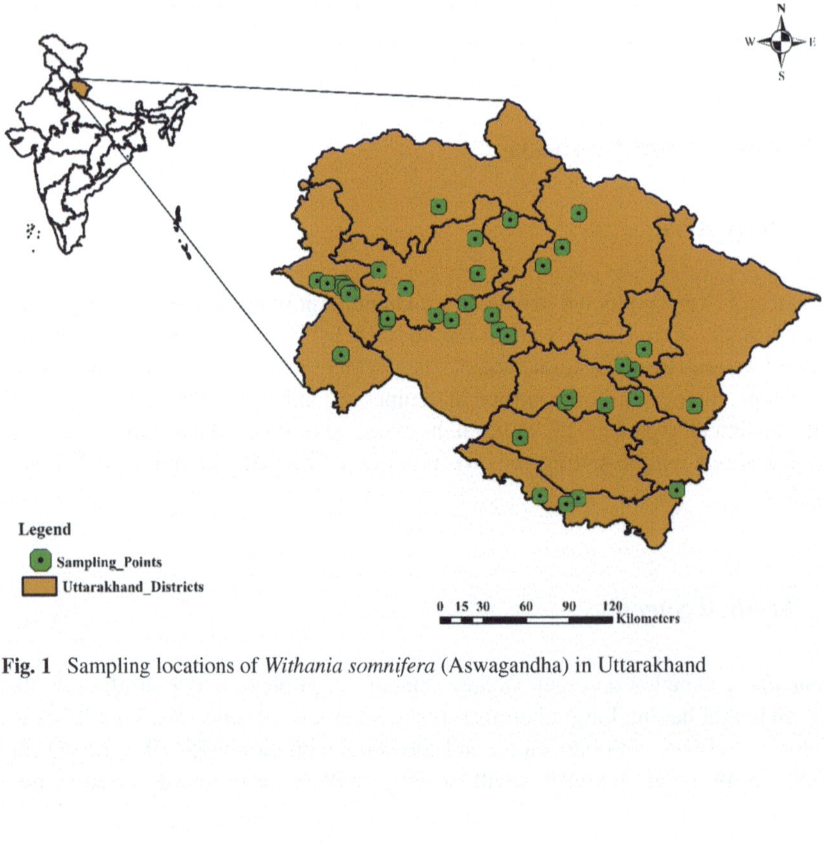

Fig. 1 Sampling locations of *Withania somnifera* (Aswagandha) in Uttarakhand

2.4 Environmental Layer

For the distribution modeling of *Withania somnifera*, we acquired a comprehensive set of environmental, topographical, land use/land cover, and soil property data with a resolution of 30 m. These datasets were sourced from various reputable providers, including WorldClim, ESRI Land Cover, and World Soil Information (ISRIC). The dataset comprises 19 bioclimatic variables, crucial for capturing climatic nuances. Additionally, topographical variables such as altitude, slope, and aspect were extracted from the Advanced Spaceborne Thermal Emission and Reflection Radiometer Global Digital Elevation Model (ASTER GDEM) database, with a spatial resolution of 30 m, using ArcGIS 10.3.

To enhance model accuracy, we applied a rigorous approach to remove highly correlated variables. This step is essential as correlated variables can negatively impact model performance and lead to inaccurate predictions, as highlighted in previous studies (Parolo et al. 2008; Merow et al. 2013; Dormann et al. 2013; Manzoor et al. 2018). The criterion for removing variables was based on the threshold correlation value of $|r| > 0.7$, aligning with the findings of various studies (Dormann et al. 2013; Manzoor et al. 2018; Sony et al. 2018; Farrell et al. 2019; Feng et al. 2019).

2.5 Model Settings and Evolution

For the potential niche modeling of *Withania somnifera*, MaxEnt, a machine leaning algorithm version 3.4.1 was employed, leveraging a dataset encompassing 43 sites in Uttarakhand. The modeling process integrated diverse environmental layers, including bioclimatic indicators and supplementary spatial data such as slope, aspect elevation, and vegetation. Transformed coordinate data in CSV format served as the model input after being converted into ASCII format to ensure compatibility with MaxEnt (Fig. 2).

The model employed 500 iterations, a convergence threshold of 0.00001, a prevalence of 0.5, 10,000 background points, and 10 percentile training presence. Additionally, 5 replicate model runs with cross-validation techniques were executed to guarantee and assess the model's reliability. MaxEnt was selected due to its reliability in accurate predictions with small sample sizes and resilience to spatial errors and sampling-biased occurrence data, as supported by previous studies (Hernandez et al. 2006; Wisz et al. 2008; Baldwin 2009; Graham et al. 2008). The individual contributions and permutation importance were analyzed using variable contribution tables and the jackknife test. Model performance was evaluated using threshold-independent metrics, including the area under the curve (AUC) of the receiver operating characteristic (ROC) curve (Hanley and McNeil 1982; Manel et al. 2001), and the true skill statistics (TSS). Logistic response options represented the probability of species presence within a range of 0–1, and the suitability ranges were determined using the 10-percentile training presence threshold rule, showcasing suitable and

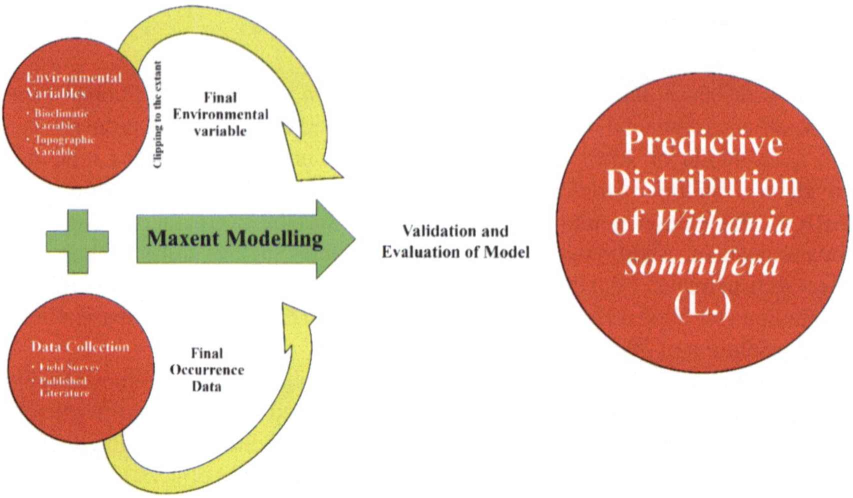

Fig. 2 Methodology adopted for the study

unsuitable areas. The average logistic outputs were imported into ArcMap 10.3 for further analysis.

3 Results

The model results unveiled a widespread distribution of *Withania somnifera* across the majority of Uttarakhand's districts, with the exception of Champawat (Fig. 3). The output of the current study highlights the heightened likelihood of *W. somnifera* occurrence in specific districts, notably Pauri, Chamoli, Udham Singh Nagar, and Dehradun. These regions emerged as hotspots, indicating a high probability of finding suitable habitats for the species. Additionally, the model projected a moderately suitable distribution in districts such as Uttarkashi, Bageshwar, Haridwar, Almora, and Tehri, suggesting a broader ecological range for *W. somnifera* across the state. This nuanced understanding of the plant's distribution is crucial for informed conservation strategies, as it not only identifies areas with high suitability for the species but also pinpoints regions with moderate suitability. Such insights provide a foundation for targeted conservation efforts, allowing for the development of region-specific plans to safeguard the natural habitats and promote the sustainable growth of this important medicinal plant.

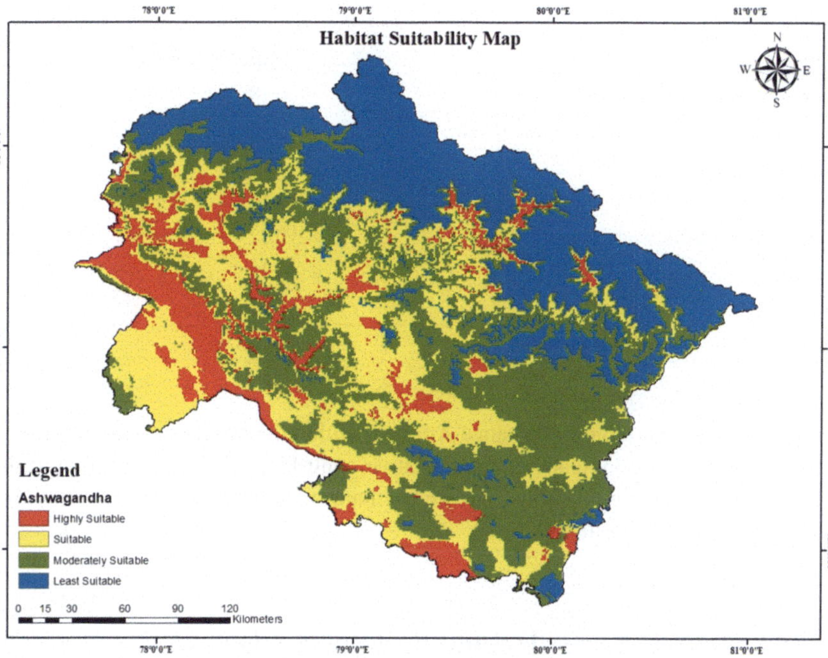

Fig. 3 Habitat suitability map of *Withania somnifera* in Uttarakhand

4 Variable Importance Analysis

Variable Importance Analysis was conducted as part of the MaxEnt modeling approach to assess the significance of different environmental factors influencing the potential distribution of *Withania somnifera* in Uttarakhand. Among the considered variables, annual mean temperature (bio1) emerged as the most critical determinant, underscoring the pronounced impact of temperature on the species' habitat suitability. Seasonal temperature variations, particularly during the warmest quarter (bio10), also played a substantial role, emphasizing the importance of specific climatic conditions for the plant's growth. Precipitation variables, including the wettest month (bio13) and wettest quarter (bio16), were identified as influential factors, highlighting the dependence of the species on adequate moisture during critical growth periods. Additionally, mean temperature during the wettest quarter (bio8), elevation, slope, and aspect were recognized as key contributors, signifying the combined influence of terrain characteristics and altitude on the distribution of *W. somnifera*. This comprehensive analysis of variable importance provides valuable insights for conservation planning and sustainable management strategies, allowing stakeholders to prioritize interventions based on the most significant environmental drivers impacting the medicinal plant's ecological niche. Table 1 presents the importance scores of environmental variables influencing the habitat suitability of *W. somnifera* in Uttarakhand,

Table 1 Variable Importance Scores for *Withania somnifera* Habitat Suitability in Uttarakhand

Variable	Importance score
Annual mean temperature (bio1)	0.18
Mean temperature of warmest quarter (bio10)	0.15
Precipitation of wettest month (bio13)	0.14
Precipitation of wettest quarter (bio16)	0.13
Mean temperature of wettest quarter (bio8)	0.12
Elevation	0.11
Slope	0.09
Aspect	0.08

as determined by Variable Importance Analysis within the MaxEnt modeling framework. The scores are indicative of the relative contribution of each variable to the overall model, providing insights into the key factors shaping the distribution of this medicinal plant species in the region.

5 Distribution Across Habitat Suitability Classes

The results of the study revealed that 17,114.56 km^2 of the area falls into the highly suitable category, indicating regions with optimal conditions for *W. somnifera*. Moderately suitable habitats, covering an area of 11,766.26 km^2, offer conditions conducive to the plant's growth but with certain limitations. The least suitable category, encompassing 16,044.9 km^2, represents areas where environmental factors may not fully support the species. This nuanced understanding of the distribution across suitability classes further informs conservation efforts and land management practices, enabling a targeted approach to safeguarding the varying habitats of *W. somnifera* in Uttarakhand.

A comparison of climatic variables related to model prediction is illustrated in Fig. 4. The environmental variable bio 16 exhibited the highest gain and contribution of 34.4% in the model. The environment variable bio 9 has the greatest impact on gain when it is deleted, whereas bio 19 has the least positive impact. This clearly shows that *Withania somnifera* distribution and spread are influenced by precipitation. It was discovered during the field studies that the plant also grows in the wild and close to agricultural fields.

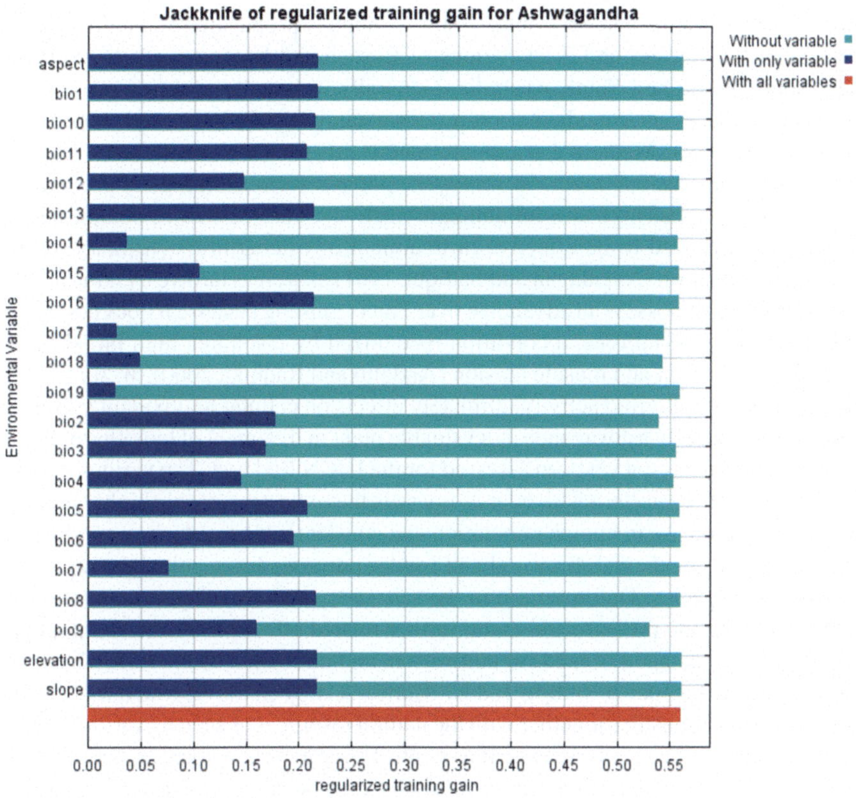

Fig. 4 Jackknife training gain plot for W. somnifera

6 Discussion

Withania somnifera is an important perennial plant species with immense therapeutic uses in traditional as well as modern system of medicine (Datta et al. 2010). Due to restorative property of roots, the species is also known as 'Winter cherry' (Winters 2006). The Indian Himalayan region (IHR), one of the richest reservoirs of biological diversity in the world, is undergoing irrational extraction of wild, medicinal herbs, thus endangering many of its high value gene stock. *Withania somnifera* is a member of solanaceae, also known for thousands of years by Ayurvedic practitioners. *Withania somnifera* root contains flavonoids, alkaloids, steroid and many active functional ingredients (Kumar et al. 2018).

The genus *Withania* is restricted and related to the old World; rather it closely belongs to the genus Physalis, the gooseberries. *Withania* possesses a natural occurrence, most probably in the drier and humid areas, spread from the Mediterranean region to throughout tropical region of Africa to South Africa and also from the Cape Verde Islands and Canary region to the Arabia and Middle East region like

India, southern China and Sri Lanka. Ashwagandha is propagated and cultivated in gardens in the warmer and drier regions of Europe and became a natural herb in New South Wales and South Australia. Generally, it is cultivated in India and in many other places as a medicinal crop (Govindaraju et al. 2003), most probably for its fleshy roots. Medicinal plants are nature gift for humanity to treat various human ailments. Among them, *W. somnifera* possess various biological properties. *W. somnifera* possess high anti-inflammatory activity. Various studies demonstrated ashwagandha proved to be effective in treatment of osteoarthritis (Kulkarni et al. 1991), inflammation (Angalagan and Sadique 1981) as well as tardive dyskinesia. *W. somnifera* though possess an advantage over Panax ginseng because it does not result in ginseng-abuse syndrome, a condition recognized by insomnia, high blood pressure, water retention and muscle tension (Andallu and Radhika 2000). *W. somnifera* may also enhance thyroid activity ultimately effects on cellular antioxidant systems. These results specify ashwaganda as a valuable plant in treatment of hypothyroidism.

W. somnifera has not yet been assessed for the IUCN Red List. The various threats are facing by *W. somnifera* which leads towards the extinction of this highly medicinal species. Increase in human population and constant unplanned over exploitation of this plant for medicinal, firewood and fodder purposes has resulted damage into the *Withania* species. Deforestation in the salt valley and Himalayas is mainly due to the demographic pressure and many further related effects like livestock population increase in demand of land for cultivation, and utilization of the forests to meet growing needs for fuel as well as fodder. Conversely, social issues such as perceptions, social relationships together with the natural resources as seen by various social groups, concerns of access to as well as control over natural resources, and matters of power relative to deforestation, have hardly ever been discussed in literature as probable driving forces to deforestation. During last many years, less rain falls and large scale deforestation in the Himalayan range has rigorously changed the environment owing to which lush green slopes have been transformed to barren land. Similar to global warming, this is causing progressive rise in temperature resulting in loss of many important medicinal plants such as *Withania* species (Ahmad et al. 2002).

Those plants with medicinal properties that have a beneficial medicinal effect on the human or animal body are referred to as medicinal plants. Uttarakhand is known for enormous biodiversity and availability of numerous aromatic and medicinal plants. According to the Uttarakhand Medicinal Plants Database (UMPDB), there are about 1127 medicinal plant species, which are widely distributed in different regions of the state of Uttarakhand (Kumar et al. 2018). *W. somnifera* is one of the important medicinal plants of the Himalayan region. All parts of Aswagandha are used for a variety of medicinal substances. Geographical information can be used to interpret a species' distribution, vulnerability, and habitat requirements. A species' conservation status might also be determined by combining data of all its known populations, information on historical range changes, and indications of its distinctive habitats. Mapping is one of the most convenient approaches to display distribution of plants through various aspects such as inherent nature of land and physiographic conditions. There is a strong inter dependent relationship between biotic and abiotic factors in an environment. Thus, species and habitat relationship modeling can be used to predict species distribution patterns thereby using parameters related topography, coil condition, microclimatic condition etc. This ecological modeling analysis clearly identifies prospective areas related to the distribution of *W. somnifera*. This may help to take initiatives for conservation of the plant related to purposes for human welfare.

7 Conclusion

From the above discussion, it can be concluded that low rainfall, over grazing, population pressure, natural and accidental fires, soil aridity, deforestation, and over exploitation of medicinal plants are the major threats to biodiversity in the Himalayan and Shiwalik range. The results of the current study suggested that the cultivation of *W. somnifera* is highly appreciated in 60% of areas in Uttarakhand including the districts of Chamoli, Dehradun, Udham Singh Nagar and Pauri. The map of the possible habitat distribution can be used in the planning of land and its management. Moreover, it would help in the selection of priority survey sites, the discovery of new populations, and the creation of conservation priority zones and resource management zones with a focus on species ecological boundaries. To further understand the additional areas where the species can develop and regenerate, distribution modelling of the related species may be useful. Thus, based on the above observation, it was concluded that plants have an important medicinal property and have a medicinal effect on the human or animal body. These products may be suggested for many nutritional diseases and disorders of the body such as diabetes, constipation, heart, and intestinal problems etc. for their regulation and controlling mechanism.

References

Adhikari BS, Babu MM, Saklani PL, Rawat GS (2007) Distribution, use pattern and prospects for conservation of medicinal shrubs in Uttaranchal State. J Mt Sci 4:155–180

Ahmad H, Ahmad A, Jan MM (2002) The medicinal plants of Salt Range. Online J Biol Sci 2(3):175–177

Ahmedullah M, Nayar MP (1986) A vanishing endemic cycad of Indian Peninsula. Nelumbo 28(1–4):169–170

Andallu B, Radhika B (2000) Hypoglycemic, diuretic and hypocholesterolemic effect of winter cherry (*Withania somnifera*) root. Indian J Exp Biol 38:607–609

Angalagan K, Sadique J (1981) Influence of an Indian medicine (Ashwagandha) on acute-phase reactants in inflammation. Indian J Exp Biol 19:245–249

Baldwin RA (2009) Use of maximum entropy modeling in wildlife research. Entropy 11(4):854–866

Barnosky AD, Matzke N, Tomiya S, Wogan GOU, Swartz B, Quental TB, Marshall C, McGuire JL, Lindsey EL, Maguire KC, Mersey B, Ferrer EA (2011) Has the Earth's sixth mass extinction already arrived. Nature 471(7336):51–57

Batugal PA (2004) Inventory and documentation of medicinal plants in 14 Asia Pacific countries. In: Medicinal plants research in Asia, vol 1: the framework and project workplans, p 3

Bhatt VP (1999) Ethnobiology of high-altitude Himalayan communities in district Chamoli: a conservation perspective. Zoos' Print Journal 14(11):137–146

Brito J, Acosta A, Lvares F, Cuzin F (2009) Biogeography and conservation of taxa from remote regions: an application of ecological-*niche* based models and GIS to North-African Canids. Biol Conserv 142(12):3020–3029

Chandra N, Joshi VC, Mishra AP (2020) Ecological status of fodder and fuelwood species in Banari Devi sacred grove of Kumaun Himalaya, Uttarakhan. Indian J Ecol 47(4):1044–1048

Chandra N, Singh G, Rai ID, Mishra AP, Kazmi MY, Pandey A, Abdo HG (2023b) Predicting distribution and range dynamics of three threatened *Cypripedium* species under climate change scenario in Western Himalaya. Forests 14(3):633. https://doi.org/10.3390/f14030633

Chandra N, Rai ID, Mishra AP, Dwivedi SK, Kotiya A, Tiwari UK, Singh G (2022) Assessing potential habitats and populations of selected medicinal herbs in Alpine areas of Uttarakhand, Western Himalaya. Indian J For 45(3):144–155. https://doi.org/10.54207/bsmps1000-2023-ITRFNL

Chandra N, Kumar A, Mishra AP, Singh G, Rawat GS (2023a) *Wigandia urens* (Ruiz & Pav.) Kunth: a new emergent invasive alien species in the Western Himalaya. Int J Environ Stud 1–14. https://doi.org/10.1080/00207233.2023.2194157

Datta AK, Das A, Bhattacharya A, Mukherjee S, Ghosh BK (2010) An overview on (*Withania somnifera*) (L.) Dunal—the 'Indian ginseng.' Med Aro Pl Sci Biotech 5:1–15

Dormann CF, Elith J, Bacher S, Buchmann C, Carl G, Carré G, Lautenbach S (2013) Collinearity: a review of methods to deal with it and a simulation study evaluating their performance. Ecography 36(1):27–46

Elith J, Leathwick J (2009) Species distribution models: ecological explanation and prediction across space and time. Annu Rev Ecol Evol Syst 40:677–697

Farrell A, Wang G, Rush SA, Martin JA, Belant JL, Butler AB, Godwin D (2019) Machine learning of large-scale spatial distributions of wild turkeys with high-dimensional environmental data. Ecol Evol 9(10):5938–5949

Feng X, Park DS, Liang Y, Pandey R, Papeş M (2019) Collinearity in ecological niche modeling: confusions and challenges. Ecol Evol 9(18):10365–10376

Gaur RD (1999) Flora of the District Garhwal, Northwest Himalaya. Transmedia

Graham CH, Elith J, Hijmans R, Guisan A, Peterson AT, Loiselle BA (2008) Evaluating the influence of spatial uncertainty in locality points for species distributional modeling. J Appl Ecol 45(1):239–247

Guisan A, Zimmermann N (2000) Predictive habitat distribution models in ecology. Ecol Model 135(2–3):147–186

Hanley JA, McNeil BJ (1982) The meaning and use of the area under a receiver operating characteristic (ROC) curve. Radiology 143(1):29–36

Hernandez PA, Graham CH, Master LL, Albert DL (2006) The effect of sample size and species characteristics on performance of different species distribution modeling methods. Ecography 29(5):773–785

Jochum G, Mudge K, Thomas R (2007) Elevated temperatures increase leaf senescence and root secondary metabolite concentrations in the understory herb (*Panax quinquefolius*) (*Araliaceae*). Am J Bot 94(5):819–826

Joshi VC, Chandra N, Sundriyal RC, Mishra AP, Abdo HG (2024) Ecological status and spatial extent of non-native shrubs *Lantana camara* L. and *Ageratina adenophora* Sprengel in the forest communities of Western Himalaya. Trees, For People 100494

Kulkarni RR, Patki PS, Jog VP (1991) Treatment of osteoarthritis with an herb mineral formulation: a double-blind, placebo-controlled, cross-over study. J Ethnopharmacol 33:91–95. https://doi.org/10.1016/0378-8741(91)90167-C

Kumar A, Kumar R, Sharma M, Kumar U, Gajula MP, Singh KP (2018) Uttarakhand medicinal plants database (UMPDB): a platform for exploring genomic, chemical, and traditional knowledge. Data 3(1):7

Manel S, Williams HC, Ormerod SJ (2001) Evaluating presence–absence models in ecology: the need to account for prevalence. J Appl Ecol 38(5):921–931

Manzoor SA, Griffiths G, Lukac M (2018) Species distribution model transferability and model grain size–finer may not always be better. Sci Rep 8(1):7168

Merow C, Smith MJ, Silander JA Jr (2013) A practical guide to MaxEnt for modeling species' distributions: what it does, and why inputs and settings matter. Ecography 36(10):1058–1069

Mishra AP, Rai ID, Pangtey D, Padalia H (2021) Vegetation characterization at community level using sentinel-2 satellite data and random forest classifier in western Himalayan Foothills, Uttarakhand. J Indian Soc Remote Sens 49:759–771. https://doi.org/10.1007/s12524-020-01253-x

Mishra AP, Kumar A, Yadav SN (2023c) Ecology and conservation of threatened medicinal plants in the Trans-Himalayan region of Nanda Devi Biosphere Reserve, Western Himalaya. Trees, for People 14:100451. https://doi.org/10.1016/j.tfp.2023.100451

Mishra AP, Chandra N, Kumar A, Sharma S, Singh G (2023a) *Mesosphaerum suaveolens* (L.) Kuntz.: a serious threat to Uttarakhand's first Ramsar site, India. Species 24:e57s1558

Mishra AP, Chandra N, Mandy JJ, Dwivedi SK, Alruzuq A, Pande CB (2023b) Analytical hierarchy process (AHP) based on the spatial assessment of an endangered alpine medicinal Herb *Aconitum heterophyllum* in the Western Himalayan environment. In: Climate change impacts on natural resources, ecosystems and agricultural systems. Springer International Publishing, Cham, pp 579–594. https://doi.org/10.1007/978-3-031-19059-9_25

Mishra AP, Chandra N, Srivastava A, Singh SK (2023b) An updated checklist of the vascular flora of Asian conservation reserve, Uttarakhand, India. Indian J For 46(1):12–21. https://doi.org/10.54207/bsmps1000-2023-7P1KSV

Mishra AP (2023) Understanding the decline of *Butea monosperma* (Lam.) Kuntze in Gorakhpur District, India: an ecological investigation. J For Geosci 1:2. https://www.esciinfo.com

Owolabi OJ, Omogbai EKI, Obasuyi O (2007) Antifungal and antibacterial activities of the ethanolic and aqueous extract of *Kigelia Africana* (Bignoniaceae) stem bark. Afr J Biotechnol 6(14):1677–1680

Parolo G, Rossi G, Ferrarini A (2008) Toward improved species niche modelling: *Arnica montana* in the Alps as a case study. J Appl Ecol 45(5):1410–1418

Phillips S, Anderson RR, Schapire R (2006) Maximum entropy modeling of species geographic distributions. Ecol Model 190(3–4):231–259

Purohit AN (2000) Medicinal plants-upgrading technology for trading the traditions. In: Proceedings in national seminar on harvesting herbs

Samant SS, Pal M (2003) Diversity and conservation status of medicinal plants in Uttaranchal State. Indian for 129(9):1090–1108

Sony RK, Sen S, Kumar S, Sen M, Jayahari KM (2018) Niche models inform the effects of climate change on the endangered Nilgiri Tahr *(Nilgiritragus hylocrius)* populations in the southern Western Ghats, India. Ecol Eng 120:355–363

Tiwari JK, Tiwari P, Rawat DS (2015) Some additions to flora of chamoli. Phytotaxonomy 15:146–151

Uniyal SK, Awasthi A, Rawat GS (2002) Current status and distribution of commercially exploited medicinal and aromatic plants in upper Gori valley, Kumaon Himalaya, Uttaranchal. Curr Sci 1246–1252

Warren DL, Glor RE, Turelli M (2008) Environmental niche equivalency versus conservatism: quantitative approaches to niche evolution. Evol 62(11):2868–2883

Winters M (2006) Ancient medicine, modern use: (*Withania somnifera*) and its potential role in integrative oncology. Altern Med Rev 11(4):269–277

Wisz MS, Hijmans RJ, Li J, Peterson AT, Graham CH, Guisan A, NCEAS Predicting Species Distributions Working Group (2008) Effects of sample size on the performance of species distribution models. Divers Distrib 14(5):763–773

Alpine and Sub-Alpine Flowering Plant Diversity Vis-á-Vis Medicinal Plants in the Sikkim Himalaya

Subhajit Lahiri and Sudhansu Sekhar Dash

Abstract The alpine region of the Sikkim Himalaya is well known as an important biodiversity rich area of the world. The biodiversity of these regions is under high anthropogenic pressure and falls within the Himalaya biodiversity hotspots as recognised by Conservation International. As a result, during the last three centuries, the entire region has drawn botanists, plant enthusiasts, and plant hunters worldwide to study and investigate its lush vegetation. Many East Himalayan species are no longer found in their natural environment. Sikkim is one of the world's most floristically diverse regions. Numerous groups of indigenous people have established native traditional wisdom there, nearly in isolation, without the assistance of sophisticated civilization. The importance of Sikkim Himalayan alpine and sub-alpine vegetation from a utilitarian point of view is immense. There are numerous publications related to several types of useful plants native to the region, including numerous medicinally important species of plants. The present communication deals with a review of the different uses of 169 plant species belonging to 47 families for the treatment of 33 major ailments among rural communities in the Sikkim Himalaya. During the different explorations of plants in the alpine and subalpine regions of the Sikkim Himalaya, the different traditional uses associated with the plants were collected from selected informants and from published literature. The traditional use patterns, parts used, modes of preparation and administration of doses were reviewed and documented in detail.

Keywords Ecosystem services · Flora · Geographical affinities · Indigenous use · Traditional knowledge

S. Lahiri
Central National Herbarium, Botanical Survey of India, Howrah, West Bengal 711103, India

S. S. Dash (✉)
Botanical Survey of India, CGO Complex, Salt Lake, Kolkata, West Bengal 700064, India
e-mail: ssdash2002@gmail.com

A. P. Mishra et al. (eds.), *Threatened Medicinal Plants in the Indian Himalayan Region*, World Sustainability Series, https://doi.org/10.1007/978-3-031-73687-2_14

1 Introduction

Sikkim is India's second smallest but most ecologically diversified Himalayan state encompassing an area of 7,096 km^2. It is distinctive in various aspects, the most notable of which is its closeness to both the Tibetan plateau and the Bay of Bengal, allowing it to have affinities with both tropical wet forests in the south and frigid desert-like environments in the north within a short distance. As part of the Eastern Himalaya Biodiversity Hotspot, the region's biodiversity is one of the richest in terms of diversity of species and varied, fragile, and complex ecosystems. The wide range of ecological habitats and environmental conditions in the state have produced all the possible types and extremes of climatic conditions that are suitable for supporting a diverse range of ecosystems. This region represents all the known Himalayan-forest ecosystems, including tropical *terai*, tropical evergreen forests, sub-Himalayan wet mixed forests, sub-tropical dense forests, wet temperate broadleaved forests, mixed Rhododendron coniferous forests, eastern oak-hemlock forests, oak-fir forests, moist alpine, dry alpine scrubs, and grasslands. Sikkim's rich and unique biodiversity represents 26 percent of all species in India. As per the current estimate, about 84% of the total land area is under forest (FSI 2019), while 5068 taxa belonging to 1491 genera and 209 families are occurring in the state (Gogoi et al. 2021). It is also hypothesised that the flora of the region is the result of the confluence of species from three major bio-geographic realms, namely the Indo-Malayan (the richest in the world), the Indo-Arctic (Eurasia), and the Afrotropical (well established). The occurrence of a high degree of endemism and a higher number of threatened plant species in the region substantiate the theory of the centre of isolation, speciation, and adaptive evolution of many plant groups such as orchids, legumes, *Pedicularis*, *Primula*, Poaceae and *Rhododendron*.

The remarkable feature of the Sikkim Himalayan region is the great deal of both species and genetic diversity within each ecosystem. However, in the changing scenario of the post- convention on biological diversity era, the perception of ecosystems has changed from species diversity to the immense ecosystem goods, ecological services, and functions they offer for overall human well-being. The mountain ranges of the state, with their beautiful landscapes, play a key role in controlling local atmospheric circulation, the hydrological cycle, and water resources, which provide a wide range of ecosystem services to the locality. Therefore, due to the rapidly expanding demand for service resources such as agricultural lands, range land for grazing, deforestation, and overexploitation for forest products, the biodiversity of these fragile ecosystems has been dwindling in their natural habitat. Besides, human-caused glitches such as habitat degradation, forest fires, and natural calamities including receding or melting glaciers, avalanches, flash floods, landslides, and stone falls worsen the disturbances in many ways and have also been severely affecting the plant diversity of the region. However, ever-increasing anthropogenic activities and overexploitation of natural resources are posing threats to this fragile ecosystem. Furthermore, research has shown that the effects of climate change are having a severe impact on the region's biodiversity (Ravindranath et al. 2012).

To analyse human plant linkages and comprehend regional human ecological links to their environment, it is necessary to record the plants and their ethnobiological values. Ethnomedicine is a traditional healthcare method that has been passed down from generation to generation for the treatment of various maladies. It is deeply intertwined with indigenous peoples' religious beliefs and practices (Lahiri and Dash 2022). Himalaya is rich in ethno medicinal important plants (EMIPs) with significant economic value and application in ayurveda or unani systems. Charaka (Charak-Samhita, 100–500 AD) and Shushruta (Shushrut-Samhita, 200–500 AD) represent the early period of Indian herbal research. While Sikkim is a densely forested region with diverse ethnic communities residing there, including Lepcha, Bhutia, Sherpa, Rai, Limboo, Nepali, etc. in remote and fringe areas have inherited a rich traditional knowledge of wild plants and rely on these natural resources for their medicinal needs. Locals refer to traditional herbal healers in Sikkim known as *Lamas* (monastery heads) or *Jakris*. However, because of increased urbanisation, traditional knowledge of wild flora is gradually fading. Several ethnobotanical studies have been conducted in Sikkim over the previous century to acquire information about and record the region's indigenous knowledge. Yet, ethnobotanical data from the alpine and subalpine areas is limited. Sikkim's alpine meadows are a good source of traditionally used medicinal herbs. Those are threatened by a variety of human-caused issues (overexploitation, grazing, trampling, i.e., road construction and uncontrolled tourism) and natural concerns, including habitat fragmentation, invasive species encroachment, climate change, and genetic drift. Therefore, the current study attempted to gather, organise, and examine information regarding the variety and usage of medicinal plants in the high-altitude terrain of Sikkim Himalaya.

2 Floristic Explorations in Sikkim Himalaya

The incredible biodiversity-rich places of the Eastern Himalaya, particularly Sikkim, have drawn many botanists and plant explorers from all over the world since David Don arrival in 1825. Though, exploitation began at least 200 years before, systematic floristic investigation in the Sikkim Himalaya did not commence until the early nineteenth century, with William Griffith's arrival in 1847. Sir Joseph Dalton Hooker (1849) began his trek in Sikkim, following the Rangeet river to its confluence with the Tista River and Tonglu peak in the Singhalila range on the Nepalese border. Sir Joseph Dalton Hooker had travelled through Dzongri and the northwestern passes of Nepal into Tibet. He hiked up the Lachen Valley to the Kongra Lama Pass, then to the Lachoong Pass. Hooker described 33 *Rhododendron* species with colour hand illustrations in *Rhododendrons* of Sikkim Himalaya (Hooker 1849, 1851). Sir Joseph Dalton Hooker and Thomas Thomson (Hooker and Thomson 1859) in their paper '*Praecursores ad Floram Indicam-Balsamineae*' deal with historic expeditions to Sikkim and Khasia Hills. The vegetation and floristic information of Sikkim on angiosperms were incorporated into the 7-volume book "*The Flora of British India*" published between 1872 and 1897, partly in Hooker and Thomson's "*Flora*

Indica" (Vol-I:1855) and Sketch of the Flora of British India. During his botanical exploration journey, he acquired roughly 3000 species in Sikkim and Bengal along with the Assam collection, as well as illustrated over 700 plants native to the region.

Following Sir Joseph Dalton Hooker expedition, numerous European botanists collected plants from the state, specially from alpine and subalpine area including Sir George King, Charles Baron Clarke, David Prain, James Sykes Gamble, Sir George Watt, Robert Pantling, George Alexander Gammie, William Wright Smith, and George H. Cave. Vast amounts of data, in the form of collections and notes, have since been collected and are now available in the Central National Herbarium, Botanical Survey of India, Calcutta, and many foreign herbariums, including the Royal Botanic Gardens, Kew; The Natural History Museum, London; Royal Botanic Garden Edinburgh, Edinburgh; University of Manchester, England; Harvard University, Cambridge etc. Smith and Cave (1911), who studied the floristic composition of alpine and subalpine flora of Zemu and Llonakh Valley in Sikkim, were followed by Smith (1911), who visited South-East Sikkim to explore the plants of this area, particularly the ridges lying between the Cho-La and the Tanka-La. Smith also visited Changu, Karponang, Nathu La, Laghep, Yak La, Dikchu Valley, Chakung Chu Valley, Kupup, Gnatong, and Dzalep La during this period. In the year 1934, Duncan visited Sikkim, particularly Thangu, Donkhya La, Yumtang, and Lachen Valley. During his visit, he photographed several *Rhododendron* species. Rolla (1963) lists the Sikkim flora, with a focus on medicinal and economic species from the north and east regions. While dealing with the Flora of Bhutan, Grierson and Long also included a record of plants from the Sikkim Himalaya and nearby locations. Dash listed the flora of 'Kabi Sacred Grove' and briefly classified numerous more plant families including Grossulariaceae, Polygonaceae, and Commelinaceae. Some legitimate studies dealing with alpine and subalpine plants published in the twenty-first century include Maity et al. (2018), Dash et al. (2021); Gogoi et al. (2021), Lahiri (2022), Lahiri et al. (2022) and Dash et al. (2023).

3 Phyto-Geographical Affinities of Sikkim Himalaya

Sikkim's alpine floristic element has an intriguing phytogeographic importance. Alpine flora of Sikkim has unique phytogeographical relationships and is said to synthesise floristic components from adjacent phytogeographical areas including Tibet, China, Bhutan, Nepal, Japan, Myanmar, Thailand, and Malaysia. The region's incredible variety is due to the continual flow of floristic components in both directions. However, a distinct distribution was noticed when compared to the flora of neighbouring regions. Based on the distribution pattern of the species throughout the Himalayas, the following categories could be summarized (Grierson and Long 1987; POWO 2023).

Taxa those are distributed from Western Himalaya to China: *Acer oblongum*, *Allium prattii*, *Anemone rupestris*, *Arisaema tortuosum*, *Astilbe rivularis*, *Clematis montana*,

Fig. 1 *Megacodon stylophorus* (C.B.Clarke) Harry Sm. distributed from the Eastern Himalaya to China

Cyananthus lobatus, *Gaultheria trichophylla*, *Juniperus recurva*, *Juniperus squamata*, *Nardostachys jatamansi*, *Paris polyphylla*, *Pleione praecox*, *Sinopodophyllum hexandrum*, *Prunus cerasoides*, *Quercus semecarpifolia*, *Ribes takare*, *Rosa sericea*, *Rubus biflorus*, *R. calycinus* and *Streptopus simplex*.

Taxa distributed from the Eastern Himalaya to China: *Cardamine griffithii*, *Cassiope selaginoides*, *Megacodon stylophorus*, *Luzula effuse* and *Papaver napaulense*.

Taxa distributed throughout the Himalaya including China* but absent from *Japan: *Parnassia nubicola*, *Rhododendron anthopogon*, *R. barbatum*, *R. campanulatum*, *Rosa macrophylla*, *Schisandra grandiflora*, *Sorbus foliolosa*, *S. microphylla*, *Spiraea canescens*, and *Tsuga dumosa*.

Taxa distributed in the Eastern Himalaya only: *Abies densa*, *Dipsacus atratus*, *Larix griffithiana*, *Primula whitei*, *Rhododendron baileyi*, *R. camelliiflorum*, *R. ciliatum*, *R. glaucophyllum*, *R. grande*, *R. lanatum*, *R. lindleyi*, *R. wallichii* and *R. wightii*.

Arctic-alpine elements: *Bistorta vivipara*, *Koenigia islandica*, *Oxyria digyna* and *Thalictrum alpinum* (Figs. 1 and 2).

4 Unique Alpine Ecosystems and EIMPs

The alpine zone of the state of Sikkim belongs to the biogeographic zone 2C-Central Himalaya and 1D-Sikkim Plateau (Rodgers and Panwar 1988; Kumar et al. 2017) which defines the extreme southern border of the Turkmenian sub-region of the Palaearctic region. Though the natural zonation of various plants is rarely separated or evident on the ground, the spatial patterns of species variety and uniqueness in

Fig. 2 *Spiraea canescens* D.Don distributed throughout the Himalaya including China and absent from Japan

eastern Himalayan flora rely heavily on altitudinal variation. The sub-alpine vegetation ranges in elevation from 2800 to 3600 m and is also known as the *Rhododendron*-Conifer Zone or cold temperate. Most of this zone is evergreen, with a *Rhododendron* and conifer mix; however, *Quercus pachyphylla* and *Q. lineata*, as well as *Acer caudatum*, *A. campbellii*, and *Betula utilis* may be found about 2700 m elevation. *Rhododendron* species such as *R. arboreum*, *R. campanulatum*, *R. grande*, *R. campylocarpum*, *R. falconeri* and *R. fulgens* become more common as one climbs higher in elevation. This forest can only be found in the east, north, and west districts of the state. *Betula utilis* grows infrequently near Lachen Valley at around 3300 m, intermingled with *Rhododendron* species. *Taxus wallichiana* thrives in sub-alpine woods above Lachung, interspersed with areas of *Tsuga dumosa* and *Abies densa*. *Tsuga dumosa* grows abundantly between 3000 and 3600 m in the Lachen Valley of North Sikkim and Tshoka valley of West Sikkim, whereas *Picea spinulosa* coexists with *Tsuga dumosa* in the Lachen and Lachung valleys. In *Abies* forest, *Arundinaria aristata* forms the dense ground cover. *Abies densa* grows gregariously between Kyongnosla, Changu and Memencho in east Sikkim, and between Samdong to Thangu in North Sikkim. Few shrubs and trees of *Rosa*, *Sorbus*, *Spirea*, and *Rubus* have also seen growing in these regions. Alpine Scrub and grasslands may be found in the north, east, and west districts of Sikkim between 3600 and 4600 m. The eastern Himalayan alpine areas are distinguished by very cold temperatures, large diurnal changes, irregular rainfall, and frequent snowfall. In the open meadows between 4500 and 5500 m or gentle mountain slopes many species of *Allium*, *Anemone*, *Bistorta*, *Cassiope*, *Delphinium*, *Potentilla*, *Primula*, *Ranunculus* are common. These general canopies of the higher altitude are usually devoid of any tree species. Above 3600 m, *Juniperus squamata*, and *J. recurva* grow on exposed sunny hill slopes around Thangu whereas in Kyongnosla it grows over 3000 m; *Ephedra gerardiana* spread in vast plains and hillocks slopes around Thangu (4200 m). *Rhododendron* thickets are a conspicuous feature of the alpine landscape, especially in the west and north districts. The alpine flora of Sikkim Himalaya, represents nearly 30% of the total flora of Sikkim (Tambe et al. 2007). The analysis of data on endemic taxa in the Himalaya revealed that the maximum representation is in the alpine region, followed by temperate and subtropical regions. Therefore, the high degree of taxonomic richness in the alpine region may be attributed to geographic isolation, habitat heterogeneity, and a varied history of migrations and evolution. Each of these eco-zones has its own biotype, defined by its floristic makeup. Chopta Valley, Dzongri, Goecha La, Green Lake, Lhonak Valley, Tso Lhamo Plateau, Memencho lake and Kyongnosla are some of the region's prominent distinctive landscapes (Figs. 3 and 4). The state is the primary catchment area for the Tista River, which flows north from Tso Lhamo Lake. The Tso Lhamo, or Sikkim plateau, is in the Trans-Himalayan region and covers only 400 km^2, making it India's smallest biogeographic province (Rawat and Tambe 2011).

Sikkim Himalayan alpine meadows are home to various high-value EIMPs such as *Aconitum heterophyllum*, *A. ferox*, *Nardostachys jatamansi*, *Dactylorhiza hatagirea*, and others. All these species have medicinal, aromatic, and other characteristics and are utilized to manufacture medicines that sell for high prices on national and worldwide markets. As a result, alpine meadows have emerged as a possible source of livelihood for those living in the difficult terrain of the Himalayas. A significant portion of EMIPs that are advantageous to humans are likewise impacted by the same climate change that puts alpine plants in danger. In recent decades, medicinal plants have been under increasing pressure because of industrialization, climate change, overexploitation for commercial purposes, and other factors. This diminishes the possibility of discovering novel pharmaceutical ingredients and has a detrimental impact on the significant segment of the world's population that relies on natural remedies. Climate change has been linked to changes in the growth pattern, distributional range, and composition of some communities, including medicinal plants, in Sikkim Himalaya.

5 Endemic Species

Sikkim's endemic species were represented by 163 taxa, including 160 narrow endemic taxa whereas165 species have been named after the state of Sikkim (Gogoi et al. 2021). A diverse range of climate fluctuations, high mountain ranges, and deep valleys provide a variety of specialised habitats for unique species. *Anaphalis hookeri*, *Astragalus lachungensis*, *Brachycaulos simplicifolius*, *Cassiope fastigata* subsp. *sennikovii*, *Codonopsis affinis*, *Gentiana springateana*, *Kuepferia pringlei*, *Ligularia kingiana* and *Lomatogonium cherukurianum* are notable alpine endemic species.

6 Alpine and Sub-Alpine Species Diversity

Knowledge of biogeographical variation in species richness is crucial for understanding the effects of climatic change on patterns of plant distribution and the conservation of biological diversity. The abundance of plant species reduces as altitude increases, although the diversity of sites within each zone varies greatly. Species richness and diversity vary along altitude gradients in mountain ecosystems, which may be used to investigate patterns in vegetation responses to changing climatic conditions. As a result, precise information on the health of these fragile ecosystems is critical for understanding the intricate nature of ecosystem services and designing an integrated conservation plan. The alpine area, on the other hand, has lower alpha species diversity because it is predominantly isolated, thinner, fragmented, and dominated by a depauperated scrub zone (Tambe and Rawat 2010). However, to date, only a little data is available on the patterns of species diversity in the alpine region of the Eastern Himalaya.

However, recent studies on rapid assessment of plant diversity in India conducted by the Forest Survey India (FSI 2019) revealed that the maximum Shannon–Wiener index for trees was 2.09 for subalpine forests. Maximum shrub diversity (H′) is 1.87 in sub-alpine forests and 0.64 in moist alpine scrub, while maximum reported herb diversity (H′) in sub-alpine forests and moist alpine scrub is 0.83 and 0.69, respectively. Recent research on the quantification of plant diversity in two landscapes of the Sikkim Himalaya undertaken by Lahiri (2022) produced highly intriguing results. Maximum tree diversity was found in alpine and subalpine forests in the west part of the state (H′ 2.46) between 3600 and 3800 m altitude, while the eastern part of the state shows the highest tree diversity (H′ 2.25) between 3200 and 3400 m altitude. Whereas the maximum shrub diversity is found in the alpine meadows of the west and east districts, it was 2.13 and 1.56, respectively. Herb species diversity (H′) varied from 3.708 in the west district to 3.596 in the alpine meadow and subalpine woodland of the east district.

The H′ values for the tree layer reported by Lahiri (2022) are comparable to the values reported from subalpine and alpine forests in Western Himalaya Forest, viz., H′: 2.09–2.29 (Bhat et al. 2020); 2.17–2.64 (Chandra et al. 2018) etc. Similarly, values reported by Lahiri (2022) were higher than those found in a similar study on Himalayan subalpine and alpine forests viz. H′: 0.73–2.62 (Rawat 2007), between 0.44 and 1.55 (Gairola et al. 2008); from 0.693 to 1.90 in tree, 0.83–2.68 in shrub and 0.863–1.705 in herb (Majumdar and Adhikari 2012) which could be attributed to differences in species composition, which imply a high degree of variation in physical landscape settings and disturbance regimes. Most of the lower alpine slopes are exposed to different disturbances, including avalanches, severe grazing, and constant flushing of nutrients (Rawat 2007). Tambe (2007) reported substantially lower values of plant species diversity value (H′:1.44–2.48) in the alpine terrain of Khangchendzonga NP (KNP) than the values obtained by Lahiri (2022) from alpine and sub alpine area of Sikkim. Some of the studies in eastern Himalaya exhibited a comparatively higher diversity index than the values reported by Lahiri (2022) such as 2.922 in trees, 3.870 in shrubs, and 4.36 in herbs (Moktan 2017). It might be owing to the forest's transitional character, which hosts species from both the upper and intermediate zones. Differences in soil micronutrients such as nitrogen and phosphorous content also influence species diversity in different landscapes. Species diversity is regulated by long-term characteristics such as community stability and evolutionary time since the variability of both micro- and macro-climate impacts diversification across various communities. Because a higher number of species with often overlapping habitats can coexist, it is feasible to suggest that diversity always leads to better sustainability.

7 Ecosystem Services

The Millennium Ecosystem Assessment's concept of 'ecosystem services' provides a valuable link between the functioning of ecosystems and their involvement in society, including the dynamics of change across place and time. Ecosystems provide provisioning services such as food, freshwater, and fuelwood; regulating services such as temperature management and pollination; cultural services; and sustaining support services such as soil development, nutrient cycling, and primary production. The Sikkim Himalayan alpine and subalpine landscapes are considered versatile because they provide an array of ecosystem services; this also plays a significant role in the relationship between loss of biodiversity and ecosystem services. This region provides goods and a variety of ecological services to millions of people, both within and beyond its physical limits. In the state of Sikkim, Alpine pasture and scrub encompass around 44% of the overall geographical area, with constant snow cover. Himalayan alpine and subalpine habitats have provided biological resources and ecosystem services that have maintained the livelihood of local Himalayan people for generations as a source of wild food, medicine, and water supply, as well as sustaining socioeconomic conditions via ecotourism. A wide range of plants from alpine meadows and subalpine forests are used for food, firewood, ethnomedicine, timber-bearing plants, fodder, spices, and other purposes. The abundance of ethnomedicine is one of the significant ecosystem services provided by alpine environments. The global market value of ethnomedicine is approximately US$120 billion (Manish 2022). These areas include a high diversity and a lot of economic trees, such as *Abies densa* (furniture, home building) and *Tsuga dumosa*, which are exploited as sources of timber and wood.

The Himalayan plants are also a source of natural dyes for the locals, who use them to colour textiles, cosmetics, skin, cave paintings, clothing, and other items. Dyes can be extracted from the root, rhizome, stem, bark, leaf, flower, fruit, seed, or complete plant, as well as resin, gum, and stem galls in a few cases. Plants such as *Berberis asiatica*, *Rhododendron arboretum*, *Urtica dioica*, and others may be found in alpine meadows and subalpine forests and used as dye. Plants and traditional local culture are inextricably linked. *Betula utilis*, *Fagopyrum esculentum*, *Rhododendron*

Thangsing, West Sikkim (3900 m asl.)

Near Goecha La, West Sikkim (4500 m asl.)

Gnathang, East Sikkim (4000 m asl.)

Kyongnosla, East Sikkim (4000 m asl.)

Fig. 3 Alpine and sub-alpine habitat of Sikkim Himalaya

anthopogon, *Saussurea obvallata*, *Prunus cerasoides*, and other sacred plants have been used to fulfil different religious ceremonies and festival celebrations.

Many plants have aromatic fragrances due to the presence of resin, turpentine, volatile oil, and other substances. The most significant '*genuine dhupi*' species is *Rhododendron anthopogon*, which produces aroma after burning due to the presence of oleoresin. *Juniperus recurva* is also a reliable source of oleoresin. Many plant species are vital resources for local populations' livelihoods, but overexploitation of firewood, fodder, lumber, food supplies, and medicinal plants has jeopardised their long-term availability in recent decades (Fig. 4).

The alpine habitat is also home to a variety of high-altitude wetlands. High-altitude lakes account for 40.79% of Sikkim's total wetland area. A total of 534 lakes comprising 3325 hectares have been established in the state, with more than 38% situated over 5000 m altitude (http://www.sikkimforest.gov.in). Among them, 215 high-altitude waterbodies were reported within protected areas (Tambe 2007). High-altitude lakes, apart from their ecological significance, play a crucial role in biodiversity, wildlife habitat, and socio-economic aspects. Chholhamu Lake (102.5 ha)

Fig. 4 Memencho Lake, East Sikkim: one of the high-altitude wetlands

in Sikkim, at 5300 m elevation, is the highest lake in India. It is the source of the river Teesta (Tista). Some of the other important wetlands in Sikkim include Gurudongmar Lake, Tso Lhamu Lake, Gyap-Zay Tso, Tsomgo Tso, Nakchok Jor Pokhori, Lampokhori, Bidan Chu Lake and Bidan Chu Lake. Wetlands are distinctive, productive ecosystems where terrestrial and aquatic habitats meet. Wetlands are essential for preserving numerous natural cycles and providing habitat for a broad variety of wetland flora and fauna. They offer food while also contributing to the flow of mountain streams and river systems. They act as a natural sponge against floods and drought and contribute to the battle against climate change. Wetlands are an essential way to serve as carbon sinks. Wetlands are incredibly productive ecosystems that offer a wide range of benefits to humanity on a global scale (Fig. 5).

8 Threats to Alpine Ecosystem and Their Component

Climate change is now considered to be one of the most persistent problems of the twenty-first century. It has a myriad of ecosystems, high floral and faunal diversity, abundant water resources, and good forest cover, with large climatic, topographical, and altitudinal variations within this small geographic area. Alpine meadows, which lie above the tree line but below the snow line, function as "sponges" by

Fig. 5 Grazing in alpine meadow of Dzongri, West Sikkim

absorbing melting snow and acting as natural water towers (Kothari et al. 2017). Alpine and subalpine areas of Sikkim harbour a wide range of specialised microhabitats that harbour substantial amounts of ethnomedicinal plants (Lahiri and Dash 2022). However, these alpine habitats are very sensitive to climate change and face tremendous anthropogenic pressure. The Alpine habitat of the Sikkim Himalaya is the source of a significant amount of highly valued medicinal plants. Though climate change threatens rare alpine plants and their habitats and could lead to the extinction of rare medicinal plants, this region also faces a high threat of terrestrial biodiversity loss due to land use change, deforestation, and infrastructure development.

The highly unscientific extraction of plants from the wild for their roots causes a serious threat to future generations as well as the microhabitat of the entire population and its associated plants. The rural environment has been quickly changing in recent years, with the beginning of development projects and the westernisation of rural society. Sikkim is strategically located in a crucial region of the country; hence, armed personnel are strongly deployed there. Human-caused habitat fragmentation and degradation have become one of the most serious threats to alpine ecosystems, as well as a major driver of species extinction. As temperatures rise and precipitation decreases, alpine meadows and plants may shift to higher elevations. This process, however, will be hampered by the lack of soils with sufficient depth for anchoring and nutrient storage at certain sites. Whereas alpine wetlands will diminish due to high evaporation, which will be worsened by settlement growth and other human

Fig. 6 Habitat degradation in alpine landscape of Kyongnosla, East Sikkim

activity, another risk in the current climate context for the alpine ecosystem is alien invasive species, which can dominate existing populations, creating dramatic alterations in ecosystems. Agricultural expansion, livestock grazing (Fig. 5), tourism, habitat degradation (Fig. 6) and other disturbances have all had a negative influence on global vegetation composition, resulting in the introduction or spread of invasive species (Pauchard and Villarroel 2002). With increasing demand and consequent pressure on medicinal plants, it is critical to document their traditional uses as well as understand their range, diversity, and availability in their natural environments. In the last decade, one of the most significant methods of medicinal plant conservation has been to provide a suitable environment for their preservation.

9 Government Initiative Towards Conservation

This sub-alpine and alpine region of the Sikkim Himalaya, known locally as the "Himal," has a distinct ecological importance. The "Himal" is also a source of high valued medicinal plants. Uncontrolled grazing, unplanned tourism, killing and trapping of wild animals, and trafficking of medicinal and aromatic plants are the primary risks for these ecosystems. Due to its remoteness and rough terrain effective protection of the "Himal" by forest workers alone is extremely difficult. To overcome these difficulties, the government of Sikkim created a workforce with the help of locals known as Himal Rakshaks. In addition to this, Government of India also created several protected areas as per IUCN guidelines. There are 8 (Eight) PAs comprising of 7 (seven) Sanctuaries and 1 (one) National Park. Non-biodegradable materials such as plastics, polybags, and biomedical/chemical waste were prohibited by legislation in 1997. Sacred peaks, caves, boulders, lakes, 'chhorten', and hot springs have been informed; climbing excursions to Mt. Khangchendzonga (8598 m) are prohibited. In 2001, the conservation of the state's unique terrestrial and aquatic ecosystems of wetlands/lakes began by forbidding commercial operations to protect the state's legacy and fragile ecosystem. Sikkim became the country's first state to form a high-level team to research glaciers and alpine ecosystems (Fig. 7).

10 Ethnomedicinal Plants

To study human plant relationships and comprehend regional human sustainable ties to their environment, it is necessary to document the plants and their ethnobiological values (Lahiri and Dash 2022). While Sikkim is a densely forested region with a diverse ethnic population that includes the Lepcha, Bhutia, Sherpa, Rai, Limboo, Nepali, and others, people in remote and fringe areas have a well-inherited rich traditional knowledge of wild plants and rely on these natural resources for food, shelter, medicines, fodder, insecticides, and other purposes. Traditional herbal healers in Sikkim are referred to as Lamas (monastery heads) or Jakris, and traditional plant use is a component of Sikkim's cultural past (Dash 2009). It is intimately linked to indigenous peoples' religious beliefs and practises (Lahiri and Dash 2022). The early phase of Indian herbal study is represented by Charaka (Charak-Samhita, 100–500 AD) and Shushruta (Shushrut-Samhita, 200–500 AD). Biswas (1956) was the first to publish Sikkim ethnobotanical studies, followed by Rolla (1963), Bennet (1983), Rai and Sharma (1994), Singh et al. (2002), Dash et al. (2003), Maity et al. (2004), Pradhan and Badola (2008), Sharma and Sharma (2010), Badola and Pradhan (2013), Dahal et al. (2017), Dahal and Borthakur (2017); Lahiri and Dash (2022) etc. *Aconitum heterophyllum*, *Nardostachys jatamansi*, *Neopicrorhiza scrophulariiflora*, and *Swertia chirayita* are frequently exported from North Sikkim (Sundriyal et al. 1998). It has been estimated that about 30,000 kg of *Nardostachys*, 10,000 kg of *Aconitum*, and 6200 kg of *Neopicrorhiza* have been going to the commercial market illegally every year (Rai and Sharma 1994). It has also been calculated that around 10,720 kg of *Aconitum heterophyllum*, 31,000 kg of *Nardostachys jatamansi*, and 3440 kg of *Swertia chirayita* had been harvested in a span of two years (Rai et al. 2000) (Figs. 8 and 9).

Rumex nepalensis, also known as "halhale" in the local language, is also available at the vegetable market, where it costs Rs. 10/kg in Namchi and Rs. 20/kg in Gangtok (Sundriyal et al. 1998). In the local market, a few people known as "Jaributy man" (herbal medicine practitioner) sell various herbal remedies. Due to the people's low socioeconomic condition and long tradition of using plant medications, rural folk collect and purchase these plants as needed. Plant uses differently by various tribe.

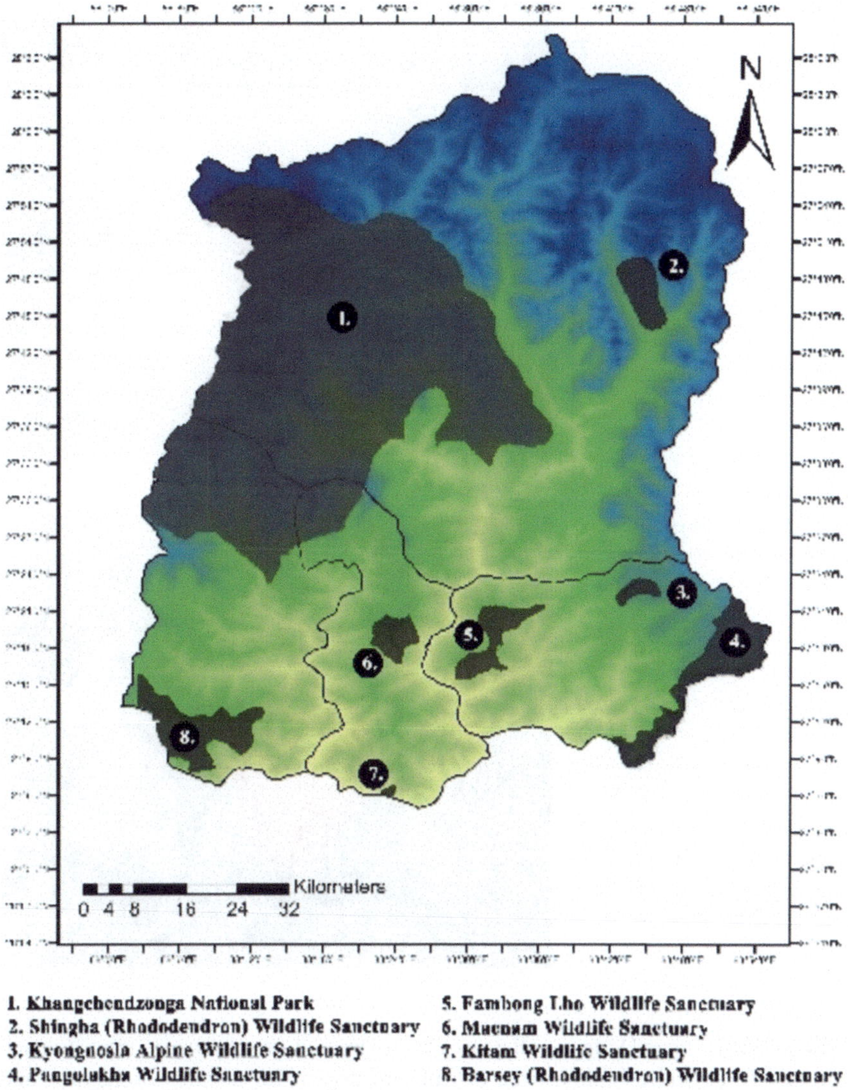

Fig. 7 Protected areas of Sikkim Himalaya

Limboo tribe from Yuksom utilises *Abies densa* leaf extract to cure asthma, bronchitis, and stomach discomfort (Badola and Pradhan 2013), however the Dzongu valley lepcha tribe exclusively uses it to treat stomach pain and fever (Pradhan and Badola 2008). The effectiveness of herbal remedies is difficult to assess. According to a study of the literature, several plants used by humans are likely to be effective; for example, excessive usage of *Caltha palustris* produces throat burning, vomiting, bloody diarrhoea, and gastrointestinal sickness (Jamloki et al. 2022).

Fig. 8 *Panax pseudoginseng* Wall

Fig. 9 *Sinopodophyllum hexandrum* (Royle) T.S.Ying

Lepcha tribe from Dzongu valley uses leaf extract of *Abies densa* to treat stomach pain and fever while limboo tribe of Yuksom uses it for asthma, bronchitis, and stomach ache (Lahiri and Dash 2022). The effectiveness of herbal remedies is difficult to assess. The leaves of *Rhododendron falconeri* and *R. barbatum* are used as a fish poison due to the presence of the glucoside ericolin (Bhattacharyya 2011), while an efficient pesticide derived from the vegetative portions of *R. thomsonii* was identified in North Sikkim (Lahiri and Dash 2022). *R. cinnabarinum* is also found in alpine areas; leaves of this *Rhododendron* are harmful to cattle and induce irritation of the eyes and face (Bhattacharyya 2011) (Figs. 10 and 11).

Field research and a thorough examination of prior work determined that the alpine and subalpine areas of Sikkim include roughly 169 species of flowering plants classified into 47 groups that have been utilised as ethnomedicine by indigenous people.

Fig. 10 *Aconitum ferox* Wall. ex Ser

Fig. 11 *Fritillaria cirrhosa* D.Don

All 169 species listed in Table 1 have their respective uses, vernacular names (if existent), and family affiliations. The field survey also revealed that the encroachment of these species' native habitats and illegal access may threaten their population. To maintain unique alpine germplasm, careful management of fragile habitats, strict restriction on overexploitation, and commercialization is essential.

Table 1 Highly used alpine and sub-alpine medicinal plants of Sikkim Himalaya

Family	Plant species	Local name	Used parts	Medicinal uses/activities	References[a]
Amaryllidaceae	*Allium prattii* C.H.Wright	Jangali piaz,	Leaves and Flower	Body aches, coughs, colds	Chhetri (2005)
	Allium wallichii Kunth	Dungdung gay	Whole plant	Altitude sickness, digestion, anaemia, cholera, diarrhoea, colds, coughs, bacterial and microbial illnesses, stomach pain, gastritis, indigestion, colic, appetite loss, intestinal cleaning	Chandra and Rawat (2015); Gewali (2008); Jamloki et al. (2022); Lahiri (2022)
Apiaceae	*Pleurospermum hookeri* C.B.Clarke	Tang-kun-dkar-po	Root	Antidote, anti-inflammatory	Chakraborty et al. (2017); Chandra and Rawat (2015); Chandra et al. (2018)
	Selinum tenuifolium (DC.) Raizada & Saxena	Bhutkesh	Roots and fruits	Headaches, fever	Kothari et al. (2017)
	Selinum wallichianum (DC.) Raizada & H.O.Saxena		Whole plant	Hand inflammation, leg, stomach, tumour	Lahiri and Dash (2022)
Araceae	*Arisaema jacquemontii* Blume	Sanp ko makai	Seed	Worm infestation, wounds, swellings, bone disorders	Bhat et al. (2020); Chhetri (2005)
Araliaceae	*Panax bipinnatifidus* Seem	Ajambari	Root, rhizome	Aphrodisiac, asthma, diabetes, gastritis, hypertension, cuts, wounds	Chandra and Rawat (2015; Lahiri and Dash (2022)
	Panax pseudoginseng Wall	Panch patay	Root	Root paste mixed with milk use to increase sexual potency and vigour	Badola and Pradhan (2013); Budha-Magar et al. (2020); Dahal and Borthakur (2017)

(continued)

Table 1 (continued)

Family	Plant species	Local name	Used parts	Medicinal uses/activities	References[a]
Asparagaceae	*Polygonatum cirrhifolium* (Wall.) Royle	Meda	Rhizome	Aphrodisiac, breathing illness, lymph disorder, tonic	Chakraborty et al. (2017); Chandra et al. (2018)
	Polygonatum hookeri Baker	Meda	Rhizome	Aphrodisiac, breathing illness, lymph disorder, tonic	Chandra and Rawat (2015)
	Polygonatum singalilense H.Hara	Meda	Flower	Wounds, regulate body temperature, body strength	Lahiri and Dash (2022); Lahiri et al. (2022)
	Polygonatum verticillatum (L.) All	Meda	Rhizome	Aphrodisiac, emollient, cardiac tonic, reduce excess joint fluid	Das et al. (2012); Lahiri and Dash (2022)
Asteraceae	*Ainsliaea latifolia* (D.Don) Sch.Bip	Meda	Root	Stomach aches, asthma, spleen enlargement	Dahal and Borthakur (2017); Dash et al. (2021b)
	Anaphalis busua (Buch.-Ham.) DC	Buki phool	Leaves	Cattle wounds	Chandra and Rawat (2015)
	Anaphalis contorta (D.Don) Hook.f	Buki phool	Leaves and head	Wound dressing, coughs	Chhetri (2005); Dahal and Borthakur (2017); Dash et al. (2021b)
	Anaphalis margaritacea (L.) Benth. & Hook.f		Leaves and flowers	Cuts	Dash et al. (2021b)
	Anaphalis triplinervis (Sims) C.B.Clarke	Buki phool	Leaves and flowers	Wounds, wound dressing	Das et al. (2012)

(continued)

Table 1 (continued)

Family	Plant species	Local name	Used parts	Medicinal uses/activities	References[a]
	Artemisia vulgaris L	Titaypati	Leaves	Antibacterial, nasal bleeding, protect the liver and spleen, mouth ulcers	Gairola et al. (2008)
	Dolomiaea macrocephala DC. ex Royle	Dhup	Whole plant	Antioxidant, indigestion, wounds, cold, flu, fever	Chandra and Rawat (2015)
	Ligularia amplexicaulis DC	Ri-sho	Shoot, leaf, root	Digestive, emetic, stomach disorder	Chakraborty et al. (2017); Chandra et al. (2018)
	Myriactis nepalensis Less	Jillethi	Leaf, flower	Fever, headache, stomach disorder	Chandra and Rawat (2015)
	Nannoglottis hookeri (C.B.Clarke ex Hook.f.) Kitam		Root	Tonic	Chandra and Rawat (2015)
	Saussurea gossypiphora D.Don	Maikopila	Whole plant	Incisions and wounds	Kothari et al. (2017)
	Saussurea obvallata (DC.) Sch.Bip	Brahma kamal	Whole plant	Bone ache, cough, cold, intestinal ailments, urinary tract disorders	Chandra and Rawat (2015)
	Saussurea simpsoniana (Fielding & Gardner) Lipsch		Inflorescence, root	Reproductive disorders, cuts, wounds	Chandra and Rawat (2015)
	Saussurea tridactyla Sch.Bip. ex Hook.f	Kapasephul	Whole plant	Fatigue, menstruation difficulties, arthritis	Kothari et al. (2017)
	Soroseris hookeriana Stebbins		Whole Plant	Fever, bone fracture	Chandra and Rawat (2015); Lahiri et al. (2022)

(continued)

Table 1 (continued)

Family	Plant species	Local name	Used parts	Medicinal uses/activities	References[a]
	Taraxacum officinale F.H.Wigg	Tukiphool	Leaves, root	Jaundice, kidney stones, gastritis, joint dislocation, wounds	Chakraborty et al. (2017); Chandra and Rawat (2015)
Berberidaceae	*Berberis angulosa* Wall. ex Hook.f. & Thomson	Karay chutro	Root	Ulcers, skin infections, conjunctivitis, urinary tract irritation, cough, cold, fever, dysentery	Chandra and Rawat (2015); Dash et al. (2021b)
	Berberis aristata DC	Chutro	Roots, stems, leaves, bark and flowers	Fever, jaundice, malaria, diarrhoea, swelling, eye disorders, rabies	Kothari et al. (2017)
	Berberis asiatica DC	Chutro	Bark, Root, Leaves and Fruit	Diarrhoea, dysentery, jaundice, fever	Badola and Pradhan (2013)
	Berberis wallichiana DC	Chitrokanra	Leaves and Fruit	Dog bite, joint pain, swells	Dahal and Borthakur (2017); Dash et al. (2021b)
	Sinopodophyllum hexandrum (Royle) T.S.Ying	Laghu patra	Tubers	Diabetes, blood purifier, vermifuge, purgative in chronic constipation	Hooker (1854); Jamloki et al. (2022); Joshi et al. (2011); Kothari et al. (2017)
Betulaceae	*Alnus nepalensis* D.Don	Sungdung kung	Bark	Antioxidant, Cancer	Gewali (2008)
	Betula utilis D.Don	Takpa, Bhojpatra	Leaf, bark	Bronchitis, cough, diuretic, wounds	Chhetri (2005); Dahal and Borthakur (2017)
Boraginaceae	*Maharanga emodi* (Wall.) DC		Flower, root	Cooling, laxative, anthelmintic, eye diseases, ear Problems, hair tonic	Chakraborty et al. (2017); Chandra and Rawat (2015); Chandra et al. (2018)

(continued)

Table 1 (continued)

Family	Plant species	Local name	Used parts	Medicinal uses/activities	References[a]
Brassicaceae	*Cardamine macrophylla* Willd	Bhotay saag	Leaves	Tonic, diuretic	Chandra and Rawat (2015)
Campanulaceae	*Codonopsis clematidea* (Schrenk) C.B.Clarke	Luptic	Whole plant	Coughs, colds, constipation, diarrhoea and gastritis	Chandra and Rawat (2015)
	Codonopsis foetens Hook.f. & Thomson	Luptic	Whole plant	Cough, cold, constipation, gastritis and diarrhoea	Chandra and Rawat (2015)
Caprifoliaceae	*Bassecoia hookeri* (C.B.Clarke) V.Mayer & Ehrend	Panri mendo	Whole plant	Fever, gout, jaundice, cough, cold, diarrhoea, dysentery	Chandra and Rawat (2015)
	Nardostachys jatamansi (D.Don) DC	Jataamansi	Root	AnMuscular cramps, edoema, paralysis, jaundice, irregular periods, low blood pressure, hysteria, piles, hair development	Budha-Magar et al. (2020); Chandra and Rawat (2015); Dahal and Borthakur (2017); Dash et al. (2021b); Dash and Misra (1997); Jamloki et al. (2022); Lahiri et al. (2022)
	Valeriana hardwickei Wall	Nakali jatamanshi	Root	Colic, epilepsy neurosis, urinary problems	Dash et al. (2003); Gairola et al. (2008)
	Viburnum nervosum D.Don		Root	Anti-inflammatory, pain reliever	Dash et al. (2021b)
Caryophyllaceae	*Silene nigrescens* (Edgew.) Majumdar		Root	Lice, dandruff, hair conditions	Biswas (1956)
Celastraceae	*Parnassia nubicola* Wall. ex Royle	Mamira	Upper part	Jaundice, fever, fatigue	Chandra and Rawat (2015)

(continued)

Table 1 (continued)

Family	Plant species	Local name	Used parts	Medicinal uses/activities	References[a]
Crassulaceae	*Rhodiola cretinii* (Raym.- Hamet) H.Ohba		Root	Breathing problems, skin diseases	Chandra and Rawat (2015); Chhetri (2005)
	Rhodiola himalensis (D.Don) S.H.Fu	Lakpaguru	Root	Respiratory issues, colds, coughs, skin infections	Chandra and Rawat (2015); Chandra et al. (2018); Lahiri et al. (2022)
Cupressaceae	*Juniperus communis* L	Vairung pate	Leaf, fruit	Asthma, bronchitis, headache	Chandra and Rawat (2015)
	Juniperus recurva Buch.- Ham. ex D.Don	Dhupi	Leaf	Fevers, skin disorders	Dash et al. (2023); Dash et al. (2021b)
Elaeagnaceae	*Hippophae tibetana* Schltdl		Fruit	Diuretic, tonic, asthma, cough, cold, hypertension	Chandra and Rawat (2015)
Ericaceae	*Cassiope fastigiata* (Wall.) D.Don	Sunthangni	Leaves and flowers	Cuts, itches and skin allergies	Chakraborty et al. (2017); Dash et al. (2021b)
	Cassiope selaginoides Hook.f. & Thomson	Sunthangni	Leaf and flowers	Acne, allergies, ringworm, itches and wounds	Chakraborty et al. (2017); Chandra and Rawat (2015)
	Gaultheria nummularioides D.Don	Shokpa	Leaves	Constipation	Chandra and Rawat (2015); Dash et al. (2021b)
	Gaultheria trichophylla Royle	Kaaligedi	Fruits and leaves	Natural antioxidant, colds, coughs, wounds	Chandra and Rawat (2015); Chandra et al. (2018); Dash et al. (2021b)

(continued)

Table 1 (continued)

Family	Plant species	Local name	Used parts	Medicinal uses/activities	References[a]
	Lyonia ovalifolia (Wall.) Drude	Angeri	Leaves	Skin diseases	Dahal and Borthakur (2017); Dash et al. (2021b)
	Rhododendron anthopogon D.Don	Sunpatey	Leaves and flower	Colds, coughs, lung disorders	Budha-Magar et al. (2020); Chandra and Rawat (2015); Lahiri and Dash (2022)
	Rhododendron arboreum Sm	Karabaka, Etok	Flower bud, petals, and leave	Diarrhoea, throat discomfort, bleeding dysentery, headache, pneumonia, slow down excessive bleeding in females	Badola and Pradhan (2013); Budha-Magar et al. (2020); Chandra and Rawat (2015); Dahal and Borthakur (2017); Dahal et al. (2017); Gairola et al. (2008); Gogoi et al. (2021)
	Rhododendron campanulatum D.Don	Nilo chimal	Leaves and stem	The juice of the leaves is used to treat coughs. Powder of dried stem and leaves used to treat colds, flu, body aches, and persistent fever, and powder of dried leaf used as snuff in rheumatism therapy	Budha-Magar et al. (2020); Gairola et al. (2008)

(continued)

Table 1 (continued)

Family	Plant species	Local name	Used parts	Medicinal uses/activities	References[a]
	R. anthopogon subsp. *hypenanthum (Balf.f.) Cullen*	Sunpati	Stem, leaf	Digestive, febrifuge, diaphoretic, cough, cold, pulmonary disorders, skin diseases	Chandra and Rawat (2015)
	Rhododendron lepidotum Wall. ex G. Don	Bhaiunakpo	Bark and Leaves	Bark juice used as purgative. Leaves and flowers paste used to purify blood. Leaves powdered used in cold and cough, fever	Bennet (1983); Dash (2009)
	Rhododendron setosum D.Don	Jhusey sunpati	Leaves and flower	Throat and muscular tissue inflammation, wounds	Chhetri (2005); Kothari et al. (2017); Lahiri and Dash (2022)
Euphorbiaceae	*Euphorbia stracheyi* Boiss	Sangmen	Whole plant	Bile problems, breathing disorder, digestive problems	Chandra and Rawat (2015)
	Euphorbia wallichii Hook.f	Thar-nu	Root	Edema, eczema, pimple, fungal infections	Chakraborty et al. (2017); Chandra et al. (2018)
Fabaceae	*Spongiocarpella nubigena* (D.Don) Yakovlev	Chyali	Whole plant	Cut, wound	Chandra and Rawat (2015)
	Thermopsis barbata Benth		Whole plant	Analgesic, cuts, wounds, dog bites, digestive, tonic, fever	Chandra and Rawat (2015)
Fumariaceae	*Dactylicapnos scandens* (D.Don) Hutch	Lahara mutu	Root tubers	Cardiac issues	Dash et al. (2021b); Forest Survey of India (2019)

(continued)

Table 1 (continued)

Family	Plant species	Local name	Used parts	Medicinal uses/activities	References[a]
Gentianaceae	*Gentiana elwesii* C.B.Clarke		Leaves	Altitude sickness	Chandra and Rawat (2015); Chandra et al. (2018); Dash et al. (2021b)
	Gentiana prolata Balf.f		Leaves	Altitude sickness	Chandra and Rawat (2015); Chandra et al. (2018); Dash et al. (2021b)
	Megacodon stylophorus (C.B.Clarke) Harry Sm		Root	Swellings, wounds	Chandra and Rawat (2015); Chandra et al. (2018); Dash et al. (2021b)
	Halenia elliptica D.Don	Tikta	Whole plant	Malaria, liver and stomach ailments	Chhetri (2005); Dash et al. (2021b); Hooker (1949–51)
	Swertia bimaculata (Siebold & Zucc.) Hook.f. & Thomson ex C.B.Clarke	Bhalay chireto	Whole plant	Fever, urinary problems, liver diseases	Chandra and Rawat (2015)
	Swertia chirayita (Roxb.) H.Karst	Chiraito	Whole plant	Fever, cold, cough, diarrhoea, stomach ache, asthma, dyspepsia, liver issues, tonic, anthelmintic, stomachic, febrifuge	Gairola et al. (2008); Jamloki et al. (2022); Lahiri (2022)
	Swertia hookeri C. B. Clarke	Lekhchiraito	Roots	Jaundice	Lahiri (2022)

(continued)

Table 1 (continued)

Family	Plant species	Local name	Used parts	Medicinal uses/activities	References[a]
	Swertia multicaulis D.Don	Sarnaguru	Whole plant	Cuts, wounds, fever, cough, cold, bodily aches, bone fractures, root paste	Jamloki et al. (2022); Lahiri (2022)
	Swertia paniculata Wall	Chiraito	Whole plant	Fever, cold, cough, diarrhoea, stomach ache, asthma, dyspepsia, liver issues, tonic, anthelmintic, stomachic, febrifuge	Chandra and Rawat (2015)
	Veratrilla baillonii Franch	Rgu-drus	Root	Antidote, jaundice, stomach disorder, Wounds	Chakraborty et al. (2017); Chandra and Rawat (2015); Chandra et al. (2018)
Geraniaceae	*Geranium nepalense* Sweet	Chunetro ghans	Whole plant	Painful urination	Dash et al. (2021b); Hooker (1872–1897)
Iridaceae	*Iris ensata* Thunb	Bhaji	Rhizome	Cold, cough	Hooker (1854)
Lamiaceae	*Eriophyton wallichii* Benth	Spang-mtshan-spuru	Whole plant	Lung infections, wounds	Dash et al. (2021b)
	Prunella vulgaris L		Whole plant	Headache, nasal congestion	Dahal and Borthakur (2017)
Liliaceae	*Fritillaria cirrhosa* D.Don		Bulb	Anaesthetic, asthma, bronchitis, stomach ache	Chandra and Rawat (2015); Hooker (1854)
	Lilium nanum Klotzsch		Whole plant	Treat poisonous bites, bone fractures, sprains, traumas	Chhetri (2005); Dash et al. (2021b); Lahiri and Dash (2022)
	Paris polyphylla Sm	Satuwa	Tuber	Stomach aches, fevers, diarrhoea, dysentery	Kothari et al. (2017)

(continued)

Table 1 (continued)

Family	Plant species	Local name	Used parts	Medicinal uses/activities	References[a]
Melanthiaceae	*Trillium govanianum* Wall. ex D.Don	Nag Chhatri	Leaves, Rhizome	Analgesic, anti-inflammatory, antifungal, indigestion, inflammation, fever	Chandra and Rawat (2015)
Nartheciaceae	*Aletris pauciflora* (Klotzsch) Hand.-Mazz	Dam-bu-ka-ra	Whole plants	Liver disorders, respiratory problems, blood infection, cuts, wounds	Chhetri (2005); Dahal and Borthakur (2017)
Orchidaceae	*Dactylorhiza hatagirea* (D.Don) Soó	Panchamle	Tubers and shoots	Bone fractures, cuts, diarrhoea, dysentery, fever, diabetes, gastritis, muscle aches, wounds	Kothari et al. (2017)
	Galearis spathulata (Lindl.) P.F.Hunt		Tuber	Cuts, wounds	Chandra and Rawat (2015); Dash et al. (2021b)
	Gymnadenia orchidis Lindl	Panch-aule	Tuber	Liver, urinary problems, gastritis, diarrhoea, cuts, and wounds, tonic, aphrodisiac	Chhetri (2005); Dash et al. (2021b); Lahiri (2022)
	Pleione hookeriana (Lindl.) Rollisson	Lasun paate	Pseudobulb	Cuts, wounds	Chandra and Rawat (2015)
	Ponerorchis chusua (D.Don) Soó	Sayno panchaunlay	Root and tuber	Cuts, diarrhoea, dysentery, stomach disorder, chronic fever, wounds	Chandra and Rawat (2015); Jamloki et al. (2022); Lahiri (2022)
	Satyrium nepalense D.Don		Tuber	Aphrodisiac, colds, coughing, fever	Badola and Pradhan (2013); Chhetri (2005)

(continued)

Table 1 (continued)

Family	Plant species	Local name	Used parts	Medicinal uses/activities	References[a]
Orobanchaceae	*Boschniakia himalaica* Hook.f. & Thomson	Not known	Seed, flower	Lung infections, tumour, blood disorders	Dash et al. (2021b); Lahiri and Dash (2022)
	Pedicularis megalantha D.Don	Lug-ru-dmar-po	Whole plant	Plant used in treatment of intestinal disorders	Lahiri and Dash (2022); Lahiri et al. (2022)
	Pedicularis siphonantha D.Don	Lug-ru-smug-po	Whole plant	Antidiarrheal, stomach disorders	Lahiri and Dash (2022); Lahiri et al. (2022)
Papaveraceae	*Meconopsis horridula* Hook.f. & Thomson	Tshər-ngon	Whole plant	Bone fractures, fever, back ache, lung, skin problems, sinusitis, liver illness, and wounds	Kothari et al. (2017); Lahiri et al. (2022)
	Meconopsis paniculata (D.Don) Prain	Ud-pal-ser-po	Whole plant	Liver and lung illnesses, aid digestion	Dash et al. (2021b); Lahiri and Dash (2022); Lahiri et al. (2022)
	Meconopsis simplicifolia (D.Don) G.Don	Ud-pel-sngon-po	Whole plant	Renal problems, antipyretic, antimalarial, respiratory difficulties, coughing, colds, blood circulation	Chhetri (2005); Dash et al. (2021b); Dash and Misra (1997); Lahiri and Dash (2022); Lahiri et al. (2022)
Pinaceae	*Abies densa* Griff	Gobre salla	Bark and Leaf	Cough, cold, fever, stomach, toothache	Chandra and Rawat (2015); Dahal and Borthakur (2017); Dash et al. (2021b); Gairola et al. (2008)

(continued)

Table 1 (continued)

Family	Plant species	Local name	Used parts	Medicinal uses/activities	References[a]
	Abies spectabilis (D.Don) Mirb	Gobre salla	Leaves	Tonic, tuberculosis, internal haemorrhage	Dahal (2019)
Plantaginaceae	*Lagotis crassifolia* Prain	Hong-ihen	Rhizome	Fever, hypertension, pneumonia, inflammation of intestines, lungs, lever	Chandra and Rawat (2015)
	Lagotis kunawurensis (Royle ex Benth.) Rupr		Whole plant	Abdominal pains, inflammation, liver tonic, fevers	Dash (2007); Dash et al. (2021b)
	Neopicrorhiza scrophulariiflora (Pennell) D.Y.Hong	Kutki	Whole plant	Antidote poison	Gewali (2008)
	Picrorhiza kurroa Royle ex Benth	Kudhi		Antiperiodic, cholagogue	Hooker (1854)
	Plantago depressa Willd		Root	Plant used as antidiarrheal	Lahiri and Dash (2022); Lahiri et al. (2022)
	Veronica ciliata subsp. *cephaloides* (Pennell) D.Y.Hong		Whole plant	Wound, ulcer, haemorrhaged	Lahiri and Dash (2022)
	Veronica himalensis D.Don		Whole plant	Wound, ulcer, haemorrhaged	Lahiri et al. (2022)
Polygonaceae	*Aconogonon molle* (D.Don) H.Hara	Thotne		Diarrhoea, dysentery	Kumar et al. (2017)
	Bistorta affinis (D.Don) Greene	Not known	Root	Pain, inflammation, stomach aches, fever	Badola and Pradhan (2013); Dahal and Borthakur (2017); Dash et al. (2021b)

(continued)

Table 1 (continued)

Family	Plant species	Local name	Used parts	Medicinal uses/activities	References[a]
	Bistorta amplexicaulis (D.Don) Greene	Raktaryaaulo	Root/ rhizome	Dysentery, diuretic, cure coughs	Chhetri (2005); Dahal and Borthakur (2017); Dash et al. (2021b)
	Bistorta vivipara (L.) Delarbre		Root	Fever, stomach discomfort	Dash et al. (2021b); Gairola et al. (2008)
	Bistorta vacciniifolia (Wall. ex Meisn.) Greene	Pulunge jhar	Whole plants	Tuberculosis, dysentery, fever	Dash et al. (2021b)
	Fagopyrum esculentum Moench	Mithayphaaper	Roots, leaves and grains	Liver problems	Hooker and Thomson (1859)
	Oxyria digyna (L.) Hill	Lug-sho	Whole plant	Appetite, fever, laxatives, smallpox, sore throat	Dash et al. (2021b)
	Rheum acuminatum Hook.f. & Thomson	Pacamchal	Stem and roots	Headaches, fevers, diarrhoea, digestive issues, rheumatism, bone fractures	Badola and Pradhan (2013); Chandra and Rawat (2015); Dahal and Borthakur (2017)
	Rheum australe D.Don	Khokim	Roots and petiole	Constipation, skin eruptions, liver diseases, internal injuries, bodily aches, diarrhoea, dysentery, ulcers, cuts, wounds	Kothari et al. (2017)
	Rheum nobile Hook.f. & Thomson	Padamchal,	Whole plants	Body aches, knee discomfort caused by water retention, cuts, wounds, laxative, diuretic, rheumatoid arthritis, heart issues, tonic after pregnancy	Chandra and Rawat (2015); Dash and Misra (1997); Jamloki et al. (2022); Lahiri and Dash (2022)

(continued)

Table 1 (continued)

Family	Plant species	Local name	Used parts	Medicinal uses/activities	References[a]
	Rumex nepalensis Sreng	Halhale	Young shoot, leaves and roots	Wounds, hair loss, hepatitis	Gairola et al. (2008); Jamloki et al. (2022)
Primulaceae	*Primula sikkimensis* Hook	Shang dril ser po	Root and flowers	Diarrhoea, fever, especially	Chandra and Rawat (2015); Lahiri and Dash (2022)
Ranunculaceae	*Aconitum dissectum* D.Don	Bikh	Root	Rheumatism, cough, asthma, fever, skin disorders, neuralgia	Dash et al. (2021a)
	Aconitum ferox Wall. ex Seringe	Kalo Bikh	Rhizome	Asthma, blood clots, food poisoning, diarrhoea, rheumatism, snake bite, stomach ache, toothache, cough, cold, bodily pain, diabetes, debility, leprosy, paralysis, spermatorrhoea, typhoid	Badola and Pradhan (2013); Jamloki et al. (2022); Kothari et al. (2017)
	Aconitum heterophyllum Wall. ex Royle	Bikh	Rhizome	Body aches, fevers, colds, coughs, nasal discharge, snakebites, food poisoning, stomach aches, antiperiodic, piles, sore throat, infertility	Gairola et al. (2008); Kothari et al. (2017)
	Aconitum heterophylloides (Brühl) Stapf	Bikhma	Root	Appetite, rheumatism, skin disorders, neuralgia	Dash et al. (2021a); Grierson and Long (1987)

(continued)

Table 1 (continued)

Family	Plant species	Local name	Used parts	Medicinal uses/activities	References[a]
	Aconitum hookeri Stapf	Bikh	Root	Poisoning symptoms, appetite, rheumatism, skin disorders, neuralgia	Bhattacharyya (2011); Dash et al. (2021a); Grierson and Long (1987)
	Aconitum naviculare (Brühl) Stapf	Bikh	Root	Appetite, rheumatism, skin disorders, neuralgia	Dash et al. (2021a); Grierson and Long (1987)
	Aconitum novoluridum Munz	Soorja banshi	Root	Boost appetite	Grierson and Long (1987)
	Aconitum palmatum D.Don	Bikhma	Root	Stomach issues, boost appetite	Chandra and Rawat (2015)
	Aconitum spicatum Stapf	Nilo Bikh	Leaves & Rhizome	Food poisoning, diarrhoea, coughing, colds, headache	Badola and Pradhan (2013); Gairola et al. (2008)
	Anemone rivularis Buch.-Ham. ex DC	Hatti paile	Whole plant	Bone fractures, wounds, headaches, coughs, fevers	Dahal and Borthakur (2017); Dahal et al. 2017
	Caltha scaposa Hook.f. & Thomson		Whole plant	Skin conditions, wounds, bone fractures, sprains	Chandra and Rawat (2015); Dash et al. (2021b)
	Caltha palustris L		Whole plant	Pain, inflammation	Dahal and Borthakur (2017); Dash et al. (2021b); Lahiri and Dash (2022)
	Clematis montana Buch.-Ham. ex DC	Pinasi Lara	Root	Colds, headaches, gout, cuts, sinusitis and wounds	Chandra and Rawat (2015); Hooker (1854)
	Thalictrum chelidonii DC	Dungri phul	Roots	Cuts, wounds	Dash et al. (2003)

(continued)

Table 1 (continued)

Family	Plant species	Local name	Used parts	Medicinal uses/activities	References[a]
	Thalictrum cultratum Wall	La- wa-sad-ma		Diphtheria and fever. Heals sores, dries serous fluids and fever	Dahal and Borthakur (2017); Dahal et al. 2017
	Thalictrum foliolosum DC	Dampatey		Cough, fever, gastritis, eye diseases, indigestion, toothache, ulcer, asthma, piles, food poisoning, jaundice, skin diseases, tonic, purgative	Chhetri (2005); Dash et al. (2003)
	Thalictrum reniforme Wall	Sngo-sprin	Whole plant	Antidote, febrifuge, gastritis, indigestion, ulcer, piles	Dahal and Borthakur (2017); Dahal et al. 2017
Rosaceae	*Fragaria nubicola* (Lindl. ex Hook.f.) Lacaita	Bhui Aiselu	Aerial parts/fruits	Boosting energy, heavy menstrual bleeding, neurological diseases, lung inflammations, chest infections	Dahal and Borthakur (2017); Dash et al. (2021b); Gewali (2008)
	Potentilla fruticosa var. *arbuscula* (D.Don) Maxim	Bajradanti	Whole plant	Antidiarrhoea, toothache, sore throats, cough, cold	Chandra and Rawat (2015); Dash et al. (2021a); Lahiri and Dash (2022)
	Potentilla lineata Trevir	Bajradanti	Root	Toothache, sore throats, cough, cold	Chandra and Rawat (2015)
	Potentilla peduncularis D.Don	Bajradanti	Whole plant	Cold, cough, fever, toothache, dysentery, sore throats	Bhattacharyya (2011); Chandra and Rawat (2015); Lahiri et al. (2022)

(continued)

Table 1 (continued)

Family	Plant species	Local name	Used parts	Medicinal uses/activities	References[a]
	Prunus cerasoides D.Don	Arupatay	Bark and Stem	Bone fracture	Dash et al. (2003); Gairola et al. (2008)
	Rubus ellipticus Sm		Young shoot and root	Stomach pain, stomach warm, headache	Gairola et al. 2008)
	Rosa sericea Lindl	Barko gulab	Flower and leaves	Wounds, headaches, liver problems, constipation, ophthalmia	Badola and Pradhan (2013); Chhetri (2005); Dash et al. (2021b); Dash et al. (2003)
	Spiraea arcuata Hook.f		Leaves	Wounds, fever, bone infections	Lahiri and Dash (2022)
Rubiaceae	*Rubia cordifolia* L	-	Root	Bladder infections, skin ailments	Gairola et al. (2008)
Rutaceae	*Skimmia laureola* (DC.) Siebold & Zucc. ex Walp		Whole plant	Smallpox	Hooker (1854)
Saxifragaceae	*Astilbe rivularis* Ham	Budho okhati		Back pain, toothache, dysentery, diarrhoea	Badola and Pradhan (2013); Jamloki et al. (2022); Joshi et al. (2011)
	Bergenia ciliata (Haw.) Sternb	Pakhanbed	Rhizome	Cracked bones, wounds, diarrhoea, dysentery, cough, cold, lung infections, piles, kidney and bladder stones	Badola and Pradhan (2013); Gairola et al. (2008); Jamloki et al. (2022)
	Bergenia purpurascens (Hook.f. & Thomson) Engl	Lekh pakhanbed	Root	Stomach-ache, pneumonia, tonsil, throat infection	Dahal and Borthakur (2017); Dash et al. (2021b)

(continued)

Table 1 (continued)

Family	Plant species	Local name	Used parts	Medicinal uses/activities	References[a]
	Saxifraga engleriana Harry Sm		Whole plant	Diuretic, gallbladder stone	Chandra and Rawat (2015)
	Saxifraga parnassifolia D.Don		Whole plant	Bile disorders, cough, cold, Jaundice, Wounds	Lahiri and Dash (2022); Lahiri et al. (2022)
Scrophulariaceae	*Hemiphragma heterophyllum* Wall	Nash jhaar, Ulakrikh	Whole plant	Pharyngitis, tonsil irritation, maintains normal blood circulation, rheumatism	Chandra and Rawat (2015); Dahal and Borthakur (2017); Dash et al. (2021b); Lahiri and Dash (2022)
Tamaricaceae	*Myricaria rosea* W.W.Sm	Ambu, Chu-shing-om-bu	Shoots	Toothache, chicken pox, fever, wounds	Budha-Magar et al. (2020); Dash et al. (2021b); Lahiri et al. (2022)
Taxaceae	*Taxus baccata* L	Chharrey salla	Leaves	Epilepsy, indigestion, rheumatism, urinary problems	Jamloki et al. (2022)
	Taxus wallichiana Zucc	Dhengre salla	Bark, leaves and fruits	Piles, muscle pain, fever, asthma, bronchitis, epilepsy, headache, diarrhoea, liver diseases	Lahiri (2022)

(continued)

Table 1 (continued)

Family	Plant species	Local name	Used parts	Medicinal uses/activities	References[a]
Thymelaeaceae	*Stellera chamaejasme* L	Decraali phool	Root	Swelling, gout, bone fracture	Dahal and Borthakur (2017); Dahal et al. 2017
Urticaceae	*Urtica dioica* L		Root, seed and shoot tips	Minor bone fractures, dislocations, diarrhoea, cough, childbirth	Gairola et al. (2008)
Violaceae	*Viola biflora* L	Ghattey gans	Roots, flowers	Antiseptic, emetic	Dash et al. (2003)

[a]Badola and Pradhan (2013); Bhattacharyya (2011); (Bennet 1983); Budha-Magar et al. (2020); (Bhat et al. 2020); Chakraborty et al. (2017); (Bhattacharyya 2011); Chandra and Rawat (2015); (Biswas 1956); Chhetri (2005); (Budha-Magar et al. 2020); Dahal and Borthakur (2017); (Chakraborty et al. 2017); Dahal (2019); (Chandra and Rawat 2015); Dahal et al. (2017); (Chandra et al. 2018); Dahal et al. (2019); (Chhetri 2005); Das et al. (2012); (Dahal 2019); Dash (2009); (Dahal and Borthakur 2017); Dash (2007); (Dahal et al. 2017); Dwivedi et al. (2019); (Das et al. 2012); (Dash 2007); Gewali (2008); (Dash 2009); Joshi et al. (2011); (Dash et al. 2023); Khare (2007); (Dash et al. 2021a); Lahiri and Dash (2022); (Dash et al. 2021b); Lepcha et al. (2011); (Dash et al. 2003); Maity et al. (2004); (Dash and Misra 1997); Panda and Mandal (2013); (Dwivedi et al. 2019); Panda (2018); (Forest Survey of India 2019); Pradhan and Badola (2008); (Gairola et al. 2008); Pradhan et al. (2021); (Gewali 2008); Rai and Das (2004); (Gogoi et al. 2021); Rastogi (2011); (Grierson and Long 1987); Rolla (1963); (Hooker 1854); Sahoo and Ansari (2009); (Hooker 1872–1897); Shankar and Rawat (2012); (Hooker 1949–51); Sherpa et al. (2015); (Hooker and Thomson 1859); Singh et al. (2002); (Jamloki et al. 2022); Sundriyal et al. (1998); (Joshi et al. 2011); Sundriyal et al. (1998); (Khare 2007); Tamang et al. (2017); (Kothari et al. 2017); Tamang et al. (2017); (Kumar et al. 2017); Tsering et al. (2017); (Lahiri 2022); Wangchuk et al. (2016); (Lahiri and Dash 2022); Wangchuk et al. (2008); (Lahiri et al. 2022).

References

Badola HK, Pradhan BK (2013) Plants used in healthcare practices by Limboo tribe in south west of Khangchendzonga Biosphere Reserve, Sikkim, India. Indian J Tradit Knowl 12(3):355–369

Bennet SSR (1983) Ethnobotanical studies in Sikkim. Indian for 109(7):477–481

Bhat JA, Kumar M, Negi AK, Todaria NP, Malik ZA, Pala NA, Kumar A, Shukla G (2020) Species diversity of woody vegetation along altitudinal gradient of the Western Himalayas

Bhattacharyya D (2011) *Rhododendron* species and their uses with special reference to Himalayas: a review. Assam Univ J 7(1):161–167

Biswas K (1956) Common medicinal plants of Darjeeling and Sikkim Himalaya. Bengal Government Press, West Bengal

Budha-Magar S, Bhandari P, Ghimire SK (2020) Ethno-medicinal survey of plants used by Magar (Kham) community, Rolpa district, Western Nepal. Ethnobot Res Appl 19:1–29. Retrieved from https://ethnobotanyjournal.org/index.php/era/article/view/1741

Chakraborty T, Saha S, Bisht NS (2017) First report on the ethnopharmacological uses of medicinal plants by Monpa tribe from Zemithang region of Arunachal Pradesh, Eastern Himalayas, India. Plants 6(13):1–12. https://doi.org/10.3390/plants6010013

Chandra S, Rawat DS (2015) Medicinal plants of the family Caryophyllaceae: a review of ethno-medicinal uses and pharmacological properties. Integr Med Res. https://doi.org/10.1016/j.imr.2015.06.004

Chandra S, Singh A, Singh CP, Nautiyal MC, Rawat LS (2018) Vascular plants distribution in relation to topography and environmental variables in alpine zone of Kedarnath Wild Life Sanctuary, West Himalaya. J Mt Sci 15(9):1936–1949. https://doi.org/10.1007/s11629-017-4738-8

Chhetri DR (2005) Ethnomedicinal plants of the Khangchendzonga National Park, Sikkim, India. Ethnobotany 17:96–103

Dahal S, Borthakur SK (2017) Medicinal plants genetic resources of Kyongnosla alpine sanctuary, Sikkim, India. Int J Curr Res 9(8):56277–56288

Dahal S, Sharma TP, Borthakur SK (2017) Database on medicinal plants of Tamze medicinal plants conservation area (MPCA) of Sikkim Himalaya, India. Nebio 8(1):45–56

Dahal S (2019). Studies on traditional medicinal plants of Sikkim. Ph.D Thesis. Gauhati University, Guwahati (AS)

Das T, Mishra SB, Saha D, Agarwal S (2012) Ethnobotanical survey of medicinal plants used by ethnic and rural people in Eastern Sikkim Himalayan Region. Afr J Basic Appl Sci 4(1):16–20

Dash SS, Maiti A, Rai SK (2003) Traditional uses of plants among the urban population of Gangtok-Sikkim. J Econ Taxon Bot 27(2):317–324

Dash S (2007) Useful plants of Kabi sacred grove, Sikkim. Nelumbo 49(1–4):79–88. https://doi.org/10.20324/nelumbo/v49/2007/73992

Dash S (2009) Traditional herbal remedies used in Sikkim, India. Nelumbo. 51:123–156. https://doi.org/10.20324/nelumbo/v51/2009/58088

Dash SS, Misra MK (1997) Medicinal plants used by the Tribal of Koraput district, Orissa. In: Mohapatra PM, Mohapatra PC (eds) Management in tribal areas: forest policy and people's participation. Concept Publishing Company, New Delhi

Dash SS, Lahiri S, Mao AA (2021) An enumeration of the flowering plants of Kyongnosla Alpine Sanctuary in eastern Sikkim, India. J Threat Taxa 13(13):20098–20117. https://doi.org/10.11609/jott.7040.13.13.20098-20117

Dash SS, Lahiri S, Chauhan AS (2023) Flora of Kyongnosla Alpine Sanctuary, Sikkim. In: Dash SS, Mao AA (eds) Botanical Survey of India, Kolkata

Dwivedi T, Kanta C, Singh LR, Sharma IP (2019) A list of some important medicinal plants with their medicinal uses from Himalayan State Uttarakhand, India. J Med Plants Stud 7(2):106–116

Forest Survey of India (2019) State of forest report. Forest Survey of India (Ministry of Environment & Forests), Dehradun

Gairola S, Rawal RS, Todaria NP (2008) Forest vegetation patterns along an altitudinal gradient in sub–alpine zone of west Himalaya, India. Afr J Plant Sci 2(6):42–48

Gewali MB (2008) Aspects of traditional medicines in Nepal, Institute of Natural Medicine, University of Toyama. Japan Global Ecol Conserv 24:e01302. https://doi.org/10.1016/j.gecco.2020.e01302

Gogoi R, Sherpa N, Franklin Benjamin JH, Agrawala DK, Rai SK, Dash SS (2021) Flora of Sikkim—a pictorial guide. Botanical Survey of India, Kolkata and Forest & Environment Department, Sikkim

Grierson AJC, Long DG (1987) Flora of Bhutan: including a record of plants from Sikkim and Darjeeling. Royal Botanic Garden

Hooker JD (1854) Himalayan Journals or Notes of a Naturlist in Bengal, the Sikkim and Nepal Himalayas, Khasia Mountains, etc. vol II. London

Hooker JD (1872–1897) The flora of British India, vol 1–7. L. Reeve & Co. Ltd, London

Hooker JD (1949–51) The Rhododendron of the Sikkim Himalaya; being and account, botanical & geographical of the Rhododendron recently discovered in the mountains of eastern Himalaya. Reeve, London

Hooker JD, Thomson T (1859) Praecursores ad Floram Indicam.-Balsamineae. J Linn Soc Bot 4(15):106–157. https://doi.org/10.1111/j.1095-8339.1859.tb01160.x

Jamloki A, Trivedi VL, Nautiyal MC, Semwal P, Cruz-Martins N (2022) Poisonous plants of the Indian Himalaya: an overview. Metabolites 12:540

Joshi K, Joshi R, Joshi AR (2011) Indigenous knowledge and uses of medicinal plants in Macchegaun, Nepal. Indian J Tradit Knowl 10(2):281–286

Khare CP (2007) Indian medicinal plants: an illustrated dictionary. Springer, New York. https://doi.org/10.1007/978-0-387-70638-2

Kothari R, Upadhya D, Theengh LT, Ghose PS, Shrestha P (2017) Climate change vulnerability assessment in snow leopard habitat. World Wildlife Fund

Kumar A, Adhikari BS, Rawat GS (2017) Biogeographic delineation of the Indian Trans-Himalaya: need for revision. Curr Sci 113(6):1032–1033

Lahiri S (2022) Studies on alpine and subalpine vascular plant diversity and community structure of two landscapes of Sikkim Himalaya. Ph.D. Thesis, University of Burdwan, West Bengal

Lahiri S, Dash SS (2022) An ethnobotanical survey of useful wild plants in Dzongri-Goecha La, Sikkim, India. Nelumbo 64(2):327–340. https://doi.org/10.20324/nelumbo/v64/2022/172609

Lahiri S, Dash S, Ghosh A (2022) An annotated checklist to the alpine and sub alpine flowering plant diversity of Dzongri-Goecha La Area, West Sikkim, India. Nelumbo 64(1):29–55. https://doi.org/10.20324/nelumbo/v64/2022/170943

Lepcha SR, Das AP (2011) Ethno-medicobotanical exploration along the international borders to Tibet autonomous region of China and the kingdom of Bhutan with special reference to the Pangolakha Wildlife Sanctuary, East Sikkim. In: Ghosh S, Das AP (eds) Recent studies in biodiversity and traditional knowledge in India. New Delhi, Gour Mahavidyalaya, pp 257–270

Maity D, Pradhan N, Chauhan AS (2004) Folk uses of some medicinal plants from North Sikkim. Indian J Tradit Knowl 3(1):66–71

Maity D, Maiti GG, Chauhan AS (2018). Vascular plants of Kanchenjunga biosphere reserve, Sikkim. Botanical Survey of India

Majumdar A, Adhikari BS (2012) Plant community composition and structure of Chenab Valley in a Part of Nanda Devi biosphere reserve. Not Sci Biol 4(3):45–56

Manish K (2022) Medicinal plants in peril due to climate change in the Himalaya. Eco Inform 68:101546

Moktan S (2017) Phytosociological characterization of vegetation in Darjeeling Hills. Ph.D. Thesis, University of North Bengal, Siliguri

Panda S (2018) Threatened ethnomedicinal plants of Singalila National Park in Darjeeling Himalaya, including their ethnic drug dosimetry: a report on an extensive field survey. In: Hartmann M, Barclay MV, Weipert J (eds) Biodiversität und Naturausstattung im Himalaya, vol 6. Naturkundemuseum Erfurt, pp 60–83. https://doi.org/10.3390/metabo12060540

Panda AK, Mandal D (2013) The folklore medicinal orchids of Sikkim. Anc Sci Life 33(2):92–96. https://doi.org/10.4103/0257-7941.139043

Pauchard A, Villarroel P (2002) Protected areas in Chile: history, current status, and challenges. Nat Areas J 22:318–330
POWO (2023) Plants of the world online. Facilitated by the royal botanic gardens, Kew. Published on the Internet. http://www.plantsoftheworldonline.org/. Retrieved 24 Nov 2023
Pradhan BK, Badola HK (2008) Ethnomedicinal plant use by Lepcha tribe of Dzongu valley, bordering Khangchendzonga biosphere reserve, in North Sikkim, India. J Ethnobiol Ethnomed 4:22. https://doi.org/10.1186/1746-4269-4-22
Pradhan DK, Ghosh J, Lepcha N, Nandi A, Banerjee D, Midday M, Naskar S, Maity D (2021) New ethnomedicinal information from lepcha community of Dzongu, Sikkim. Explor Anim Med Res 11(2):179–187. https://doi.org/10.52635/eamr/11.2.179-187
Rai LK, Prasad P, Sharma E (2000) Conservation threats to some important medicinal plants of the Sikkim Himalaya. Biol Cons 93:27–33
Rai U, Das AP (2004) Diversity of Tree species in the foot hill region of Darjeeling. In: Pandey AK, Wen J, Dogra VV (eds) Plant taxonomy advances and relevance. New Delhi, pp 81–93
Rai LK, Sharma E (1994) Medicinal plants of Sikkim Himalaya: status, uses and potential. Bishen Singh Mahendrapal Singh, Dehradun
Rastogi S (2011) A review of aconite (Vatsanabha) usage in Ayurvedic formulations: traditional views and their references. Spatula DD 1:233–244. https://doi.org/10.5455/spatula.20111220112858
Ravindranath NH, Sharma N, Sagadevan A, Jayaraman M, Munsi M (2012) Climate change and its impacts on forests of Sikkim. In: Arrawatia ML, Tambe S (eds) Climate change in Sikkim—patterns, impacts and initiatives. Information and Public Relations Department, Government of Sikkim, Gangtok, pp 125–140
Rawat GS (2007) Alpine vegetation of the western Himalaya: species diversity, community structure, dynamics and aspects of conservation. Ph.D. thesis. Botany Department, Kumaun University
Rawat GS, Tambe S (2011) Sikkim Himalaya: unique features of biogeography and ecology. In: Arrawatia ML, Tambe S (eds) Biodiversity of Sikkim. Information and public relations department, Government of Sikkim, pp. 1–12
Rodgers WA, Panwar SH (1988) Planning a wildlife protected area network in India, vol I&II. Wildlife Institute of India
Rolla SR (1963) A botanical tour in the Sikkim State, Eastern Himalayas. Nelumbo 5(2):166–205. https://doi.org/10.20324/nelumbo/v5/1963/76417
Sahoo A, Ansari A (2009) Less known ethnic uses of plants of South Sikkim. Nelumbo 51:219–222. https://doi.org/10.20324/nelumbo/v51/2009/58144
Shankar R, Rawat M (2012) Conservation of traditional medicinal practices and pharmaceutically important medicinal plants in Mizoram, India. J Biol Life Sci 4(2):95–102
Sharma TK, Sharma S (2010) Medicinal plants of Sikkim, vol 1. Sharma and Sharma Publication, pp 1–6
Sherpa MT, Mathur A, Das S (2015) Medicinal plants and traditional medicine system of Sikkim: a review. World J Pharm Pharm Sci. 4(2):161–184
Singh HB, Prasad P, Rai LK (2002) Folk medicinal plants in the Sikkim Himalayas of India. Asian Folklore Stud 61(2):295–310. https://doi.org/10.2307/1178975
Smith WW (1911) A new Gentian and two new Swertias from the East Himalaya. J Asiat Soc Bengal 7(3):77–79
Smith WW, Cave GH (1911) The vegetation of Zemu and Lhonak valley of Sikkim. Rec Bot Surv India 4(1):141–260
Sundriyal M, Sundriyal RC, Sharma E, Purohit AN (1998) Wild edibles and other useful plants from the Sikkim Himalaya, India. Oecologia Montana 7:43–54
Tamang M, Pal K, Rai SK, Kalam A, Ahmad SR (2017) Ethnobotanical survey of threatened medicinal plants of West Sikkim. Int J Botany Stud 2(6):116–125

Tambe S (2007) Ecology and management of the alpine landscape in the Khangchendzonga national park, Sikkim Himalaya. Ph.D. thesis. Forest Research Institute, University Dehra Dun, Uttarakhand

Tambe S, Rawat GS (2010) The alpine vegetation of the Khangchendzonga landscape, Sikkim Himalaya. Mt Res Dev 30(3):266–274. https://doi.org/10.1659/MRD-JOURNAL-D-09-00058.1

Tsering J, Tam N, Tag H, Gogoi BJ, Apang O (2017) Medicinal orchids of Arunachal Pradesh: a review. Bull Arunachal for Res 32(1&2):1–16

Wangchuk P, Samten U, Thinley J, Afaq SH (2008) High altitude plants used in Bhutanese traditional medicine (g.so-ba-rig-pa). Ethnobotany 20:54–64

Wangchuk P, Namgay K, Gayleg K, Dorji Y (2016) Medicinal plants of Dagala region in Bhutan: their diversity, distribution, uses and economic potential. J Ethnobiol Ethnomed 12(28):1–18. https://doi.org/10.1186/s13002-016-0098-7

Indigenous Medicinal Plant Utilization and Conservation Imperatives: An Ethnobotanical Case Study from Arunachal Pradesh, India

Abhimanyu Chaturvedi, Tridipa Biswas, and Shasikant Chaturvedi

Abstract The present study elucidates the indigenous medicinal plant utilization by various tribes in 11 blocks across the districts of Tirap, Longding, Changlang, and East Siang in Arunachal Pradesh and accentuates the significant role of local communities in the conservation of medicinal plants. The region is ethnobotanically rich, with abundant availability and knowledge of medicinal plants. The study was conducted by collecting information on local medicinal plants in consultation with traditional healers and knowledge holders of local tribal communities, with the aim of documenting the traditional knowledge on the use of medicinal plants for the human well-being benefit of mankind. The findings of this research highlight the significant reliance of tribal communities in Arunachal Pradesh on plant-based remedies for their healthcare needs. The common medicinal plants used by Minyong, Mishing, Padam, Nocte, Tangsa, Tutsa, and Wancho tribes were studied, and around 57 species under 40 families were identified and documented. Majority of them are herbaceous, followed by trees, shrubs, and whole plants. Study shows that increasing demand for medicinal plants is leading to deforestation and overexploitation, further reducing the diversity of this valuable resource in the region. Thus, the study high lights the significance of ethnobotanical research, emphasizing the necessity for conservation efforts and the documentation of traditional medicinal knowledge to ensure the continued availability and utilization of these resources for the benefit of future generations. This chapter provides valuable insights into the rich ethnobotanical heritage of Arunachal Pradesh, which provides the potential for further research and conservation initiatives to safeguard this traditional medicinal knowledge and associated medicinal plants.

A. Chaturvedi
Krishi Vigyan Kendra, Deomali, Tirap 792129, Arunachal Pradesh, India

T. Biswas (✉)
G.B. Pant National Institute of Himalayan Environment, North-East Regional Centre, Chandranagar, Itanagar 791113, Arunachal Pradesh, India
e-mail: tridipa.b@gbpihed.nic.in

S. Chaturvedi
Krishi Vigyan Kendra, Yachuli, Lower Subansiri 791120, Arunachal Pradesh, India

A. P. Mishra et al. (eds.), *Threatened Medicinal Plants in the Indian Himalayan Region*, World Sustainability Series, https://doi.org/10.1007/978-3-031-73687-2_15

Keywords Conservation · Medicinal plants · Ethnobotany · Tribe · Traditional knowledge

1 Introduction

Medicinal plants are worldwide recognized as valuable sources of herbal products (Mishra et al. 2023a; Chandra et al. 2023a), playing vital roles in disease prevention and treatment (Sofowora et al. 2013). They have been used for centuries in various cultures and are an integral part of traditional healthcare systems (Namsa et al. 2011; Sofowora et al. 2013; Rahman et al. 2022). Unfortunately, these plants are rapidly disappearing (Mishra 2023) due to factors such as climate change, acculturation, deforestation, overutilization, and habitat destruction (Barata et al. 2016; Xia et al. 2022; Theodoridis et al. 2023). Therefore, it is essential to conserve and sustainably use medicinal plants to ensure the preservation of ethnomedicinal knowledge and their availability for future generations (Kalita and Khan 2013; Gajurel et al. 2015; Barata et al. 2016; Chen et al. 2016; Sharma 2018; Hu et al. 2020; Huang et al. 2022; Kibonde 2020; Kumar et al. 2023). The conservation of medicinal plants involves different measures like collection, propagation, evaluation, disease identification, storage, and distribution. Conservation of medicinal plants and their genetic resources can be undertaken by in-situ and ex-situ conservation (Chandra et al. 2022). Ethnobotanical studies are crucial for documenting the indigenous knowledge of medicinal plants (Chandra et al. 2020; Hu et al. 2020; Huang et al. 2022) and implementing conservation measures (Mishra et al. 2023d).

Local communities, particularly those associated with traditional knowledge and practices, play a significant role in the conservation of medicinal plants (Kibonde 2020; Chandra et al. 2023b). They have inherited their knowledge through generations and rely on these plants for their healthcare and livelihoods (Namsa et al. 2011; Sahu et al. 2023). The use of medicinal plants in human life has been considered to be safe, with minimal side effects, and in harmony with nature (Van Wyk and Prinsloo 2020). However, the demand for medicinal plants is increasing due to globalization (Chen et al. 2016), leading to further depletion of plant diversity (Volenzo and Odiyo 2020). Urgent efforts are needed to prevent the loss of valuable traditional knowledge and conserve medicinal plants, which are crucial for the well-being of indigenous communities (Ray and Saini 2022; Rawat et al. 2021; Mishra et al. 2023c).

Located in northeastern region of India, Arunachal Pradesh is known for its rich biodiversity and abundant population of medicinal and aromatic plants (Kalita and Khan 2013; Gajurel et al. 2015; Mishra et al. 2021, 2023b). Plant diversity in Arunachal Pradesh include angiosperms 5000 species, gymnosperms 39, ferns 400, bamboo 89, canes 18, medicinal plants 500, wild and edible plants 400, orchids 650 (Sharma 2018). The tribal communities in Arunachal Pradesh heavily depend on forests for their sustenance, and medicinal plants serve as a vital resource for their healthcare and livelihoods (Ramya 2014; Singh et al. 2021). These tribal communities have their own unique ethnomedicinal practices, acquired through close observation of nature. However, the traditional knowledge and practices related to medicinal plants in Arunachal Pradesh have not been extensively studied (Shankar et al. 1993; Singh et al. 1995, 2021; Rawat et al. 1997; Rawat and Shankar 2002), and the conservation efforts in the region are limited to tropical plant species.

The research chapter aims to explore traditional knowledge of medicinal plants among local communities and tribal groups in several blocks across the districts of Tirap, Longding, Changlang, and East Siang in Arunachal Pradesh. The local communities of the state, living in fragile ecosystems, have developed location specific knowledge to meet out their daily food, ethno-medicines and other livelihood needs including cultural practices. Such traditional knowledge is nurtured by indigenous institutions, and that have enabled communities to conserve local biodiversity having crucial ecological significance. Given the value of traditional knowledge in tribal communities, the major objectives of this study are to (i) investigate the ethnomedicinal practices, which locals use to treat and cure a variety of illnesses; and (ii) highlight the socio-cultural, ecological and diversity aspects of the local plant species that are used. The findings of this study will contribute to the development of sustainable conservation strategies for medicinal plants in the surveyed blocks, ensuring the preservation of valuable traditional knowledge for the well-being of the indigenous communities. This study also highlights the significance of ethnobotanical research and provide insights for guiding sustainable practices in the region.

2 Study Area

Arunachal Pradesh, a state located in northeast India is home to a rich diversity of medicinal plants. It extends between 26° 28′–29° 30′ N latitude and 91° 30′–97° 30′ E longitude, covering an area of 83,743 km^2. The survey was conducted during August 2023 in several blocks in Arunachal Pradesh (Fig. 1), including Lazu, Dadam, Khonsa and Soha blocks in the district of Tirap, Kanubari block in the district of Longding, Khimiyang, Miao and Yatdam blocks in the district of Changlang, and Pasighat, Mebo, Ruksin blocks in the district of East Siang.

The survey focused on the interplay between ethnographic diversity, rich biodiversity, and the vital use of medicinal plants. The study area is home to 12 tribes and approximately 46 ethnic communities (Fig. 2), all belonging to scheduled tribes. The chosen blocks were purposively selected based on the ethnicity and heavy dependence of the local communities on plant resources for their health management and livelihood needs. The immense biodiversity in these areas reflects the habitats associated with a wide altitudinal range of medicinal plants. The ethnobotanical research has been conducted in the region to highlight the importance of traditional knowledge and the need for conservation measures.

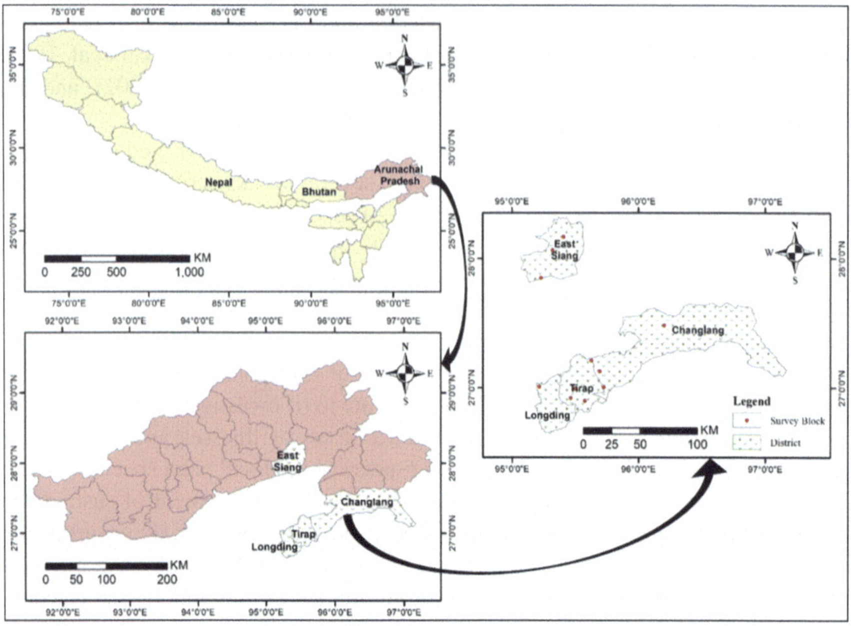

Fig. 1 Map showing intensive study area

Fig. 2 Interaction with ethnic communities

3 Materials and Methods

The study utilized the snowball sampling method, which is a non-probability sampling technique. This approach was necessary as the target population consisting of local communities and tribal groups in the surveyed blocks of Arunachal Pradesh was difficult to access and hidden. The researchers had limited control over the sampling process. The main purpose was to explore the key traditional healers and traditional knowledge holders (TKHs) within this population. In this method, an initial group of respondents who met the study criteria were selected and they were requested to provide the names of other potential respondents. The total number of knowledge holders chosen for the study was 16, and the selection was based on their status as traditional healers and traditional knowledge holders. The oral consent of knowledge holders was obtained to disclose their ethnomedicinal knowledge to the general public, ensuring that the potential subjects' privacy was not violated. Data was collected through personal interviews and participant observations with the selected knowledge holders. Personal interviews were conducted using an unstructured schedule and data were recorded qualitatively to allow for deeper meaning and narrations to be explained. The information gathered on ethnomedicinal plants was analyzed by at least two knowledge holders, ensuring that the data collected accurately represented the target population's traditional knowledge and use of medicinal plants. Transect walks were also conducted to verify some selected plant species used in ethnomedicine by these knowledge holders, adding an additional layer of validity to the study. The overall aim of the study was to raise awareness among other stakeholders and encourage learning and conservation practices. However, this research does not promote or endorse any practices that could potentially be used for commercial purposes by third parties.

4 Results and Discussion

A total of 57 species of medicinal plants used by the local communities in the surveyed blocks of Arunachal Pradesh were identified (Table 1). The study emphasized the reliance of tribal communities on plant-based remedies for healthcare. The study found that the areas surveyed were ethnobotanically rich with abundant availability and knowledge of medicinal plants. Most of these ethnobotanical plants were categorized as herbs (47%), followed by trees (32%), shrubs (19%), and whole plants (2%) belonging to 40 families. Several parts of individual plant species were used as medicine. The traditional knowledge of medicinal plants varied among the different tribes, reflecting their distinct cultural practices and beliefs. The indigenous tribes in the study region have a rich knowledge of medicinal plants, with each tribe utilizing a variety of herbs, shrubs, and trees for their traditional medicine. For instance, the Minyong tribe in Pasighat block of East Siang district relies on one herb, one shrub, and two tree species for medicinal purposes, while the Mishing tribe in the Mebo block of East Siang district employs two herbs and two shrubs. The Nocte tribe in Dadam, Khonsa, and Soha blocks of Tirap district utilizes five herbs, five shrubs, and six tree species for medicinal practices, and the Padam tribe in the Ruksin block of East Siang district employs two herbs and one shrub in their traditional medicine. The Tangsa tribe, spanning Yatdam and Miao blocks in Changlang district relies on three herbs, two shrubs, and six tree species for their medicinal needs, while the Tutsa tribe in Lazu block of Tirap district and Khimiyang block of Changlang district utilizes 10 herbs, three shrubs, and one whole plant in their traditional medicine. Lastly, the Wancho tribe, located in the Kanubari block of Longding district, employs four herbs and one tree species in their medicinal practices. This variation highlights the rich diversity of traditional medicinal practices in the region. The recorded uses of medicinal plants included treatments for various diseases and disorders, such as rheumatism, diabetes, malaria, typhoid, cold and cough, asthma, blood pressure, jaundice, and others. The study findings demonstrate the significant reliance of tribal communities in Arunachal Pradesh on plant-based remedies for their healthcare needs. The absence of modern infrastructure and limited access to healthcare facilities have contributed to the continued use of traditional medicinal practices. The local communities rely on their traditional knowledge of medicinal plants as a primary healthcare system.

The present study found that increasing demand for medicinal plants is leading to deforestation and overexploitation, which further reduced the diversity of this valuable resource in this region. The unsustainable harvesting practices and the destruction of habitats pose a threat to the availability and sustainability of medicinal plants in the region. The traditional practice of burning the forest for cultivation has further contributed to the destruction of medicinal plants, leading to the loss of medicinal plant species and disrupting the delicate balance of the ecosystem. Urgent ethnobotanical studies and conservation measures are necessary to preserve indigenous knowledge and medicinal plant diversity in the surveyed blocks of Arunachal Pradesh. These studies can help identify sustainable harvesting practices, promote the cultivation of medicinal plants, and raise awareness about the importance of conserving these valuable resources. This information can be used for the conservation and utilization of indigenous knowledge and can serve as a model for low-cost healthcare (Kalita and Khan 2013; Ramya 2014).

5 Strategies for Medicinal Plant Conservation

Due to increasing demand for medicinal plants in the study area, medicinal plant resources are rapidly depleting. The present study highlights the urgent need for medicinal plant conservation methods and initiatives to reduce the loss of wild plant diversity and meet the growing demand for medicinal plants. Conservation is the practice and a crucial approach to protecting and managing genetic diversity, species, and ecosystems to maintain the balance of natural and ecological systems. It involves various strategies such as collection, propagation, assessment, disease detection and eradication, storage, and dispersal, with the ultimate aim of preventing the extinction of any species or plant varieties. Conservation methods for medicinal plants and their genetic diversity can be achieved through in-situ and ex-situ conservation methods.

Table 1 Ethnobotanical documentation of medicinal plants utilized by indigenous tribal communities in surveyed blocks of Arunachal Pradesh, India

Sr. no.	Local name	Common name	Scientific name	Family	Medicinal uses	Habit	Tribe name	Block Name	District Name
1	Bish	Aconite/Indian Vish	*Aconitum ferox*	Ranunculaceae	Antipyretic, rheumatism and skin diseases	Herb	Tutsa	Lazu	Tirap
2	Daruga/Boch	Sweet flag	*Acorus calamus*	Araceae	Carminative, general tonic, diarrhea and dysentery	Herb	Tutsa	Lazu	Tirap
3	Dilap	Hooker chives	*Allium hookerii*	Liliaceae	High blood pressure	Herb	Tutsa	Lazu	Tirap
4	Singer	Cheesel wood	*Alstonia scholaris*	Apocynaceae	Diabetes, malaria and typhoid	Herb	Tutsa	Lazu	Tirap
5	Kuhumkanta	Mexican poppy	*Argemone maxicana*	Papaveraceae	Urinary tract infection and liver disorders	Herb	Tutsa	Lazu	Tirap
6	Hinger	Indian birthwort	*Aristolochia indica*	Aristolochiaceae	Skin disease and snake bite	Whole plant	Tutsa	Lazu	Tirap
7	Kalmeg	Bitter weed	*Andographis paniculata*	Acanthaceae	General tonic, fever, worms, dysentery and liver disorder	Herb	Nocte	Dadam	Tirap
8	Kordoi	Star fruit	*Averrhoea carambola*	Averrhoaceae	Jaundice and anemia	Tree	Nocte	Dadam	Tirap
9	Ogjok	Sadabahar	*Bauhinia variegata*	Caesalpiniaceae	Piles and dysentery	Shrub	Nocte	Dadam	Tirap

(continued)

Table 1 (continued)

Sr. no.	Local name	Common name	Scientific name	Family	Medicinal uses	Habit	Tribe name	Block Name	District Name
10	Brahmi	Brahmi	*Bacopa monneri*	Scrophulariaceae	General tonic and skin diseases	Herb	Nocte	Dadam	Tirap
11	Bhang	Indian hemp	*Canabis sativa*	Cannabaceae	Anthelminthic, bronchitis and piles	Herb	Nocte	Khonsa	Tirap
12	Bengi	Kassod tree	*Cassia siamea*	Fabaceae	Intestinal worms and scabies	Tree	Nocte	Khonsa	Tirap
13	Chinchayong	Dal chini	*Cinnamomum bejolghota*	Lauraceae	Liver disorders and fever	Tree	Nocte	Khonsa	Tirap
14	Dhuvephool	Periwinkle	*Catharanthus roseus*	Apocynaceae	Malaria and jaundice	Herb	Nocte	Khonsa	Tirap
15	Kiling-kiro	Asiatic pennywort	*Centella asiatica*	Apiaceae	Cough and asthma	Herb	Nocte	Khonsa	Tirap
16	Tejpat	Malabar leaf	*Cinnamomum tamala*	Lauraceae	Carminative, colic pain and diarrhea	Tree	Nocte	Khonsa	Tirap
17	Kaginemu	Fingered citron	*Citrus medica*	Rutaceae	Gastric and tumors	Tree	Nocte	Soha	Tirap
18	Pathinebu	Sour orange	*Citrus aurantium*	Rutaceae	Laxative and urinary tract infections	Tree	Nocte	Soha	Tirap
19	Bhati	Glory bower	*Clerodendrum colebrookianum*	Verbenaceae	Blood pressure, cough, cold and liver disorders	Shrub	Nocte	Soha	Tirap

(continued)

Table 1 (continued)

Sr. no.	Local name	Common name	Scientific name	Family	Medicinal uses	Habit	Tribe name	Block Name	District Name
20	Toku	Blue flowered glory tree	*Clerodendrum serratum*	Verbenaceae	Fever	Shrub	Nocte	Soha	Tirap
21	Mishimiteeta	Gloden thread herb	*Coptis teeta*	Ranunculaceae	Cough, cold, diarrhea, dysentery and liver disorder	Shrub	Nocte	Soha	Tirap
22	Kewpachla	Crepe ginger	*Costus speciosus*	Costaceae	Piles and ear diseases	Shrub	Nocte	Soha	Tirap
23	Autenga	Elephant apple	*Dillenia indica*	Dilleniaceae	Hair tonic and dysentery	Tree	Tutsa	Khimiyang	Changlang
24	Tami	Finger millet	*Eleusine coracana*	Poaceae	Blood pressure, liver disorders and asthma	Herb	Tutsa	Khimiyang	Changlang
25	Hanna	Perrenial Buckwheat	*Fagopyrum dibotrys*	Polygonaceae	Fracture and skin diseases	Herb	Tutsa	Khimiyang	Changlang
26	Gnomu	Buckwheat	*Fagopyrum esculentum*	Polygonaceae	Rheumatic pain and fever	Herb	Tutsa	Khimiyang	Changlang
27	Kewpachla	Spiral ginger	*Costus variegata*	Costaceae	Ear diseases	Herb	Tutsa	Khimiyang	Changlang
28	Kaw	Amlavetasa	*Garcinia pedunculata*	Clussiaceae	Diarrhea and dysentery	Tree	Tutsa	Khimiyang	Changlang
29	Dikang	Dekang tree	*Gymnocladus burmanicus*	Fabaceae	Skin problems	Tree	Tutsa	Khimiyang	Changlang
30	Banhaldi	Perfume ginger	*Hedychium spicatum*	Zingiberaceae	Asthma and blood dysentery	Herb	Tutsa	Khimiyang	Changlang

(continued)

Table 1 (continued)

Sr. no.	Local name	Common name	Scientific name	Family	Medicinal uses	Habit	Tribe name	Block Name	District Name
31	Kutki	Chaulmugra	*Hydnocarpus kurzii*	Flacourtiaceae	Skin diseases and general tonic	Tree	Tangsa	Yatdam	Changlang
32	Domlishi	Star anise	*Illicium griffithii*	Magnoliaceae	Stimulant and carminative	Tree	Tangsa	Yatdam	Changlang
33	Akrot	Persian walnut	*Juglans regia*	Juglandaceae	Anthelminthic and rheumatisms	Tree	Tangsa	Yatdam	Changlang
34	Tayer	Mountain pepper	*Litsea cubeba*	Lauraceae	Headache, stomach trouble and fever	Tree	Tangsa	Yatdam	Changlang
35	Ekauna	Thumbai	*Leucas aspera*	Lamiaceae	Asthma and bronchitis	Herb	Tangsa	Yatdam	Changlang
36	Myoka	Curry leaf tree	*Murraya koenigii*	Rutaceae	Asthma and bronchitis	Tree	Tangsa	Yatdam	Changlang
37	Jatamansi	Spikenard	*Nardostachys jatamansi*	Velerianaceae	Female disorders, urinary tract infection and indigestion	Herb	Tangsa	Yatdam	Changlang
38	Puko	Night blooming jasmine	*Nyctanthes arbortristis*	Oleaceae	Malaria	Shrub	Tangsa	Yatdam	Changlang
39	Bhatghila	Indian trumpet flower	*Oroxylum indicum*	Bignoniaceae	Hair tonic, anti-helminthic and carminative	Tree	Tangsa	Miao	Changlang
40	Dhundi	Sleeping beauty	*Oxalis corniculata*	Oxalidsaceae	Wormicide and liver tonic	Herb	Tangsa	Miao	Changlang

(continued)

Table 1 (continued)

Sr. no.	Local name	Common name	Scientific name	Family	Medicinal uses	Habit	Tribe name	Block Name	District Name
41	Mewang	Naudin	*Osbeckia glauca*	Melastomaceae	Liver disorders	Shrub	Tangsa	Miao	Changlang
42	Nyopo	Chinese knotweed	*Polygonum chinense*	Polygonaceae	Malarial fever	Herb	Wancho	Kanubari	Longding
43	Yeppe	Chinese fever vine	*Paederiafoetida*	Rubiaceae	Jaundice	Herb	Wancho	Kanubari	Longding
44	Gulachin	Nose gay	*Plumeria rubra*	Apocynaceae	Cough and bronchitis	Tree	Wancho	Kanubari	Longding
45	Konya	Dock leaf knotweed	*Aconogonum rumicifolium*	Polygonaceae	Pain killer and liver disorders	Herb	Wancho	Kanubari	Longding
46	Mehung	Himalayan May apple	*Sinopodophyllum hexandurm*	Podophylaceae	Stimulant, purgative and stomach disorders	Herb	Wancho	Kanubari	Longding
47	Knokbua	Five fingers	*Potentilla lineata*	Rosaceae	Toothache and diarrhea	Herb	Minyong	Pasighat	East Siang
48	Zingnangsing	Burans	*Rhododendron arboreum*	Ericaceae	Diarrhea and dysentery	Tree	Minyong	Pasighat	East Siang
49	Manjista	India madder	*Rubia cordifolia*	Rubiaceae	Vomiting and female disorders	Tree	Minyong	Pasighat	East Siang
50	Koppi	Pea egg plant	*Solanum torvum*	Solanaceae	Malaria and diabetics	Shrub	Minyong	Pasighat	East Siang
51	Koppir	Dutch egg plant	*Solanum khasianum*	Solanaceae	Malaria and diabetics	Shrub	Mishing	Mebo	East Siang

(continued)

Table 1 (continued)

Sr. no.	Local name	Common name	Scientific name	Family	Medicinal uses	Habit	Tribe name	Block Name	District Name
52	Chirata	Bhunimba	*Swertia chirata*	Gentianaceae	General tonic, stomachic, chronic fever, bronchial and asthma	Shrub	Mishing	Mebo	East Siang
53	Minhang	Toothache plant	*Spilanthes oleracea*	Asteraceae	Sinusitis and asthma	Herb	Mishing	Mebo	East Siang
54	Kinnu	Goat weed	*Scoparia dulcis*	Scropulariaceae	Cough	Herb	Mishing	Mebo	East Siang
55	Hogunilota	Guduchi	*Tinospora cordifolia*	Menispermace	Anemia and urinary troubles	Herb	Padam	Ruksin	East Siang
56	Gagra	Mejenga	*Zanthoxylum oxyphylum*	Rutaceae	Liver tonic	Shrub	Padam	Ruksin	East Siang
57	Kekir	Ginger	*Zingiber siangensis*	Zingiberaceae	Jaundice, typhoid and dysentery	Herb	Padam	Ruksin	East Siang

6 Dependency and Role of Local Communities in Medicinal Plant Conservation

The study documented notable observations pertaining to the dependency and role of local communities in medicinal plant conservation across various blocks within the districts of Tirap, Longding, Changlang, and East Siang in Arunachal Pradesh.

6.1 Medicinal Plants as a Source of Food

The ethnic and indigenous communities residing in surveyed blocks have played a vital role in the conservation and utilization of a diverse array of medicinal plants, many of which serve dual purposes as both food and traditional medicines. Notably, millet has held a central place in the diets of numerous ethnic groups for generations within this region. The year 2023 is designated as the "Year of Millets", which significantly increased the popularity of millets among the local communities. Which extends beyond dietary consumption to an appreciation for their medicinal properties, particularly in managing conditions like diabetes. This shift underscores the increasing recognition of millet as not only a sustenance source but also a therapeutic resource. The preservation of these traditional food systems by indigenous groups holds paramount importance, as it not only contributes to the conservation of regional biodiversity but also advances the cause of food security and public health in this ecologically rich and culturally diverse region (Shankar et al. 1993).

6.2 Medicinal Plants with Religious Significance

Tribal communities have a profound connection with nature, worshipping trees and flowers as embodiments of deities. In their belief systems, Gods and Goddesses reside within these botanical beings. Consequently, local communities become inadvertent conservators of numerous medicinal plant species, as their spiritual reverence inherently protects these valuable resources. Studies and surveys consistently highlight the profound influence of plants and flowers on the lives, traditions, and healthcare practices of these communities, showcasing the intricate relationship between culture, spirituality, and biodiversity conservation (Singh et al. 1995).

6.3 Medicinal Plants Used as Antidotes

Local communities play a crucial role in conserving specific plants valued for their efficacy as antidotes against venomous bites and stings. For instance, *Aristolochia indica* (Indian birthwort) plant is extensively utilized for treating snakebites and scorpion stings in the Tirap district. The preservation of such plants by these communities not only safeguards valuable traditional knowledge but also contributes significantly to both healthcare practices and biodiversity conservation. This intersection of traditional wisdom and conservation underscores the critical role of indigenous knowledge in preserving both human well-being and natural resources (Rawat et al. 1997).

6.4 Medicinal Plants for Edible Purposes

The multifunctional utility of numerous plant species, where different parts like seeds for food, roots for sustenance, and leaves, stems, and roots for medicinal applications, underscores the rich tradition of local communities conserving medicinal plants since ancient times within surveyed regions. The adeptness of these communities in preserving and utilizing medicinal plants underscores their critical role in safeguarding traditional knowledge and maintaining the ecological balance of surveyed blocks. Such harmonious coexistence and utilization of nature's bounty by local communities not only contribute to the preservation of medicinal plant diversity but also signify the enduring relationship between humans and their environment. These practices provide invaluable insights into sustainable resource management, ensuring the continued availability of these vital botanical resources for future generations.

6.5 Medicinal Plants for Orthopedic Treatment

Medicinal plants have been a valuable resource for orthopedic treatment in the surveyed blocks. Various plant parts including roots, stems, leaves and bark of plants are used, which are known for their effectiveness in treating bone fractures. Notably, *Fagopyrum dibotrys*, commonly known as *Periwinkle*, is an example of a prominent medicinal plant widely used for this purpose, especially in Changlang district. The application of traditional knowledge in harnessing the therapeutic properties of these medicinal plants underscores the rich tradition of indigenous healing practices in these regions, providing a potential avenue for further exploration and integration into contemporary healthcare practices for orthopedic care and fracture management. Several studies have documented the use of medicinal plants for the treatment of bone fractures by these tribal communities of Arunachal Pradesh, highlighting the importance of traditional knowledge in conserving biodiversity and promoting human health.

6.6 *Medicinal Plants for Traditional Medicine*

Local communities inhabiting the surveyed blocks have played a crucial role in the conservation of medicinal plants since time immemorial, as these botanical resources have been integral to their healthcare practices. Across generations, indigenous knowledge has guided the utilization of numerous plant species for a wide range of ailments, including fever, diarrhea, skin issues, and weakness. This profound understanding and sustainable utilization of medicinal flora underscore the rich tradition of local healing practices, intertwining human health with the natural environment. By preserving and perpetuating this traditional wisdom, these communities not only contribute significantly to the conservation of biodiversity but also promote the well-being of their communities. Their role in safeguarding these invaluable resources reflects an enduring relationship between humans and the ecosystems they inhabit, highlighting the essential synergy between traditional ecological knowledge and the preservation of our planet's botanical diversity.

7 Conclusion

This ethnobotanical study provides valuable insights into the traditional knowledge and use of medicinal plants among local communities and tribal groups in Arunachal Pradesh. Medicinal plants have played an enduring role in human history, providing sustenance, traditional remedies, and contributions to modern medicine. The preservation of indigenous knowledge and conservation efforts by local communities are essential for ensuring the sustainable use of these botanical resources. The preservation of traditional knowledge regarding the medicinal uses of plants is not only a matter of cultural significance but also essential for the well-being of future generations and the sustainability of our natural resources. As this valuable knowledge is orally transmitted and vulnerable to being lost over time, documenting it is essential. The local communities in the surveyed regions of Arunachal Pradesh are integral to the conservation of medicinal plants, owing to their deep-rooted traditions, spirituality, and indigenous knowledge. This research underscores the urgency of conserving both this indigenous knowledge and the diversity of medicinal plants. Urgent ethnobotanical studies and conservation measures are necessary to promote sustainable harvesting practices, cultivate medicinal plants, and raise awareness about the significance of conserving these valuable resources. Furthermore, fostering collaboration between traditional wisdom and modern scientific approaches holds great promise for developing holistic and sustainable conservation strategies. By combining the insights of indigenous communities with cutting-edge research, we can enhance our understanding of these plants' ecological roles, medicinal properties, and cultivation methods, ultimately promoting their responsible use and preservation. By taking proactive steps to document, conserve, and collaborate, we can ensure that these precious gifts of nature continue to benefit humanity for generations to come while maintaining the delicate balance of our natural ecosystems. The findings of this study can serve as a model for low-cost healthcare and contribute to the conservation and utilization of indigenous knowledge in the surveyed blocks of Arunachal Pradesh.

References

Barata AM, Rocha F, Lopes V, Carvalho AM (2016) Conservation and sustainable uses of medicinal and aromatic plants genetic resources on the worldwide for human welfare. Ind Crops Prod 88:8–11

Chandra N, Joshi VC, Mishra AP (2020) Ecological status of fodder and fuelwood species in Banari Devi sacred grove of Kumaun Himalaya, Uttarakhand. Indian J Ecol 47(4):1044–1048

Chandra N, Rai ID, Mishra AP, Dwivedi SK, Kotiya A, Tiwari UK, Singh G (2022) Assessing potential habitats and populations of selected medicinal herbs in Alpine areas of Uttarakhand, Western Himalaya. Indian J For 45(3):144–155. https://doi.org/10.54207/bsmps1000-2023-ITRFNL

Chandra N, Kumar A, Mishra AP, Singh G, Rawat GS (2023a) *Wigandia urens* (Ruiz & Pav.) Kunth: a new emergent invasive alien species in the Western Himalaya. Int J Environ Stud 1–14. https://doi.org/10.1080/00207233.2023.2194157

Chandra N, Singh G, Rai ID, Mishra AP, Kazmi MY, Pandey A et al (2023b) Predicting distribution and range dynamics of three threatened *Cypripedium* species under climate change scenario in Western Himalaya. Forests 14(3):633. https://doi.org/10.3390/f14030633

Chen SL, Yu H, Luo HM, Wu Q, Li CF, Steinmetz A (2016) Conservation and sustainable use of medicinal plants: problems, progress, and prospects. Chin Med 11(37):1–10

Gajurel PR, Ronald Kh, Buragohain R, Rethy P, Singh B, Potsangbam S (2015) On the present status of distribution and threats of high value medicinal plants in the higher altitude forests of the Indian Eastern Himalaya. J Threat Taxa 7(6):7243–7252

Hu R, Lin C, Xu W, Liu Y, Long C (2020) Ethnobotanical study on medicinal plants used by Mulam people in Guangxi, China. J Ethnobiol Ethnomed 16(40):1–50

Huang SS, Huang CH, Ko CY, Chen TY, Cheng YC, Chao J (2022) An ethnobotanical study of medicinal plants in Kinmen. Front Pharmacol 12:1–15

Kalita J, Khan ML (2013) Medicinal plants from the high altitudes of the western part of Arunachal Pradesh, India and their trade. Int J Conserv Sci 4(3):337–346

Kibonde SF (2020) Indigenous knowledge and conservation of medicinal plants in Rungwe District, Tanzania. Open Access Library J 7(7):1–10

Kumar A, Adhikari BS, Mishra AP (2023) Status of threatened medicinal and Biosphere Reserve, Uttarakhand. A report (TR No./2022/35) submitted to International Union for Conservation of Nature—India, p 103

Mishra AP (2023) Understanding the decline of *Butea monosperma* (Lam.) Kuntze in Gorakhpur District, India: an ecological investigation. J For Geosci 1:2. https://www.esciinfo.com

Mishra AP, Rai ID, Pangtey D, Padalia H (2021) Vegetation characterization at community level using sentinel-2 satellite data and random forest classifier in western Himalayan Foothills, Uttarakhand. J Indian Soc Remote Sens 49:759–771. https://doi.org/10.1007/s12524-020-01253-x

Mishra AP, Kumar A, Yadav SN (2023a) Ecology and conservation of threatened medicinal plants in the Trans-Himalayan region of Nanda Devi Biosphere Reserve, Western Himalaya. Trees For People 14:100451. https://doi.org/10.1016/j.tfp.2023.100451

Mishra AP, Chandra N, Kumar A, Sharma S, Singh G (2023b). *Mesosphaerum suaveolens* (L.) Kuntz.: a serious threat to Uttarakhand's first Ramsar site, India. Species 24:e57s1558

Mishra AP, Chandra N, Mandy JJ, Dwivedi SK, Alruzuq A, Pande CB (2023c) Analytical Hierarchy Process (AHP) based on the spatial assessment of an endangered alpine medicinal herb *Aconitum heterophyllum* in the Western Himalayan Environment. In: Climate change impacts on natural resources, ecosystems and agricultural systems. Springer International Publishing, Cham, pp 579–594. https://doi.org/10.1007/978-3-031-19059-9_25

Mishra AP, Chandra N, Srivastava A, Singh SK (2023d) An updated checklist of the vascular flora of Asan Conservation Reserve, Uttarakhand, India. Indian J For 46(1):12–21. https://doi.org/10.54207/bsmps1000-2023-7P1KSV

Namsa ND, Mandal M, Tangjang S, Mandal SC (2011) Ethnobotany of the Monpaethnic group at Arunachal Pradesh, India. J Ethnobiol Ethnomed 7(31):1–14

Rahman MH, Roy B, Chowdhury GM, Hasan A, Saimun MSR (2022) Medicinal plant sources and traditional healthcare practices of forest-dependent communities in and around Chunati Wildlife Sanctuary in southeastern Bangladesh. Environ Sustain 5:207–241

Ramya T (2014) Forest as a source of livelihood for the tribes in Kurung Kumey District, Arunachal Pradesh. North East India Hum Landsc 99–114

Rawat MS, Shankar R (2002) Strategies for conservation and cultivation of medicinal plants in Arunachal Pradesh. Himal Biosph Reserv 4(1–2):75–78

Rawat MS, Singh VK, Shankar R (1997) Cultivation of medicinal plants in social forestry programme in Arunachal Pradesh. Bull Medico-Ethno Bot Res 17(3–4):169–174

Rawat N, Mishra AP, Chandra N, Painuly V, Negi GS, Bisht MPS, Purohit S (2021) Maxent modeling for predicting the potential distribution of *Valeriana jatamansi* jones. in Chakrata forest division of Garhwal Himalaya. Indian J Ecol 48(2):367–372

Ray S, Saini MK (2022) Impending threats to the plants with medicinal value in the Eastern Himalayas region: an analysis on the alternatives to its non-availability. Phytomed Plus 2:1–6

Sahu H, Mishra AP, Jakhwal P, Purohit P (2023) Threats and challenges to sustainable agriculture and rural development in India: implications for agricultural extension. J Glob Agric Ecol 15(3):1–10. https://doi.org/10.56557/jogae/2023/v15i38470

Shankar R, Singh VK, Rawat MS (1993) Medicinal plants from Dibang valley (Arunachal Pradesh): social forestry and afforestation. Bull Medico-Ethno Bot Res 14(3–4):144–149

Sharma E (2018). Biodiversities in the Himalayas—trends, perception and impacts of climate change. In: *Al* IMBC-Plenary Session 1: climate change and its implications for mountain, pp 1–13

Singh VK, Shankar R, Rawat MS (1995) Prospects of medicinal plants cultivation in Arunachal Pradesh. Bull Medico-Ethno Bot Res 17(3–4):133–137

Singh RK, Lego YJ, Sureja AK, Srivastava RC, Hazarika BN (2021) People and plant: learning with Adi community on ethnomedicinal practices and conservation in Arunachal Pradesh, India. Indian J Tradit Knowl 20(1):74–82

Sofowora A, Ogunbodede E, Onayade A (2013) The role and place of medicinal plants in the strategies for disease prevention. Afr J Tradit Complement Altern Med 10(5):210–229

Theodoridis S, Drakou EG, Hickler T, Thines M, Nogues-Bravo D (2023) Evaluating natural medicinal resources and their exposure to global change. Lancet Planet Health 7(2):155–163

Van Wyk AS, Prinsloo G (2020) Health, safety and quality concerns of plant-based traditional medicines and herbal remedies. S Afr J Bot 133:54–62

Volenzo T, Odiyo J (2020) Integrating endemic medicinal plants into the global value chains: the ecological degradation challenges and opportunities. Heliyon 6:1–11

Xia C, Huang Y, Qi Y, Yang X, Xue T, Hu R, Deng H, Bussmann RW, Yu S (2022) Developing long-term conservation priority planning for medicinal plants in China by combining conservation status with diversity hotspot analyses and climate change prediction. BMC Biol 20(89):1–20

The manufacturer's authorised representative in the EU is Springer Nature Customer Service Centre GmbH, Europaplatz 3, 69115 Heidelberg, Germany. If you have any concerns regarding our products, please contact ProductSafety@springernature.com

Printed and bound by CPI Group (UK) Ltd, Croydon, CR0 4YY
01/10/2026
02225914-0010